Advances in Intelligent Systems and Computing

Volume 671

Series editor

Janusz Kacprzyk, Polish Academy of Sciences, Warsaw, Poland
e-mail: kacprzyk@ibspan.waw.pl

The series "Advances in Intelligent Systems and Computing" contains publications on theory, applications, and design methods of Intelligent Systems and Intelligent Computing. Virtually all disciplines such as engineering, natural sciences, computer and information science, ICT, economics, business, e-commerce, environment, healthcare, life science are covered. The list of topics spans all the areas of modern intelligent systems and computing.

The publications within "Advances in Intelligent Systems and Computing" are primarily textbooks and proceedings of important conferences, symposia and congresses. They cover significant recent developments in the field, both of a foundational and applicable character. An important characteristic feature of the series is the short publication time and world-wide distribution. This permits a rapid and broad dissemination of research results.

More information about this series at http://www.springer.com/series/11156

Rahul Kher · Nikhil Gondaliya
Mukesh Bhesaniya · Latif Ladid
Mohammed Atiquzzaman
Editors

Proceedings of the International Conference on Intelligent Systems and Signal Processing

ISSP 2017

Springer

Editors
Rahul Kher
Department of Electronics
and Communication Engineering
G.H. Patel College of Engineering
and Technology
Vallabh Vidyanagar, Gujarat
India

Nikhil Gondaliya
Department of Information Technology
G.H. Patel College of Engineering
and Technology
Vallabh Vidyanagar, Gujarat
India

Mukesh Bhesaniya
Department of Electrical Engineering
G.H. Patel College of Engineering
and Technology
Vallabh Vidyanagar, Gujarat
India

Latif Ladid
SnT
University of Luxembourg
Esch-sur-Alzette
Luxembourg

Mohammed Atiquzzaman
University of Oklahoma
Norman, OK
USA

ISSN 2194-5357 ISSN 2194-5365 (electronic)
Advances in Intelligent Systems and Computing
ISBN 978-981-10-6976-5 ISBN 978-981-10-6977-2 (eBook)
https://doi.org/10.1007/978-981-10-6977-2

Library of Congress Control Number: 2017956309

Printed on acid-free paper

This Springer imprint is published by Springer Nature
The registered company is Springer Nature Singapore Pte Ltd.
The registered company address is: 152 Beach Road, #21-01/04 Gateway East, Singapore 189721, Singapore

Preface

International Conference on Intelligent Systems and Signal Processing (ISSP 2017) aims to spread awareness in the research and academic community regarding the cutting-edge technological advancements revolutionizing the world. The main emphasis of this conference is on dissemination of information, experience, and research results on the current topics of interest through in-depth discussions and participation of researchers from all over the world. The objective is to provide a platform to the scientists, research scholars, and industrialist to interact and exchange ideas in a number of research areas. This will facilitate communication among researchers in different fields of Electrical Engineering, Electronics and Communication Engineering as well as Information Technology. The conference aims to address the following major research areas: Signal Processing and Communication Engineering; Applied Electronics and Emerging Technologies; Computer Vision and Machine Learning; Big Data and Cloud Computing; Power Electronics and Power Systems; and Automation, Control and Electrical Drives.

ISSP 2017 is facilitated with Pre-conference Workshops on March 23, 2017. Pre-conference Workshops aim to discuss theoretical and practical aspects of emerging technologies by renowned technocrats in the field of Communication and Information Technology and Electrical Engineering.

Vallabh Vidyanagar, India — Rahul Kher
Vallabh Vidyanagar, India — Nikhil Gondaliya
Vallabh Vidyanagar, India — Mukesh Bhesaniya
Esch-sur-Alzette, Luxembourg — Latif Ladid
Norman, USA — Mohammed Atiquzzaman

About the Book

The book provides insights into International Conference on Intelligent Systems and Signal Processing (ISSP 2017) held at G.H. Patel College of Engineering and Technology, Gujarat, India, during March 24–25, 2017. The book comprises contributions by the research scholars and academicians covering the topics in Signal Processing and Communication Engineering, Applied Electronics and Emerging Technologies, Computer Vision and Machine Learning, Big Data and Cloud Computing, and Advanced Intelligent Power Electronics and Drives Systems. The main emphasis of the book is on dissemination of information, experience, and research results on the current topics of interest through in-depth discussions and contribution of researchers from all over the world. The book is useful for research community, academicians, industrialists, and postgraduate students across the globe.

About the Book

Contents

About the Editors

Dr. Rahul Kher received B.E. (Electronics) from Sardar Patel University in 1997, M.Tech. (Electrical Engineering) from Indian Institute of Technology Roorkee in 2006, and Ph.D. (Electronics and Communication Engineering) from Sardar Patel University in 2014. He has a teaching and research experience of over 18 years. His research interests include Biomedical Signal and Image Processing, Medical Image Analysis, and Healthcare Monitoring Systems. He has published four books and more than 60 research papers in various international journals and conferences. He is a Senior Member of IEEE and was the founder Secretary of Signal Processing Society (SPS) Chapter of IEEE Gujarat Section during 2013–2015. He has been on the Reviewer Panel/TPC member of many international journals and conferences including the IEEE Communication Society Magazine; Journal of Biomedical Signal Processing and Control (Elsevier); Journal of Computer Networks (Elsevier); International Journal of Advanced Intelligence Paradigms (InderScience); Biomedical Engineering: Application, Basis and Communication (World Scientific); 1st Global IoT Summit (GIoTS '17), Geneva, Switzerland; 3rd Global IoT Innovation Forum, Barcelona, Spain; 3rd Annual Int. Conf. on Wireless Comm. and Sensor Networks (WCSN 2016); 2016 IEEE World Forum on Internet of Things (WF-IoT), Virginia, USA; and many more.

Dr. Nikhil Gondaliya is working as Associate Professor and Head in the Department of Information Technology. He obtained Doctor of Philosophy in Computer Engineering from Gujarat Technological University and master's degree in Computer Engineering from Sardar Patel University. He is having about 14 years of experience in teaching at undergraduate level and 6 years in postgraduate level. He has published more than 20 papers in various national/international conferences and journals. He has guided more than 40 projects at undergraduate level and more than 12 dissertations at postgraduate level. His areas of interest are Ad hoc Wireless Networks, Data Mining, and Internet of Things (IoT).

Prof. Mukesh Bhesaniya has received the B.E. degree in Electrical Engineering from Sardar Patel University, Gujarat, India, in 2000 and the M.Tech. degree in Electrical Engineering from the Indian Institute of Technology Kanpur, India, in

2009, and is currently pursuing Ph.D. degree in Electrical Engineering from Indian Institute of Technology Bombay, Mumbai, India. He has received 'Academic Excellence Award' for the best academic performance during masters in Electrical Engineering from Indian Institute of Technology Kanpur. In 2003, he joined G.H. Patel College of Engineering and Technology, where he is currently an Associate Professor in the Department of Electrical Engineering. He has published/presented number of research papers at national and international journals/conferences, and he has also filed one national patent and one international patent for invention in power electronic converters. He is a reviewer of IEEE Transactions on Power Delivery. His research interests include Power Electronic Converters, HVDC, and Flexible AC Transmission Systems.

Dr. Latif Ladid is a Senior Researcher at Interdisciplinary Centre for Security, Reliability and Trust, SnT, Université du Luxembourg. He works on multiple European Commission Next Generation Technologies IST Projects: 6WINIT, First Pioneer IPv6 Research Project; 6WINIT, Euro6IX, Eurov6, NGNi, Project initiator of SEINIT, and SecurIST. Dr. Ladid initiated the new EU project u-2010 to research Emergency & Disaster and Crisis Management, re-launched the Public Safety Communication Forum, supported the new IPv6++ EU Research Project called EFIPSANS and the new safety & Security Project using IPv6 called Secricom, and co-initiated the new EU Coordination of the European Future Internet Forum for Member States called CeFIMS. He holds the following positions: President, IPv6 FORUM; Chair, European IPv6 Task Force; Emeritus Trustee, Internet Society; Board Member, IPv6 Ready & Enabled Logos Program; and Board Member, World Summit Award. Dr. Ladid is also a member of 3GPP PCG, 3GPP2 PCG; Vice Chair, IEEE ComSoc EntNET; member of UN Strategy Council; member of IEC Executive Committee; Board Member of AW2I; Board Member of Nii Quaynor Institute for Research in Africa; and member of the Future Internet Forum EU Member States, representing Luxembourg.

Dr. Mohammed Atiquzzaman is a Professor in Computer Science Department at The University of Oklahoma, USA. There, he teaches courses in Data Networks and Computer Architecture. His research interests and publications are in Next-Generation Computer Networks, Wireless and Mobile Networks, Satellite Networks, Switching and Routing, Optical Communications, and Multimedia over networks. Many of the current research activities are supported by the National Science Foundation (NSF), National Aeronautics and Space Administration (NASA), and the US Air Force. He serves as the Editor-in-Chief of Journal of Network and Computer Applications and Vehicular Communications; associate editor of IEEE Communications Magazine, Journal of Wireless and Optical Communications, International Journal of Communication Systems, International Journal of Sensor Networks, International Journal of Communication Networks and Distributed Systems, and Journal of Real-Time Image Processing.

Hybrid Compression Method Using Compressive Sensing (CS) Theory for Various Biometric Data and Biomedical Data

Rohit Thanki, Vedvyas Dwivedi and Komal Borisagar

Abstract A hybrid compression method based on compressive sensing (CS) theory proposed for various biometric data and biomedical data in this paper. The data compression method is designed using CS theory, discrete cosine transform (DCT), discrete wavelet transform (DWT), and singular value decomposition (SVD). In this method, first DCT and then DWT are applied on data to get hybrid coefficients of data (which is a combination of wavelet coefficients and frequency coefficients). Then, hybrid measurement is produced by application of CS theory on these coefficients. The SVD is applied to the hybrid measurement of data to get S matrix, U matrix, and V matrix of the data. Then, low-rank approximation is applied on these matrices to get compressed matrices. The inverse SVD is applied on these compressed matrices to get a compressed hybrid measurement of the data. After getting compressed hybrid measurement at the receiver side, CS theory reconstruction is applied on it to get compressed hybrid coefficients of the data. Then, first inverse DWT and then inverse DCT are applied on compressed hybrid coefficients to get reconstructed and compressed data. This method is tested and analyzed for various biometric and biomedical data such as speech signal and ECG signal. The experimental results show that this method is equally performed for all types of biometric and biomedical data. This method is also tested and analyzed for color biometric images such as face image and biomedical image such as magnetic resonance imaging (MRI) image.

Keywords Biomedical · Biometric · CS theory · Data compression
ECG signal · Face image · Magnetic resonance imaging (MRI) image
Speech signal

R. Thanki (✉) · V. Dwivedi
C. U. Shah University, Wadhwan City, Gujarat, India
e-mail: rohitthanki9@gmail.com

K. Borisagar
Atmiya Institute of Technology and Science, Rajkot, Gujarat, India

R. Kher et al. (eds.), *Proceedings of the International Conference on Intelligent Systems and Signal Processing*, Advances in Intelligent Systems and Computing 671,
https://doi.org/10.1007/978-981-10-6977-2_1

1 Introduction

The research on biometric and the biometric system is increased nowadays because biometric characteristics are used for personal identification and verification. The meaning of biometric is "life to measure" and is divided into two types such as behavioral characteristics and physiological characteristics. So nowadays, a biometric characteristic of a person is acquired and stored at various organizations and institutions for easy verification. The biometric characteristics such as fingerprint, face, speech, iris, retina, and signature are widely used for person verification [1]. There is a lot of biometric data available at system database. One of the examples is fingerprint and iris data of Aadhaar project of India. In medical science, the biomedical data such as images and signals are used for better health treatment. These biomedical data are observed physiological activities of a human organ. The biomedical data such computed tomography (CT) image, magnetic resonance imaging (MRI) image, ECG signal, and EEG signal are used for treatment related to a heart problem and brain problem, respectively. There is a lot of biomedical data available at large hospital. This creates problem related to storage of biometric data and biomedical data at system database of biometric system and storage of hospital, respectively.

So in this paper, we have discussed new compression method based on compressive sensing (CS) [2–4] theory for various biometric data and biomedical data. This method is applied and analyzed on biometric data like speech signal and color face image and biomedical data like ECG signal and color MRI image. The rest of paper is organized as follows: In Sect. 1.1, information of related work is given. The information of compressive sensing (CS) theory is given in Sect. 2. The information of implementation of proposed signal compression method is given in Sect. 3. The results and discussion of proposed compression method for biometric data and biomedical data are given in Sect. 4. Finally, the conclusion of the paper is given in Sect. 5.

1.1 Related Work

The many researchers are proposed various compression methods using CS theory for biometric data and biomedical data. Authors in [5] represented CS theory for audio signal compression. Authors gathered audio data from multiple sensors, and then CS theory is applied on it to get compressed audio data. Authors in [6] proposed DCT and CS theory-based compression method for the audio signal. Authors have applied CS theory on DCT coefficients of the audio signal and generate a compressed audio signal. Authors in [7, 8] proposed wavelet transform and CS theory-based compression method for ECG signal. In these methods, CS theory is applied on wavelet coefficients of ECG signal to get compressed ECG signal. Authors in [9] proposed K-SVD adaptive dictionary and CS-based compression method for the speech signal. In this method, K-SVD algorithm and CS theory process are combined to get compressed speech signal. Authors in [10] proposed wavelet transform and CS-based compression method for CT image and MRI image.

After reviewing research papers, it is clearly indicated that most methods are designed and proposed using single transform. These methods are applied to one type of data or information. So in this paper, various signal processing transforms and CS theory-based compression method are proposed and analyzed for various types of biometric data and biomedical data. In this paper, first DCT and after that DWT is applied on data to get hybrid coefficients of the data. The CS measurement of the data is generated using compressive sensing (CS) theory process. Then, SVD and low-rank approximation are applied to CS measurement of the data to get compressed CS measurement of the data. The compressed hybrid coefficients of the data are recovered from compressed CS measurement using CS recovery algorithm such as OMP algorithm [4]. Finally, IDWT and then IDCT are applied on compressed hybrid coefficients to get a reconstructed and compressed data. The complete process for compression of the data is given in Sect. 4.

2 Information on Compressive Sensing (CS) Theory

A signal S can become sparse signal when only a few nonzero elements are presented in the signal. The signal S can be converted into a sparse signal by applying signal transform basis matrix on it. The signal has x nonzero coefficients (sparse coefficients) which are represented as follows:

$$x = \Psi \times S \times \Psi' \tag{1}$$

where x is the sparse coefficients of signal and Ψ is the transform basis matrix.

The CS measurement of the signal using compressive sensing (CS) theory can be represented by using following equation [2, 3]:

$$y = A \times x \tag{2}$$

where y is the CS measurement with dimension $M \times N$, A is known as measurement matrix with dimension $M \times N$, x is known as sparse coefficients with dimension $N \times N$. Here, the value of M is deciding the size of CS measurement which is providing compression to the signal.

To reconstruct a signal from CS measurement, various CS recovery algorithms are available in the literature [2–4]. A greedy algorithm such as orthogonal matching pursuit (OMP) is used in this paper. This algorithm is proposed and designed by Tropp and Gilbert [4]. It is used in this paper for the extraction of sparse coefficients from CS measurement. This algorithm is defined by three basic steps which are matching, orthogonal projection and residual update and recovering one nonzero sparse coefficient in each iteration. It can be mathematically explained by using below equation:

$$x^* = \mathrm{OMP}(y, A) \tag{3}$$

where x^* is extracted sparse coefficients which provide the extracted from the CS measurement y.

3 Proposed Hybrid Compression Method

The most of the existing data compression method based on CS theory are designed using single transform. So in this proposed method, different transforms such as DCT, DWT, and SVD are used along with CS theory for data compression method. The framework for proposed data compression method is shown in Fig. 1, and the mathematical steps for signal compression are given below. This proposed data compression method is analogous to conventional compression method [11]. In the proposed method, source encoding and source decoding are performed by CS theory, while channel encoding and channel decoding are performed by SVD-based low-rank approximation method.

- Take data and calculate the size of data. If data is signal, then it is converted into the matrix for easy processing in MATLAB software. The discrete cosine transform (DCT) is applied to data to get DCT coefficients of the data.

$$D = \mathrm{DCT}(S) \tag{4}$$

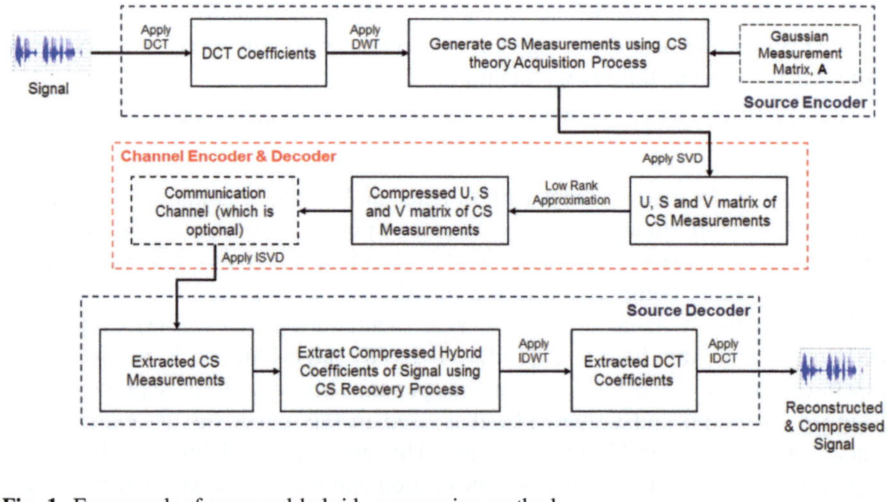

Fig. 1 Framework of proposed hybrid compression method

where D is DCT coefficients of the data and S is any biometric data or biomedical data.

- The wavelet basis matrix and its inverse basis matrix are multiplied with DCT coefficients to get hybrid coefficients (wavelet coefficients of DCT coefficients) of the data. The wavelet basis matrix is generated using the wavmat function in MATLAB [12–14].

$$W = \Psi \times D \times \Psi' \tag{5}$$

where W is hybrid coefficients (wavelet coefficients of DCT coefficients) of the data and Ψ is a wavelet basis matrix.

- The CS measurement of the data is generated using hybrid coefficients and Gaussian measurement matrix. The Gaussian measurement matrix is generated using zero mean and one variance.

$$y = A \times W \tag{6}$$

where y is a CS measurement of the data and A is Gaussian measurement matrix with a size of $M \times N$.

- Apply singular value decomposition (SVD) on CS measurement of data to get U matrix, S matrix, and V matrix of CS measurement of the data.

$$[U, S, V] = \mathrm{SVD}(y) \tag{7}$$

- The low-rank approximation using M value of measurement matrix A is applied on U matrix, S matrix, and V matrix to get compressed matrices.

$$\begin{aligned} U^* &= (:, 1, M) \\ S^* &= (1 : M, 1 : M) \\ V^* &= (:, 1 : M) \end{aligned} \tag{8}$$

where U^*, S^*, and V^* are compressed U matrix, S matrix, and V matrix of CS measurement of the data, respectively.

- Inverse SVD is applied on compressed matrices to get compressed CS measurement of the data.

$$y^* = U^* \times S^* \times V^* \tag{9}$$

where y^* is compressed CS measurement of the data.

- After getting compressed CS measurement, compressed hybrid coefficients of the data are recovered by CS recovery algorithm [4].

$$W^* = \text{OMP}(y^*, A) \tag{10}$$

where $W*$ is compressed hybrid coefficients (wavelet coefficients of DCT coefficients) of the data and OMP is orthogonal matching pursuit.

- Then, inverse wavelet basis matrix and its original basis matrix are multiplied with compressed hybrid coefficients of the data to get compressed DCT coefficients of the data.

$$D^* = \Psi' \times W^* \times \Psi \tag{11}$$

where $D*$ is compressed DCT coefficients of data.

- Inverse DCT is applied on compressed DCT coefficients to get a reconstructed and compressed data. If the data is signal, then matrix is converted into a vector to get a reconstructed and compressed data.

$$S^* = \text{IDCT}(D^*) \tag{12}$$

where $S*$ is a reconstructed and compressed data.

4 Results and Discussion

The testing and analysis of proposed hybrid compression method are performed using different biometric data such as face image and speech signal and biomedical data such as MRI image and ECG signal. The test biometric data and biomedical data are discussed in Sect. 4.1. The quality measures for proposed compression method are given in Sect. 4.2. The performance of proposed compression method for speech signal and ECG signal is discussed in Sect. 4.3. The performance of proposed compression method for color face image and color MRI image is discussed in Sect. 4.4.

4.1 Test Biometric Data and Biomedical Data

The performance of any data compression method varies with different types of data. Therefore, in this paper, biometric data such as face image and speech signal and biomedical data such as MRI image and ECG signal are taken for analysis of proposed compression method. In Fig. 2, speech signal has a size of 62,500 samples and it is taken from the Freesound Web site [15]; ECG signal has a size of 3600 samples, and it is taken from the Physionet Web site [16]. The size of the face and MRI image is 256 × 256 pixels. The face image is taken from FEI face database [17, 18]. The FEI face database is Brazilian face database for research which is

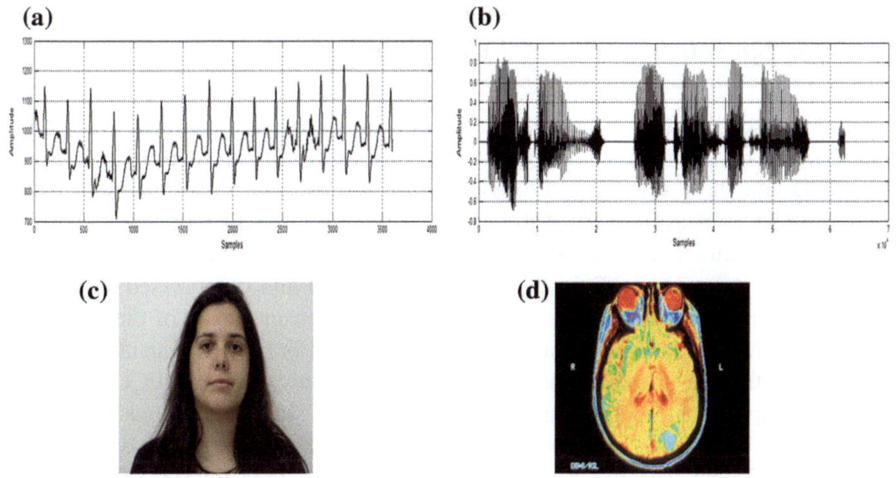

Fig. 2 Test biometric data and biomedical data **a** ECG signal. **b** Speech signal. **c** Face image [17, 18]. **d** MRI image [19]

created by Artificial Intelligence Laboratory of FEI in Sao Bernardo do Campo, Sao Paulo, Brazil [19]. The MRI image is taken from the on line course on Probability Theory and Mathematical Statistics of Pennsylvania State University, example of Lesson 31 [21]. The performance of proposed method is carried out for different CS measurements which are given in Sects. 4.2 and 4.3. In this method, wavelet basis matrix is generated using Haar wavelet.

4.2 Quality Measures for Proposed Hybrid Compression Method

The quality of the compressed signal is measured by root-mean-square error (RMSE) [7], signal-to-noise ratio (SNR) [20], and correlation coefficients (CoC) [8]. The quality of the compressed image is measured by RMSE, peak signal-to-noise ratio (PSNR) [20], and structural similarity index measure (SSIM) [21]. The high value of CoC and SSIM indicated that compressed signal and the compressed color image are more similar to the original signal and original color image, respectively. The value of CoC and SSIM lies in 0–1. The compression ratio of proposed compression method is calculated using below equation:

$$\mathrm{CR} = \frac{\mathrm{Size}_{\mathrm{Original\_Data}}}{\mathrm{Size}_{\mathrm{CS\_Measurement}}} \tag{13}$$

4.3 Performance Analysis of Proposed Compression Method for Speech Signal and ECG Signal

In the proposed method, hybrid coefficients of the signal are generated using DCT and DWT. The CS theory is applied on it to get CS measurement of the signal. The SVD and the low-rank approximation are applied to CS measurement to get compressed CS measurement of the signal. The compressed sparse coefficients of the signal are extracted using OMP algorithm. Finally, IDCT and IDWT are applied on compressed sparse coefficients to get reconstructed and compressed signal. For analysis of proposed method, different sizes of CS measurements of the signal are generated. Then, reconstructed and compressed signal is generated using these CS measurements. The proposed method is tested and analyzed using ECG biometric signal and speech biometric signal. Figures 3 and 4 show the reconstructed and compressed ECG signal and speech signal, respectively. These signals are generated using different sizes of CS measurements. Table 1 shows the RMSE, SNR, and

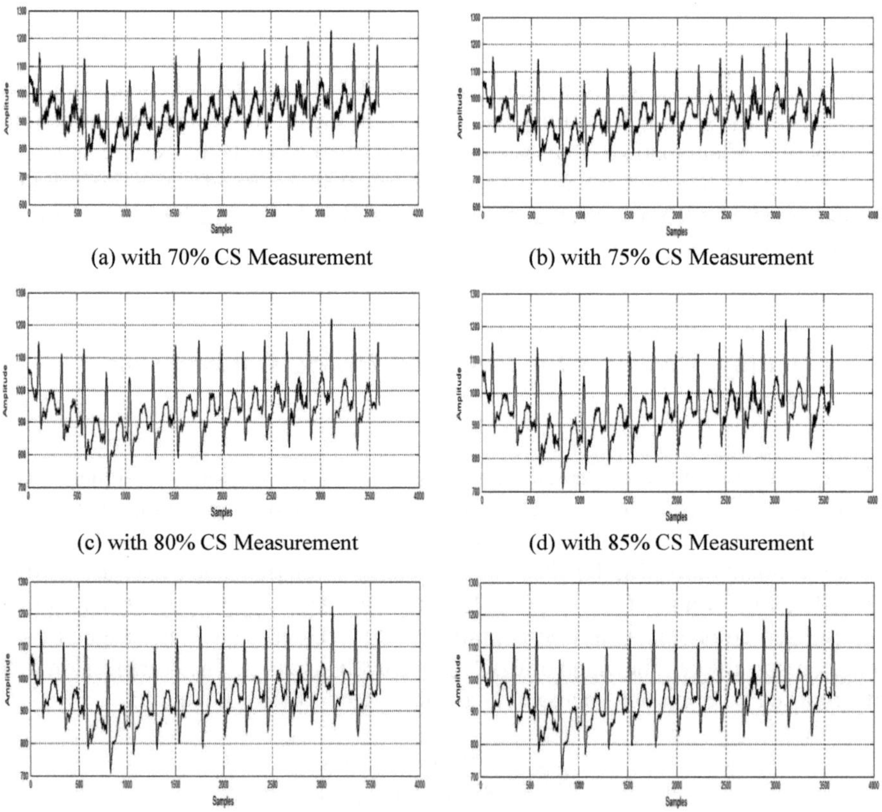

(a) with 70% CS Measurement
(b) with 75% CS Measurement
(c) with 80% CS Measurement
(d) with 85% CS Measurement
(e) with 90% CS Measurement
(f) with 95% CS Measurement

Fig. 3 Reconstructed and compressed ECG signal for different sizes of CS measurements

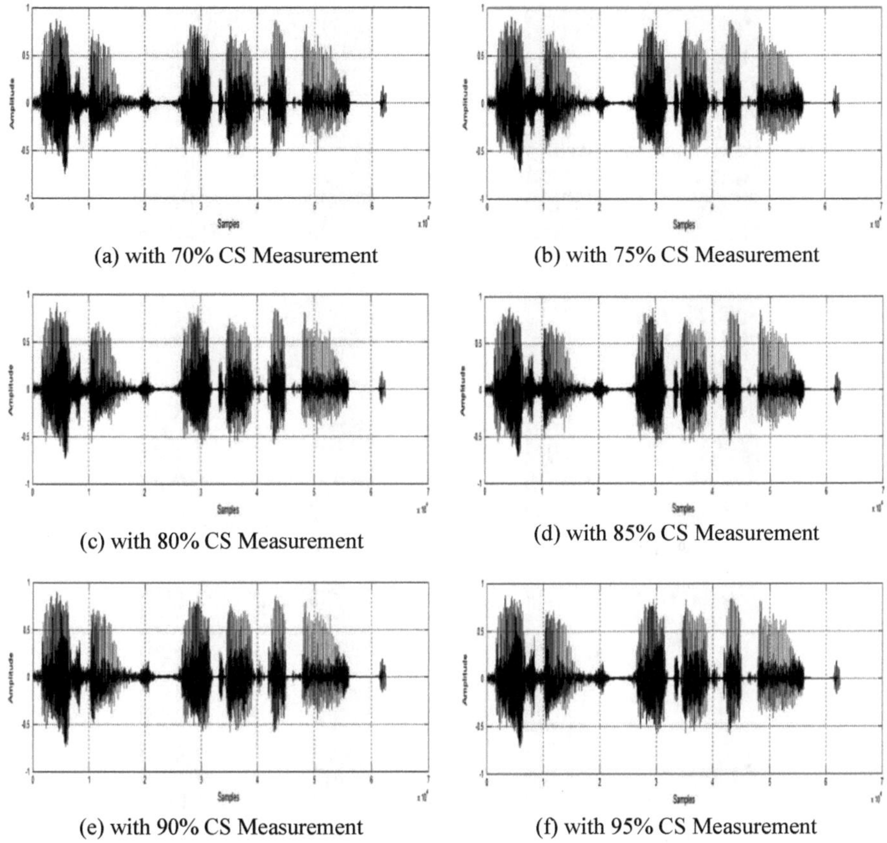

Fig. 4 Reconstructed and compressed speech signal for different sizes of CS measurements

Table 1 Quality measures using different sizes of CS measurements for ECG signal and speech signal

Quality measure for ECG signal				Quality measure for speech signal				CR
Different sizes of CS measurement	RMSE	SNR (dB)	CoC	Different sizes of CS measurement	RMSE	SNR (dB)	CoC	
43,750 (70%)	1.18	37.4	0.985	2520 (70%)	3.33	12.2	0.967	1.43
47,000 (75%)	2.13	37.8	0.990	2700 (75%)	2.96	12.6	0.971	1.33
50,000 (80%)	1.84	41.4	0.993	2880 (80%)	1.97	12.7	0.973	1.25
53,500 (85%)	1.57	41.9	0.995	3060 (85%)	2.09	13.0	0.975	1.17
56,250 (90%)	1.21	43.6	0.997	3240 (90%)	2.88	13.4	0.976	1.11
59,500 (95%)	1.48	44.9	0.998	3420 (95%)	1.49	13.5	0.978	1.05

CoC values for the different sizes of CS measurement for the ECG signal and speech signal. The results in Table 1 show that when the size of CS measurement is increased, the quality of compressed ECG signal and speech signal is also

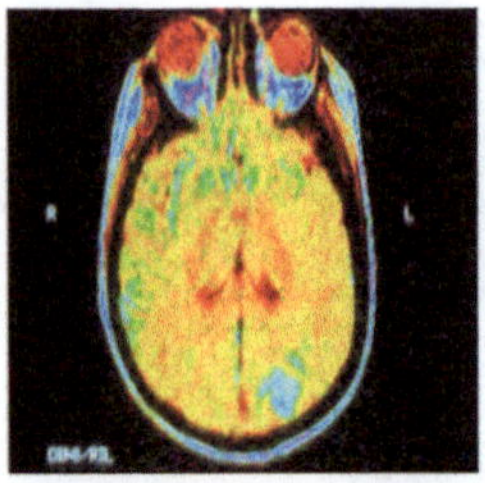

(a) with 70% CS Measurement

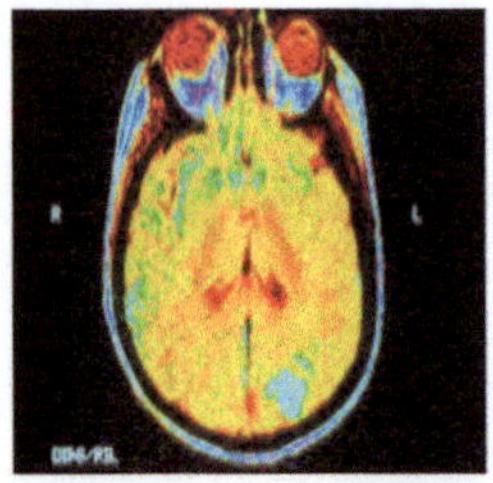

(b) with 75% CS Measurement

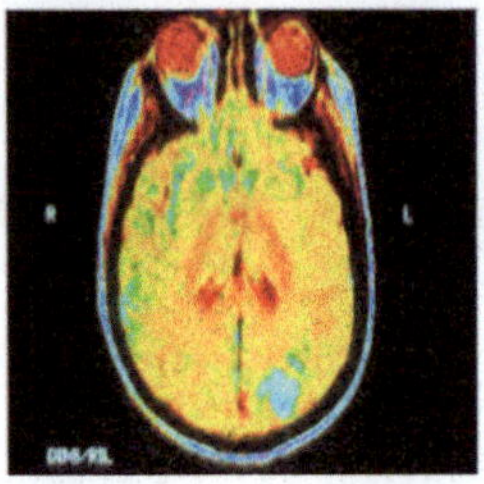

(c) with 80% CS Measurement

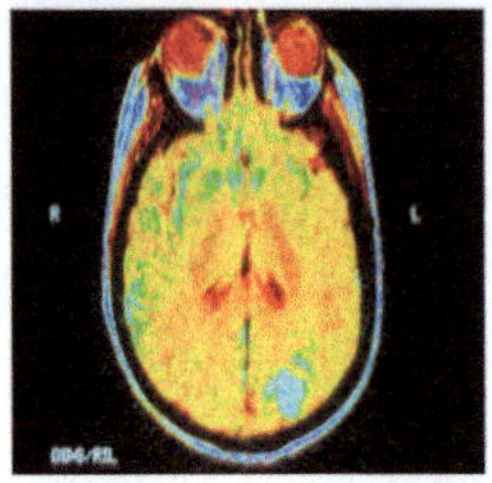

(d) with 85% CS Measurement

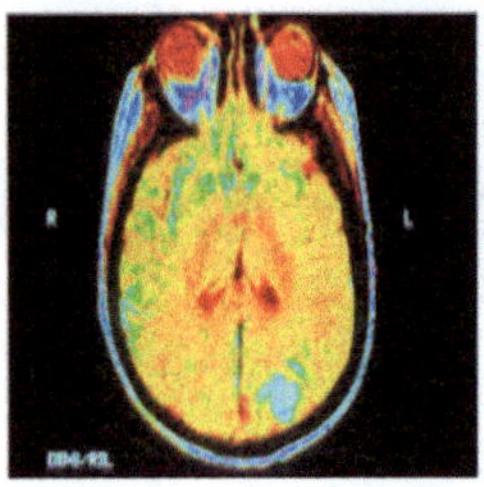

(e) with 90% CS Measurement

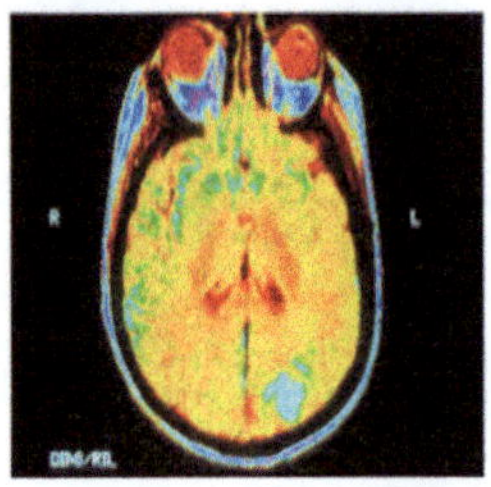

(f) with 95% CS Measurement

Fig. 5 Reconstructed and compressed Color MRI Image for different sizes of CS measurements

increased. The compression ratio is decreased when the size of CS measurement is increased.

4.4 *Performance Analysis of Proposed Compression Method for Color Face Image and Color MRI Image*

The proposed method is also tested different types of data such as color face biometric image and color MRI biomedical image. When this method is applied to the color image, the first color image is decomposed in three different channels like R channel, B channel, and G channel. The proposed method is applied to the individual channel to generated compressed data of R channel, B channel, and G channel. Then, reconstructed and compressed color image is generated using compressed R channel, B channel, and G channel. Figures 5 and 6 show the

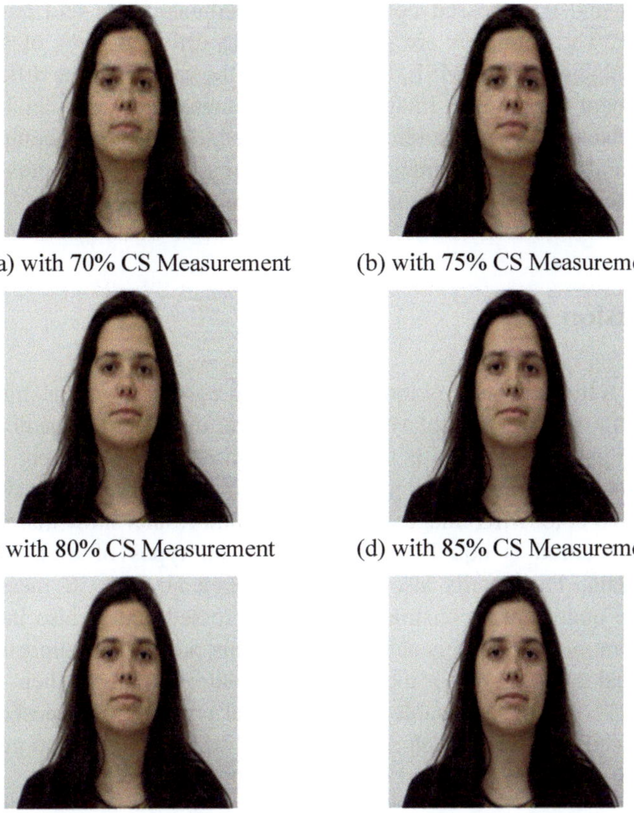

(a) with 70% CS Measurement (b) with 75% CS Measurement

(c) with 80% CS Measurement (d) with 85% CS Measurement

(e) with 90% CS Measurement (f) with 95% CS Measurement

Fig. 6 Reconstructed and compressed Color Face Image for different sizes of CS measurements

Table 2 Quality measures using different sizes of CS measurements for color MRI image and color face image

Different sizes of CS measurement	Quality measure for color MRI image			Quality measure for color face image			CR
	RMSE	PSNR (dB)	SSIM	RMSE	PSNR (dB)	SSIM	
137,472 (70%)	5.64	33.1	0.867	2.82	39.1	0.957	1.43
147,456 (75%)	5.69	33.2	0.882	2.78	39.2	0.965	1.33
157,440 (80%)	5.56	33.2	0.896	2.74	39.4	0.971	1.25
167,424 (85%)	5.53	33.3	0.906	2.71	39.5	0.974	1.17
177,408 (90%)	5.51	33.3	0.915	2.68	39.6	0.978	1.11
187,392 (95%)	5.48	33.4	0.927	2.66	39.7	0.982	1.05

reconstructed and compressed color MRI biomedical image and color face biometric image. These images are generated using different sizes of CS measurements. Table 2 shows the RMSE, PSNR, and SSIM values for the different sizes of CS measurement for the face image and MRI image. The results in Table 2 show that when the size of CS measurement is increased, the quality of compressed color face image and color MRI image is also increased. The compression ratio is decreased when the size of CS measurement is increased.

5 Conclusion

In this paper, a hybrid compression method using compressive sensing (CS) theory is proposed for compression of various biometric data and biomedical data. This proposed compression method is tested and analyzed using different types of biometric data such as speech signal and color face image and biomedical data such as ECG signal and color MRI image. The experimental results show that this proposed compression method is equally performed for all types of biometric data and biomedical data. The results also show that when size of CS measurements is increased, the quality of reconstructed and compressed data is also increased. The compression ratio up to 1.43 is achieved using this proposed compression method. But biomedical data is used for the better treatment of a patient; then, compression ratio up to 1.11 can be applicable for biomedical data using proposed method. In future, advanced transforms such as curvelet and ridgelet are used to generate space coefficients of biometric data and biomedical data.

Statement on Consent The author(s) have obtained consent from the individuals for publication who participated in this study. If any issue arises hereafter, then the authors will be solely responsible.

References

1. Jain, A., Kumar, A.: Biometric Recognition – An Overview, Second Generation Biometrics. The Ethical, Legal and Social Context, E. Mordini and D. Tzovaras (Eds.), Springer, pp. 49–79 (2012).
2. Donoho, D.: Compressed Sensing. IEEE Transaction on Information Theory, 52(4), pp. 1289–1306 (2006).
3. Candes, E.: Compressive Sampling. In: Proceedings of the International Congress of Mathematicians, 3, pp. 1433–1452, IEEE, Madrid, Spain (2006).
4. Tropp, J., Gilbert, A.: Signal Recovery from Random Measurements via Orthogonal Matching Pursuit. IEEE Transactions on Information Theory, 53(12), pp. 4655–4666 (2007).
5. Griffin, A., Tsakalides, P.: Compressed Sensing of Audio Signals using Multiple Sensors. In: Proceedings of IEEE 16th European Signal Processing Conference, pp. 1–5, IEEE (2008).

6. Moreno-Alvarado, R., Martinez-Garcia, M.: DCT-Compressive Sampling of Frequency Sparse Audio Signals. In: Proceedings of the World Congress on Engineering, 2, pp. 6–8, IEEE (2011).
7. Abo-zahhad, M. Hussein, A., Mohamed, M.: Compression of ECG Signal based on Compressive Sensing and the Extraction of Significant Features. International Journal of Communications, Network and System Sciences, 8, pp. 97–117 (2015).
8. Mishra, A., Thakkar, F., Modi, C., Kher, R.: ECG Signal Compression using Compressive Sensing and Wavelet Transform. In: Proceedings of 2012 Annual International Conference on the IEEE Engineering in Medicine and Biology Society, pp. 3404–3407, IEEE (2012).
9. Zhou, Y., Heming, Z.: Speech Signal Compressed Sensing based on KSVD Adaptive Dictionary. Journal of Theoretical and Applied Information Technology, 48(2), pp. 1237–1243 (2013).
10. Sevak, M., Thakkar, F., Kher, R., Modi, C.: CT Image Compression using Compressive Sensing and Wavelet Transform. In: Proceedings of 2012 International Conference on Communication Systems and Network Technologies (CSNT), pp. 138–142, IEEE, Rajkot (2012).
11. Gonzales, R., Woods, R.: Digital Image Processing. pp. 422, Prentice Hall Inc., New Jersey (2002).
12. Vidakovic, B.: Statistical Modelling by Wavelets. pp. 115–116, Wiley (1999).
13. Yan, J.: Wavelet Matrix. The University of Victoria, Canada (2009).
14. Claude, F., Navarro, G.: The Wavelet Matrix. In: International Symposium on String Processing and Information Retrieval, pp. 167–179, Springer Berlin Heidelberg (2012).
15. For Speech Signal: https://freesound.org.
16. For ECG Signal: http://www.physionet.org/physiobank/database/nstdb/.
17. FEI Face Database (2005, 2006). Available: http://fei.edu.br/~cet/facedatabase.html.
18. Thomaz, C., Giraldi, G.: A New Ranking Method for Principal Components Analysis and its Application to Face. Image and Vision Computing, 28(6), pp. 902–913 (2010).
19. For Color MRI Image: https://onlinecourses.science.psu.edu/stat414/node/202.
20. Mrak, M., Grgic, S., Grgic, M.: Picture Quality Measures in Image Compression Systems. In: Proceedings of the IEEE Region 8 EUROCON 2003. The computer as a Tool, 1, pp. 233–236, IEEE, Slovenia (2003).
21. Wang, Z., Simoncelli, E., Bovik, A.: Multiscale Structural Similarity for Image Quality Assessment. In: Proceedings of the 37th IEEE Asilomar Conference on Signals, Systems and Computers, 2, pp. 1398–1402, IEEE, Pacific Grove (2003).

Design, Implementation and Testing of Signal Processing Module for an Impact Ionization Dust Detector

J. P. Pabari, A. D. Jadav and U. Y. Kakaiya

Abstract The study of dust at high altitudes on the Mars has been in literature for a long time and a few sources are expected, which can contribute to it. Recent MAVEN observations found the dust from about 150 to 1000 km height from the Mars surface and it is expected to be interplanetary in nature. An in situ measurement of dust at orbital altitude is needed to study the dust and a few attempts were made in past in this direction. An impact ionization dust detector, called Mars Orbit Dust Experiment (MODEX) has been proposed to find the source, abundance, flux and distribution of the dust at Mars by future orbiter. A prototype model of the dust detector is under development at PRL. The detector principle, its electronics design, implementation, simulation and testing are presented in this article. Also, back end processing and derivation of the signal parameters are included along with user interface for the analysis. The signal processing module could be useful for detection of a charge pulse generated due to dust impact.

Keywords Dust · Hyper velocity · Impact · Mars · Planet

1 Introduction and Motivation

It is known that dust devils prevailing near the Mars surface play vital role in the background opacity on Mars and it could be one possible source of dust in Mars atmosphere. However, no physical process can explain the transport of surface dust to high altitudes. Other possible source of dust at orbital altitudes could be escaping

J. P. Pabari (✉)
Physical Research Laboratory, Ahmedabad, India
e-mail: jayesh@prl.res.in

A. D. Jadav
VVPEC, Rajkot, India

U. Y. Kakaiya
DDU, Nadiad, India

R. Kher et al. (eds.), *Proceedings of the International Conference on Intelligent Systems and Signal Processing*, Advances in Intelligent Systems and Computing 671,
https://doi.org/10.1007/978-981-10-6977-2_2

dust particles from the natural satellites of the Mars, viz. Phobos and Deimos. Continuous bombardment of dust on Phobos and Deimos can cause the secondary ejecta, which can easily escape the parent body due to very low escape velocity (10 m/s for Phobos and 6 m/s for Deimos [1, 2]). Escaping dust is predicted to form dust rings within orbits of Phobos and Deimos and therefore also around Mars. However, no such rings have been detected to the present day [3]. The dimensions of the dust torus could be as large as ~5 Mars radii in both the planes, i.e., horizontal and vertical planes, as found in literature [4]. The remaining possible source of high altitude dust could be interplanetary in nature. Igenbergs et al. [5] have reported a Mars Dust Counter (MDC) on the Nozomi mission, which detected ~100 particles, several of interplanetary, during the cruise phase. A PADME mission with a dust experiment has been proposed by Lee et al. [6]. Recently, MAVEN [7] observed the dust at altitudes ~150 to 1000 km by a Langmuir Probe and Wave (LPW) instrument and it is expected to be interplanetary in nature. A Langmuir probe is primarily used to study plasma parameters and a dedicated dust detector is needed to understand the source of the dust particles, unambiguously. To study origin, abundance, distribution and seasonal variation of Martian dust, a Mars Orbit Dust Experiment (MODEX) has been proposed [8]. A prototype design of an impact ionization dust detector has been initiated at PRL.

The remaining part of the article is organized as follows. Section 2 presents the system block diagram and principle of detection. Section 3 shows design of front end electronics, analytical derivation, simulation results and noise figure computation. Section 4 gives implementation of the electronics and its practical testing results. Section 5 presents Graphical User Interface (GUI) and signal analysis and, paper ends with summary.

2 Impact Ionization Dust Detector

The high altitude dust at Mars is expected to have hypervelocity, i.e., velocity greater than 1 km/s. Whenever such a hypervelocity dust particle makes an impact on a metal (gold) target dust detector, an impact plasma is created. The plasma plume travels within the detector box and reach near the collection plates of the detector. A positively biased plate will capture the plasma electrons, while a negatively biased plate will attract the plasma ions. The background noise is separated from the impact charge, using a neutral channel. A block diagram of the dust detector is shown in Fig. 1a and a photograph of the detector box is depicted in Fig. 1b. The detector cross-section is 150 mm × 150 mm and the height is ~200 mm in the initial detector design with payload mass of about 1.5 kg. The charge pulse generated by the impact plasma is processed by the electronics to arrive at particle velocity and mass, using the rise time (t_r) and the maximum charge (Q) of the impact pulse [5]

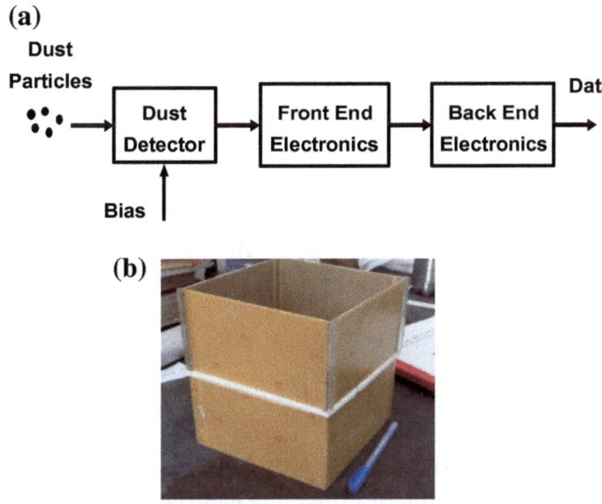

Fig. 1 **a** Block diagram of dust detector (top) and **b** a snapshot of detector box (bottom)

$$t_r = c_g v^{\eta} \tag{1}$$

and

$$\pm\frac{Q}{m} = c_r v^{\beta} \tag{2}$$

respectively. In the above, c_g, c_r, η and β are constants to be determined by calibration experiments [9]. The mass-velocity distribution is useful to understand the possible source of incoming dust particles [5].

3 Front End Electronics Design and Simulation

The front end electronics consists of a Charge Sensitive PreAmplifier (CSPA) to convert the impact charge into a voltage signal and a pulse shaper to shape the pulse for further measurements. The rise time of incoming pulse could be few microseconds to few tens of microseconds which is preserved in the charge sensitive preamplifier. The pulse fall time is decided by feedback components of the CSPA and typically, it could be a few hundred microseconds. The large fall time is desirable for further processing by subsequent stages. The pulse shaper comprises of a CR stage and double RC stage $(RC)^2$ for making the pulse into semi Gaussian shape. A compensating resistor is connected in parallel with the capacitance of the CR stage for pole-zero cancellation. The feedback components of CR stage are helpful for better shaping of the pulse. To accommodate possibly large dynamic range of the input signal, a logarithmic compression is achieved using a diode in the CR stage. The RC stage time constant should be comparable to the time constant of

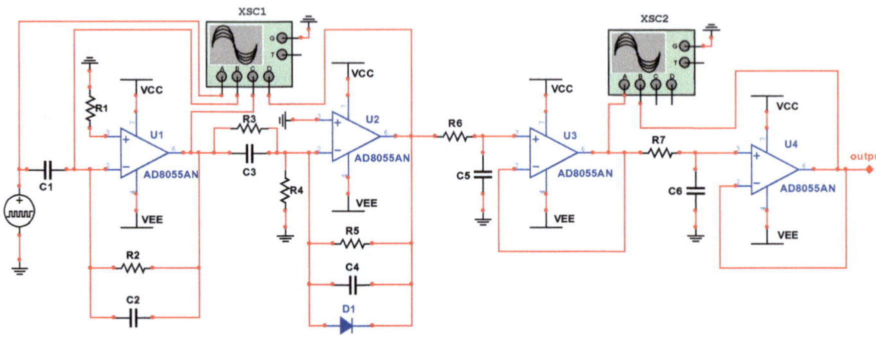

Fig. 2 Preliminary circuit design of front end electronics for dust detector

CR stage. A double integrator stage is used to achieve an improved pulse shape. A circuit schematic of front end electronics is depicted in Fig. 2 along with the other simulation units like signal source and oscilloscope.

In practice, the actual signal is due the dust impact on detector surface and its repetition rate depends on the arrival of dust particles to the detector. MAVEN measured the dust rate about 0.001 hits/s at Mars [7]. This dust impact rate could be larger if there is any comet passing near the Mars. To simulate this scenario, we have used a pulse source in the software and carried out optimization of the circuit. The input signal used for simulation is in the form of an impulse train, which can be represented by

$$v_{\text{in}}(t) = \sum_{k=-\infty}^{k=\infty} \delta(t - kT) \tag{3}$$

where k is an integer and T is pulse repetition time period which depends on the incoming dust flux. The output of charge sensitive preamplifier is given by

$$v_{\text{csp}}(t) = \frac{-1}{c_{\text{f}}} \mathrm{e}^{-t/\tau_1} \cdot v_{\text{in}}(t) \tag{4}$$

where $\tau_1 = C_{\text{f}} R_{\text{f}}$ the output of CR cum logarithmic stage is given by

$$v_2(t) = 2\tau_2 - \frac{c_3}{c_4} \sqrt{2\, R_{\text{S}} I_0 - v_{\text{csp}}(t)} \tag{5}$$

where I_0 = reverse saturation current of diode and $\tau_2 = C_4 R_{\text{S}} = C_{\text{S}} R_6 = C_6 R_7$. The output of RC^2 stage is given by

$$v_3(t) = \tau_2^2 \mathrm{e}^{-2\tau_2 t} v_2(t) \tag{6}$$

The final output of the front end electronics is given by combination of Eqs. (3)–(6) and representing the output in terms of input voltage. We have simulated

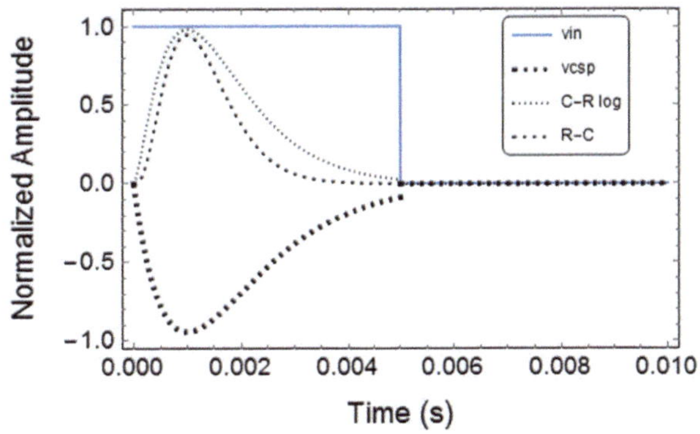

Fig. 3 Theoretical pulse response of various stages of front end electronics

theoretical pulse responses of all stages in the circuit and the results are shown in Fig. 3. In addition, circuit simulation has been carried out for the practical circuit and the results are depicted in Figs. 4 and 5. One can observe from the theoretical and simulation results that the pulse responses are satisfactory.

Furthermore, noise analysis was carried out to understand the noise figure of the circuit given in Fig. 2. The noise analysis results are depicted in Fig. 6 where the asterisk plot shows the input noise spectrum while triangle plot shows the output noise spectrum. From the results in Figs. 4, 5 and 6, we found the circuit noise figure of the detector electronics as 1.59 dB.

Further, we have varied the feedback components of the CSPA to understand their effect on the noise figure. Figure 7 show variation in Noise Figure (in dB) with respect to the feedback resistance and capacitance of CSPA. It has been found that the noise figure remains almost constant for smaller values (a few GΩ) of the feedback resistance, while it monotonically reduces for higher values of the feedback resistance. Also, the noise figure is found to decrease with increasing feedback capacitance. This is due to increase in the noise for the rise in total input capacitance of CSPA. In noise analysis, the feedback capacitance adds to the total input capacitance [10] and hence, the feedback capacitance should be kept small (a few pF typically). Due to increase in the output SNR, the overall noise figure is reduced.

4 Front End Electronics Implementation and Testing

The front end electronics has been implemented on a PCB as shown in Fig. 8. The testing was carried out using the charge pulse generated by triboelectricity. The negative charge was applied at the input and again negative signal is produced at the output after double inversion in the circuit. Further, there is no signal inversion and we get negative pulse as final output. The result of the PCB testing is depicted in Fig. 9 from which one can observe that the pulse shape is reasonably good which may be used by back end processing to derive its parameters.

Fig. 4 Input pulse, CSPA output and CR (logarithmic) stage output

5 GUI and Signal Analysis

The pulse produced at the output of front end electronics is given to a Digital Phosphor Oscilloscope (DPO) and data were transferred to a computer for analysis. We made a GUI using MATLAB for the analysis of signal captured, as depicted in Fig. 10. The parameters of interest are the pulse rise time and pulse peak value. These two parameters are used in Eqs. (1) and (2) after overall calibration to retrieve the physical quantity of interest. Using the rise time and peak value of the charge pulse, one can derive the impact velocity and mass of incoming dust particle. The mass-velocity distribution is finally used for finding the source of dust particle.

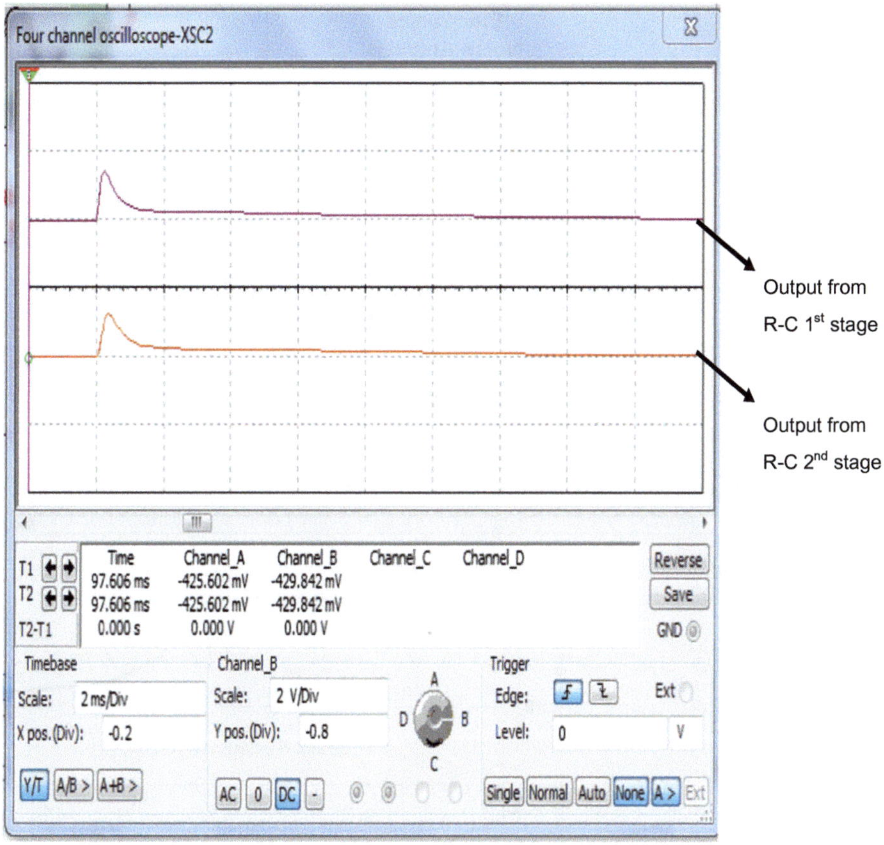

Fig. 5 Output of first and second RC stages

6 Summary

The front end electronics of dust detector has been designed and simulated for a dust detector. The circuit has been optimized and pulse responses of various stages are found satisfactory. Also, the noise analysis has been carried out to understand noise figure of the circuit and we found that the circuit noise figure is 1.59 dB. Furthermore, the circuit has been implemented on a PCB and its practical testing was carried out. The pulse data were analysed using the MATLAB GUI, which can be used to derive the parameters of incoming dust particle.

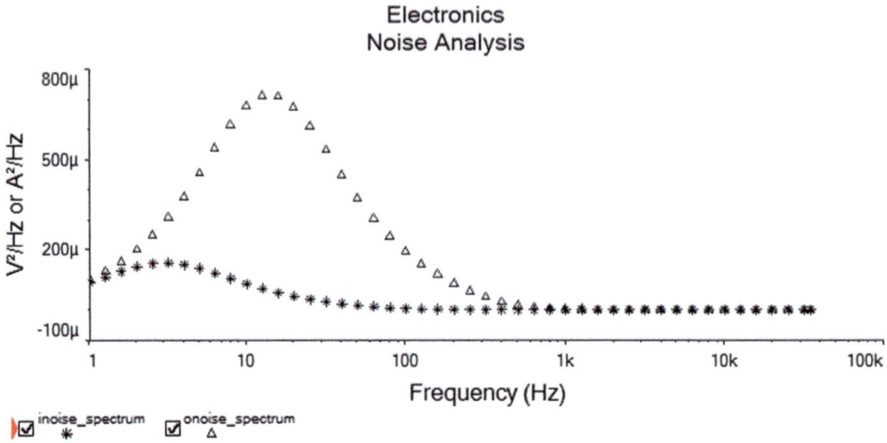

Fig. 6 Input and output noise spectrum

Fig. 7 Variation in noise figure with feedback parameters

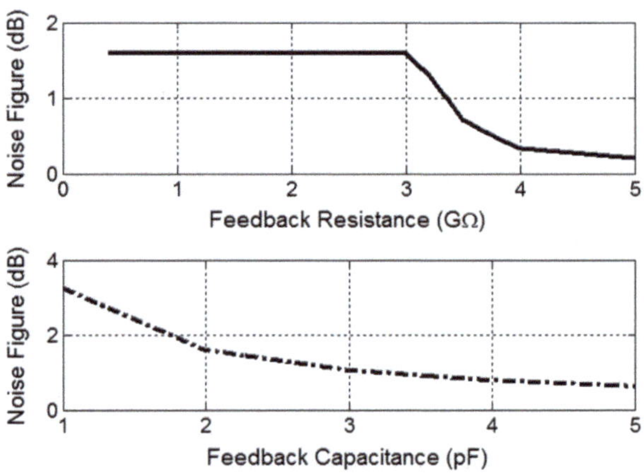

Fig. 8 Snapshot of front end electronics PCB with components mounted on it

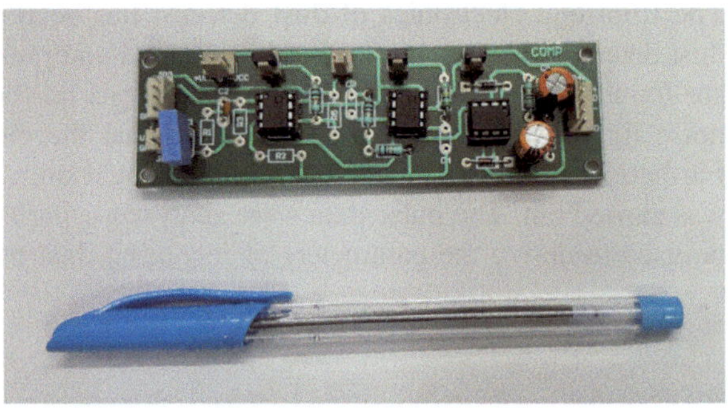

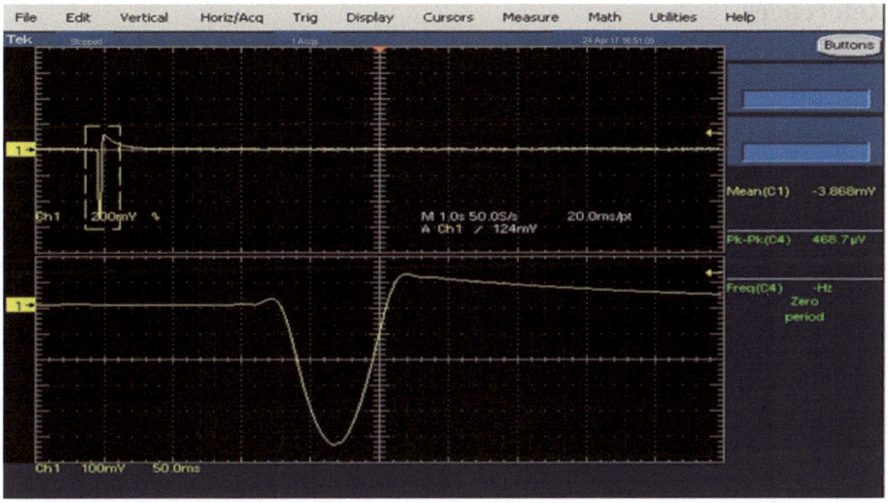

Fig. 9 Result of PCB testing

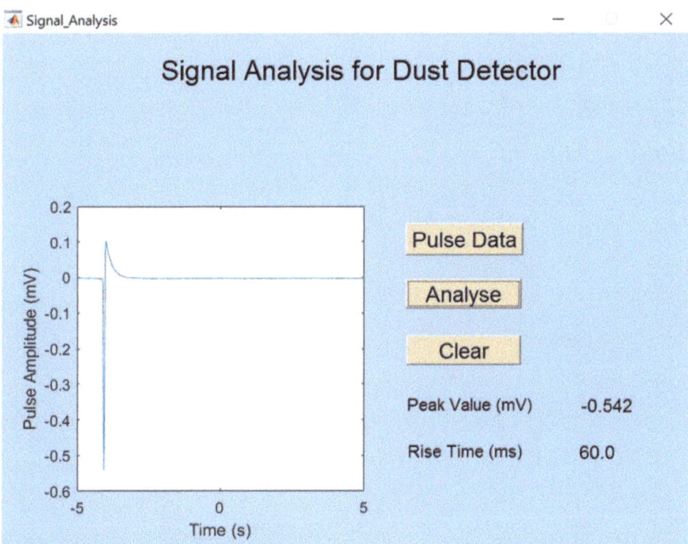

Fig. 10 MATLAB GUI for signal analysis

References

1. Anonymous a, http://solarviews.com/eng/phobos.htm, last accessed 2016/6/16.
2. Anonymous b, http://solarviews.com/eng/deimos.htm, last accessed 2016/6/16.
3. Oberst, J., Zakharov, A. and Schulz, R.: Why Study Phobos and Deimos? An Introduction to the Special Issue, Planetary and Space Science, 102, 1 (2014).
4. Krivov, A. V. and Hamilton, D. P.: Martian dust belts: waiting for discovery, Icarus, 128, 335–353 (1997).
5. Igenbergs, E. et al.: Mars Dust Counter, Earth Planets Space, 50, pp. 241–245 (1998).
6. Lee, P. et al.: Phobos And Deimos & Mars Environment (PADME): A LADEE-Derived Mission to Explore Mars's Moons and the Martian Orbital Environment, 45th LPSC, #2288 (2014).
7. Andersson, L. et al.: Dust observations at orbital altitudes surrounding Mars, Science, 350 (6261), aad0398 (2015).
8. Pabari, J. P., Bhalodi, P. J. and Patel, D. K.: Mars Orbit Dust Experiment (MODEX) for future Mars orbiter, 47th Lunar and Planetary Science Conference, #1419, The Woodlands, Texas (2016).
9. Igenbergs, E. et al.: Mars Dust Counter on Board ISAS PLANET-B, Proc. 20th Int. Sym. Space Tech. and Space, pp. 1222–1229, Gifu, Japan (1996).
10. Spieler, H.: Front-End Electronics and Signal Processing, Physics Division, Lawrence Berkeley National Laboratory, Berkeley, CA (2002).

Automatic Fire Detection Using Combination of Color Cue and Flame Flicker

Ripal Patel, Kashyap Mandaliya, Pushkar Shelar, Rushi Savani and Chirag I. Patel

Abstract This paper presents the novel algorithm for automatic fire detection from still images and video sequences. Proposed technique has been using the color cue and flame flicker for detecting fire. This paper proposes a combination of two algorithms to detect fire from video clips. Firstly, the algorithm defines the method to detect fire in static images which can be called as color feature technique. Secondly, the algorithm defines to detect the fire in video sequences, which can be called as flicker technique. Furthermore, the final result is generated from the combination of both the results to get the exact fire output eliminating noise regions. Hence, in results, exact fire region is detected which satisfies both algorithms. Lastly, the buzzer is attached to our system; it will ring when the fire is detected by our algorithm. Results have been evaluated for still images from benchmark dataset. The overall performance of the proposed technique is appropriate.

Keywords Flame detection · Color · Flicker detection · Fire detection in video

1 Introduction

With the development of the economy, the number of skyscrapers is increasing. By traditional methods, major property damage and huge loss happen in these places and result into very bad social impact which is not good for the economic growth of a country. We are designing a module which can be able to detect the fire and smoke on early stage, and also at the same time fast processing can be there so the

R. Patel (✉) · K. Mandaliya · P. Shelar · R. Savani
Electronics & Communication Department, BVM Engineering College,
Vallabh Vidyanagar 388120, Gujarat, India
e-mail: ripal.patel@bvmengineering.ac.in

C. I. Patel
Computer Science & Engineering Department, Institute of Technology,
Nirma University, Ahmedabad 382481, Gujarat, India

R. Kher et al. (eds.), *Proceedings of the International Conference on Intelligent Systems and Signal Processing*, Advances in Intelligent Systems and Computing 671,
https://doi.org/10.1007/978-981-10-6977-2_3

fire rescue operation is done in very less time resulting in less damage. Video-based fire and smoke detection can be able to detect pinpoint fire from long distances as soon as the fire starts and allowing the fire to be dealt with before it gets out of control. Furthermore, cameras can cover very large areas, and video-based fire detection even has the potential to be placed on mobile platforms planes.

2 Literature Survey

Töreyin et al. [1] proposed a completely unique technique to notice fireplace and/or flames in the period by the process of the video information. A common method to detect the fire is based on some color features or motion features. However, the author proposed the method which works based on flame and fire flicker. This phenomenon is evaluated using temporal wavelet analysis. Periodic behavior of flame flicker is being monitored and analyzed in the wavelet domain. Different types of features like the nonuniform boundary of the fire-colored region and color variation in flame have been considered in this approach. Another technique has been proposed by Liu and Ahuja [2] for vision-based fire detection. The detection of fire region has been evaluated in three various domains. Firstly, frequency domain analysis has been performed using probability density distribution. Secondly, Fourier representation of the shape of flames has been achieved. Thirdly, temporal changes have been designed to build temporal signature.

Borges and Izquierdo [3] proposed low-level feature representation of fire regions. Fire flicker and random behavior of fire are key characteristics of fire to analyze, and based on this phenomenon, various low-level features like color, area size, and surface coarseness are considered. Another approach is proposed for an automatic system for fire detection [4] in the sequences of videos. The method used is the information of color and motion taken from the sequences of videos to determine the location of the fire. In earlier vision-based methods of detection of fire, this method is useful for more regions due to the insensitiveness of the motion of the camera. Two basic methods are not proper to the earlier algorithms for the recognizing the fire in global camera motion's presence or the site motion and the recognizing the fire in the movies and entertainment to the possibility of the use in rating system which is automatic.

3 Proposed Approach

Block diagram of the proposed video-based fire detection approach is shown in Fig. 1. Specifically, it consists of five major steps of proposed algorithm.

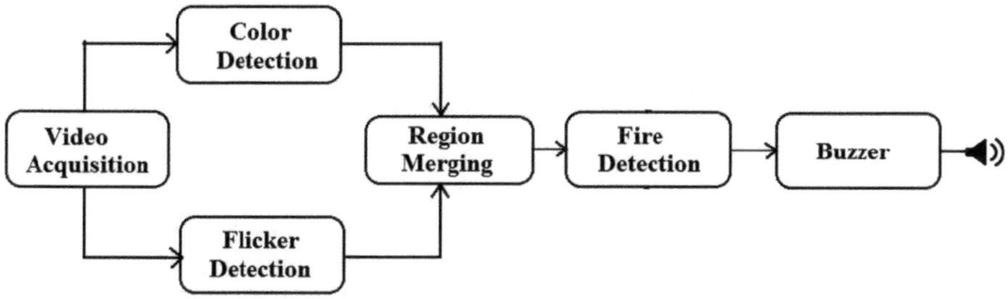

Fig. 1 Block diagram of proposed fire detection system

3.1 Color Detection

In this paper, following algorithm is evaluated for fire detection using color features. The first step of our algorithm is to find color values of moving pixels which is compared with predetermined color distribution. The images are then converted to RGB color range Matrix.

- Selection yellow color range of photograph:
  In this method, conversion of the true color image into yellow color component is taken place.
- Count to how much a binary inside the image.
  Comparison of the yellow pixel value is performed using summation.

3.2 Flicker Detection

The flicker detection algorithm has been designed based on two properties. Firstly, it is observed that the fire is flickered between frequency ranges 1–10 Hz. Secondly, fire is typically the strongest source of a light; hence, the intensity of luminance tends toward maximum value. In proposed system, to track moving object we had used the time derivative of the luminance. The time derivative of the static object is zero, and on the other side, the time derivative of moving object is nonzero. To estimate the tendency of fire to flicker around a region we had used cumulative time derivative method. Cumulative strength and luminance weight are incorporated in above two mentioned properties.

Flicker detection algorithm [5] Each frame of video clip has been converted from RGB color space to YUV color space.

- Time derivative of luminance (Y) (i.e., D_i) has been approximated using following equation:

$$D_i(x, y) = |Y\_{i}(x, y) - Y\_{i-1}(x, y)| \quad (1)$$

- Cumulative time derivative (CTD) computation

$$\mathrm{CT}_i(x, y) = \alpha \mathrm{CT}_{i-1}(x, y) + (1 - \alpha) w_i(x, y) D_i(x, y) \quad (2)$$

where α is the cumulative strength, $w_i(x, y)$ is luminance weight, $\mathrm{CT}_i(x, y)$ is equal to zero. As more frames in video, more accurate the cumulative time derivative information it carries. Then after, resetting the pixel in normalized $\mathrm{CT}_i(x, y)$ is used (Eq. 2). Convert $\mathrm{CT}_i(x, y)$ to 8-bit luminance range.
- Setting all pixels in CT_N, which have less than nonzero values in CT_N to 0's by keeping strong move regions. CT_N is final image resulting flame flicker.

3.3 Region Merging

The observation has been made that the fire in the fire regions is clustered together. Hence, the merging operation has been evaluated on fire regions and results in the elimination of another noisy region which contains no fire or any motion. Therefore, region merging of fire areas from resulted images has been evaluated using AND process of the output of color feature and flicker process.

3.4 Fire Detection and Buzzer

The final fire is obtained by the combination of color and flicker algorithm. After then, the buzzer is attached at the last of our system. The main function of buzzer here is to ring as soon as the fire is detected by our algorithms.

4 Experimental Setup

Proposed approach is evaluated on the system platform of 64-bit operating system processor—INTEL(R) CORE(TM)i5CPU using MATLAB version—R2012a having frequency of 2.67 GHz.

5 Dataset

In proposed system, the experiments are performed into two parts: (1) static images. (2) video sequences. The proposed system of detection of fire is tested with static images using proposed color featuring algorithm, which are taken from various sources. Later on, merging of color and flicker feature is applied on video. The proposed approach is evaluated on six different types of videos from [5], and they include various types of fire sources like nature like sun, matchstick, fire caught in the building, the fire in the chimney.

6 Results and Discussions

In Fig. 2, all pictures are taken in daylight and the results which are shown above in which some pictures contain the fire and some are not containing the fire. The result of proposed color feature algorithm, applied to the still images, is shown in Fig. 2.

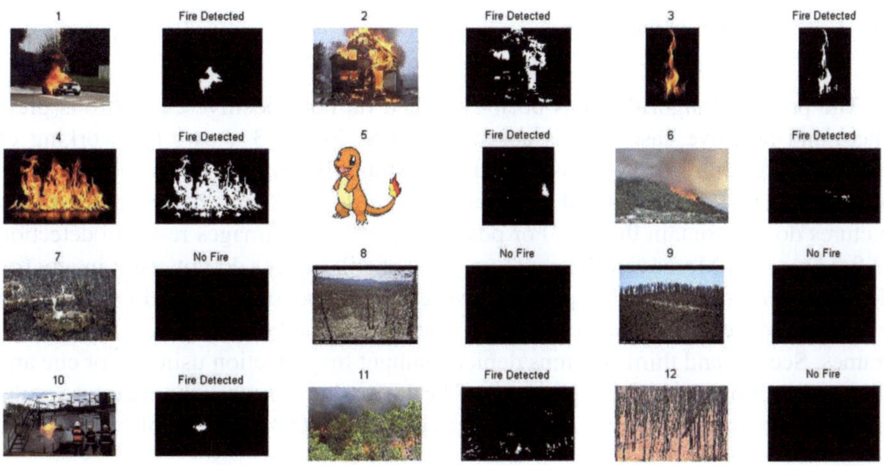

Fig. 2 Results for static image using color as cue

Fig. 3 Positive cases of fire images

The proposed algorithm has been evaluated on both positive case (fire is presence) and negative case (fire is absence) of fire. Figure 3 shows the working of proposed algorithm on the positive image. Similarly, Fig. 4 shows the result for negative images. In Fig. 4, all pictures taken in daylight and the results in which pictures do not contain the fire. For positive cases of still images result in detection of fire properly. Moreover, the negative cases of still images do not result in any fire detection feature. Figure 5 represents the results of proposed method evaluated on video sequences. The first column of the result displays snapshot from video frames. Second and third columns depict resultant fire detection using color cue and flicker detection method, respectively. The last column shows the merged results from both the binary images using proposed merging technique. Table 1 compares

Fig. 4 Negative cases of fire images

the fire detection results by applying our proposed algorithm for all six videos. The list of videos containing fire is video 073-1, video 073-2, video 073-8, video 073-9, and non-fire videos are video 073-6 and video SUN. For video 073-1, the fire recognition rate is 83.1% which is good. On the other side, specifically, our system detects sun as a fire (video 073-6) as we are using color feature technique to detect fire. Our system correctly detects non-fire video frames as shown in Table 1. However, another system is false to detect non-fire video frames.

Fig. 5 **a** Original video sequence. **b** Color output. **c** Flicker output. **d** Merged image

Table 1 Detailed results of fire video dataset

Sr. no.	Total no. of frames	Frames with fire	False +ve	False −ve	Recognition rate (%)	Description
Video 073-1	592	542	0	50	83.1	A burning house
Video 073-2	42	42	0	0	100	Fire in chimney
Video 073-6	83	0	0	83	100	A settling sun
Video 073-8	33	33	0	0	100	Burning building
Video 073-9	83	83	0	0	100	Burning paper
Video SUN	236	0	193	43	42.37	Sun in afternoon

7 Conclusion

The system proposed the automatic fire detection system based on vision-based technique for still images and video sequences. For the still image technique, the results are superior as shown in the figure at detecting fire, and for the video sequences technique, our system is consistent and reliable at detecting all forms of fire. The proposed technique is a fusion of two features, which are color features and flame flicker features. To avoid any aliasing in the decision of flame detection, we have proposed region merging technique. Experimental results show the outstanding performance of the proposed technique. In future, frequency domain analysis of fire flicker can be considered to improve the outcome of the proposed system.

References

1. Töreyin, B.U., Dedeoğlu, Y., Güdükbay, U. and Cetin, A.E., Computer vision based method for real-time fire and flame detection. Pattern recognition letters, 27(1), pp. 49–58 (2006).
2. Liu, Che-Bin, and Narendra Ahuja. "Vision based fire detection." Pattern Recognition, 2004. ICPR 2004. Proceedings of the 17th International Conference on. Vol. 4. IEEE, (2004).
3. Borges, Paulo Vinicius Koerich, and Ebroul Izquierdo. "A probabilistic approach for vision-based fire detection in videos." IEEE transactions on circuits and systems for video technology 20, no. 5: 721–731(2010).
4. Walter Phillips III, Mubarak Shah, and Niels da Vitoria Lobo, Flame Recognition in Video, Pattern Recognition Letters, Vol. 23 (1–3), pp. 319–327, January, (2002).
5. Qi, Xiaojun, and Jessica Ebert. "A computer vision based method for fire detection in color videos." International journal of imaging 2, no. S09: 22–34 (2009).
6. Patel, Chirag I., Sanjay Garg, Tanish Zaveri, Asim Banerjee, and Ripal Patel. "Human action recognition using fusion of features for unconstrained video sequences." Computers & Electrical Engineering (2016).

Performance Analysis of Supervised & Unsupervised Techniques for Brain Tumor Detection and Segmentation from MR Images

Brijesha D. Rao and Mukesh M. Goswami

Abstract Brain tumor detection and segmentation from the magnetic resonance images (MRI) is a difficult task as in the MR brain images, various tissues such as white matter, gray matter, and cerebrospinal fluid have complicated structures that make it difficult to segment the tumor. An automated system for brain tumor detection and segmentation will help the patients for proper treatment planning. Also, it will improve the diagnosis and reduce the diagnostic time. Segmentation of brain tumor MR images is the most difficult task as the tumor varies in terms of size, shape, location, and texture. In this paper, we discuss various supervised and unsupervised techniques for brain tumor detection and segmentation such as K-nearest neighbor (K-NN), *K*-means clustering, and using morphological operators. We also review the results obtained.

Keywords Brain tumor · Magnetic resonance image (MRI) · Feature extraction Supervised and unsupervised techniques

1 Introduction

Brain tumor is the extra cells growing in the brain, which form a mass of tissue. Survey says that 7.6 million people die among 12.7 million cancerous people in a year [1]. Gliomas are the primary brain tumor which found more frequent in adults [2]. The survival rate for high-grade tumor is almost two years while several years for low-grade tumors. If the automated system for brain tumor detection and segmentation is available, then it may help to minimize the rate of diagnosis of brain tumor abnormality which helps the patients for proper treatment planning.

B. D. Rao (✉) · M. M. Goswami
Information Technology, Dharmsinh Desai University, Nadiad, India
e-mail: brijesha_056102@yahoo.co.in

M. M. Goswami
e-mail: mukesh.goswami@gmail.com

R. Kher et al. (eds.), *Proceedings of the International Conference on Intelligent Systems and Signal Processing*, Advances in Intelligent Systems and Computing 671,
https://doi.org/10.1007/978-981-10-6977-2_4

MRI is a widely used technique for tumor analysis as it provides better visualization of the internal structure in detail. It also provides better contrast between the different soft tissues of the body. The MRI scan of the patient can be taken from three separate angles such as axial, sagittal, and coronal plane [3]. T1, T2, T1c (Contrast Enhanced), fluid attenuation inversion recovery (FLAIR), etc., are mostly used MRI sequences for brain tumor analysis which provides different anatomical and functional information of the brain tumor MRI images.

The brain tumor segmentation process contains the steps such as preprocessing, extraction of features from MR images, and segmentation using supervised or unsupervised techniques. Here, we will deliberate different methods for supervised and unsupervised learning which are useful for brain tumor detection from the MRI images.

2 Methodology

Figure 1 shows the generalized process for brain tumor detection.

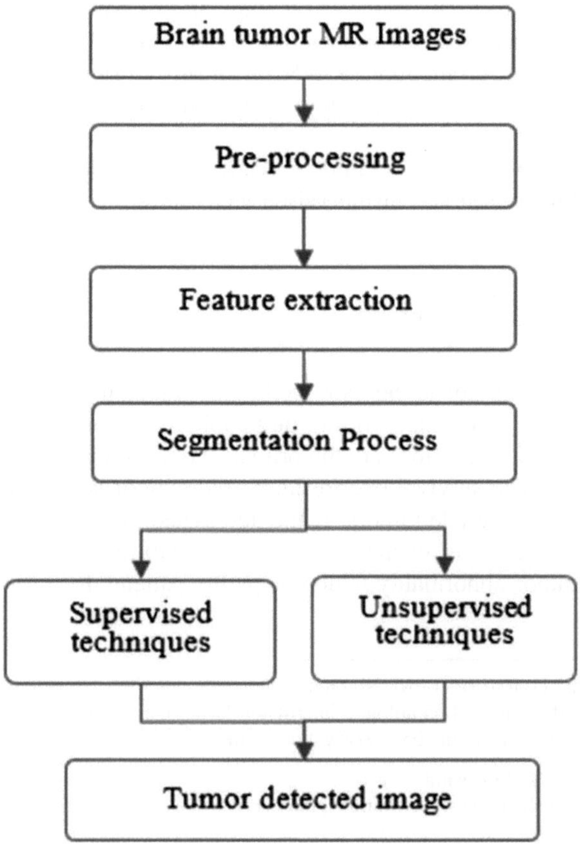

Fig. 1 Workflow of the tumor detection system

A. *Image Acquisition and Preprocessing*

In order to scale the current state-of-the-art in automated brain tumor segmentation and to compare it between different methods, Multimodal BRATS challenge is organized in conjunction with the MICCAI 2015 conference. For this purpose, they prepared and made available a unique dataset of MR scans of low-grade and high-grade glioma patients which is analyzed by several human experts [4]. In this research work, we are using BRATS 2015 dataset of different high-grade and low-grade glioma patients. The dataset contains training data of 220 HGG and 54 LGG images with ground truth and also the testing data of 110 HGG and LGG images [5]. Basically, the dataset contains 3D images in .mha format which is a metafile used in medical domain to represent more information. Here, we will work on axial view (with respect to the z-axis) of the image. The size of the 2D image is 240×240.

As a preprocessing step, we will normalize the image to grayscale range from 0 to 255. Mostly, we assumed that our images are already preprocessed and no need of noise removal, skull-stripped removal, image enhancement, etc. After normalization of the images, features will be extracted.

B. *Feature Extraction*

For the detection of the tumor region, our initial goal is to extract relevant features from the grayscale images. Characteristics obtained from the features are key to differentiate the tissue of the tumor region with the normal tissue of the brain. Texture-based features are mainly used features for tumor detection purpose. The first-order intensity-based statistical features such as mean, variance, standard deviation, skewness, kurtosis [6], and second-order statistical features using GLCM can be used for segmentation purpose [6]. In the proposed work, the first-order statistical features are computed from the Flair and T2 MR images, considering the window size of the 3×3 neighborhood where the center pixel is replaced by the function applied to it.

C. *Segmentation*

By segmenting the image, it divides the image into subregions. In tumor detection, we have to separate the normal tissue and abnormal tissue as a different region. Various supervised and unsupervised techniques can be used for segmentation process. Support vector machine (SVM), K-nearest neighbor (K-NN), artificial neural network (ANN), decision tree (DT), etc., are the various supervised learning technique which uses labeled information for classification. In unsupervised learning, label information is not available, so clustering using fuzzy *c*-means, *k*-means, mixture model, as well as gray-level segmentation technique using morphological and watershed segmentation can be applied for segmentation.

3 Supervised and Unsupervised Techniques for Tumor Detection

For automatic brain tumor detection, we need to train the model using the supervised and unsupervised machine learning techniques. As discussed above, many supervised and unsupervised techniques are available but, for our initial experiments, we have tested supervised and unsupervised machine learning techniques using K-NN and *K*-means, respectively. We have also used the morphological operator for our experiments. These techniques are described below.

A. *Supervised Technique*

The flow of brain tumor detection using supervised learning technique is shown in Fig. 2.

1. *K-Nearest Neighbor (K-NN)*:

K-NN is a simple classifier that stores all available training samples, and the testing samples can be classified based on a similarity measure, i.e., using distance functions [7]. The test case is classified using majority vote taken from the neighbors, and finally, the class label is assigned based on the class most common among all its K-nearest neighbors. For $K = 1$, the class of its nearest neighbor is assigned. To provide the optimal value of K, one should properly analyze the data. A large number of K can be more precise as it reduces the overall noise. K-NN-based techniques are used in [8–10].

B. *Unsupervised Technique*

In the unsupervised learning technique, we do not have a data with class label. So on the basis of similarity or dissimilarity, the clusters are formed. Here, we discuss *K*-means clustering and morphological operators, which work as an unsupervised learning technique for brain tumor detection.

1. *K-means clustering*

K-means clustering is used for solving low-level segmentation task. First, the images are taken as an input. Next, the points that have similar color are grouped together. Euclidean distance calculates the distance between the clusters. After

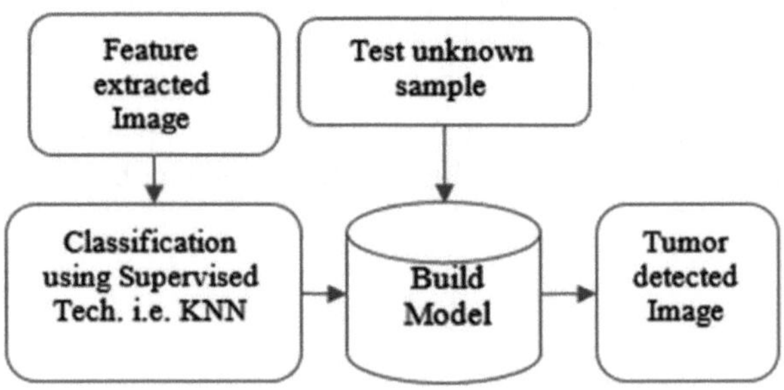

Fig. 2 Supervised technology for tumor detection

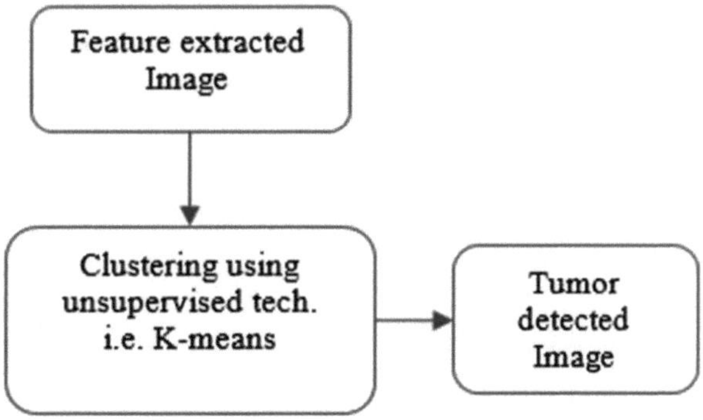

Fig. 3 Unsupervised technology for tumor detection

clustering, the mean of clustering is taken. Fuzzy c-means requires more computation time than *K*-means. The flow of unsupervised learning technology using *k*-means is shown in Fig. 3. This technique is used in [11, 12].

2. *Using Morphological operators*

Morphological image processing is a technique which uses a number of nonlinear operations associated with the shape in an image. For the processing of binary images, morphological operations are suited because it relies on the ordering of pixel values; morphological operations can also be applied to grayscale images because the light transfer functions used in that are unknown so their absolute pixel values are of less interest. Morphological techniques use various structuring element such as disk, oval, square. The structuring element is placed at all possible locations in the image and compared with the corresponding neighborhood which can be "fits" or can be "hits" the neighborhood [13]. The morphological operators such as erode (for shrinking), dilate (for filling and expanding) reconstruct, and regional max are mostly applied to the brain tumor MRI images which separate the foreground and background objects and help to find out the tumor region. Figure 4 shows the flow for tumor segmentation using morphological operator. This technique is used in [7, 12, 14].

4 Performance Measures

The performance measures will be computed using positive predictive value (PPV), sensitivity, and Dice score which are described here.

- **Positive Predictive Value**: PPV is also known as precision value. It is a measure that gives the probability of positive prediction being correct with respect to all positive prediction made by the system. The high value of precision gives good accuracy. It can be computed using the given formula.

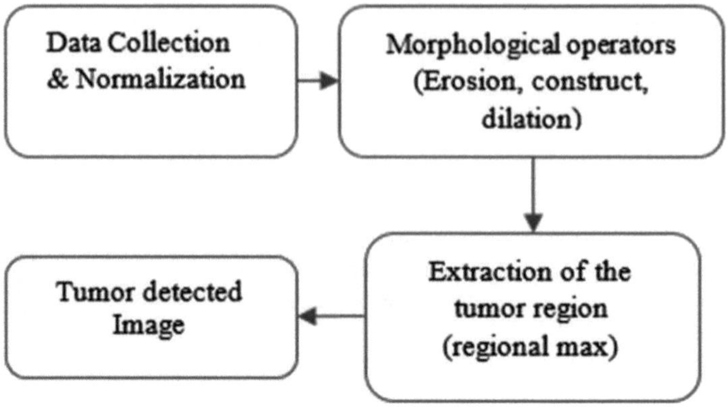

Fig. 4 Tumor detection using morphological operators

$$\text{PPV} = \frac{\text{No. of True Positive (TP)}}{(\text{No. of True Positive (TP)} + \text{No. of False Positive (FP)})}$$

- **Sensitivity**: It is also known as recall or the hit rate. It gives the probability of a positive case being identified properly by a system. To compute the sensitivity, we can use the following formula.

$$\text{Sensitivity} = \frac{\text{TP}}{\text{TP} + \text{FN}}$$

- **Dice**: It normalizes the number of true positives to the average size of the two segmented areas. It is also known as *F*-score.

$$\text{Dice Score} = \frac{2.\text{TP}}{(2.\text{TP} + \text{FN} + \text{FP})}$$

5 Experimental Outcomes and Analysis

For our primary experiments, we have used the MRI images of BRATS dataset which is already labeled so we can use it for the supervised machine learning technique, and same images have been used for the unsupervised techniques for brain tumor detection.

A. *Outcomes of K-NN, K-means, and morphological operators*

Figure 5 shows the result of tumor detection using K-NN.

Figure 6 illustrates the result of tumor detection using *K*-means clustering technique

Figure 7 shows the output for tumor detection using morphological operators.

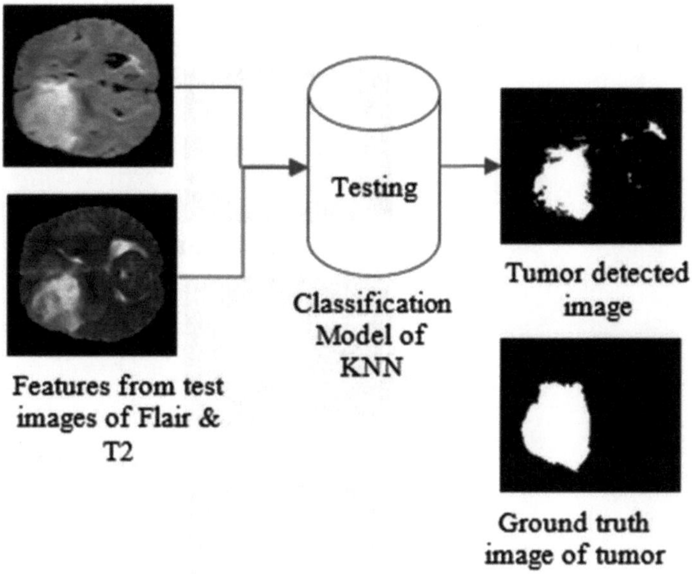

Fig. 5 Brain tumor detection using K-NN

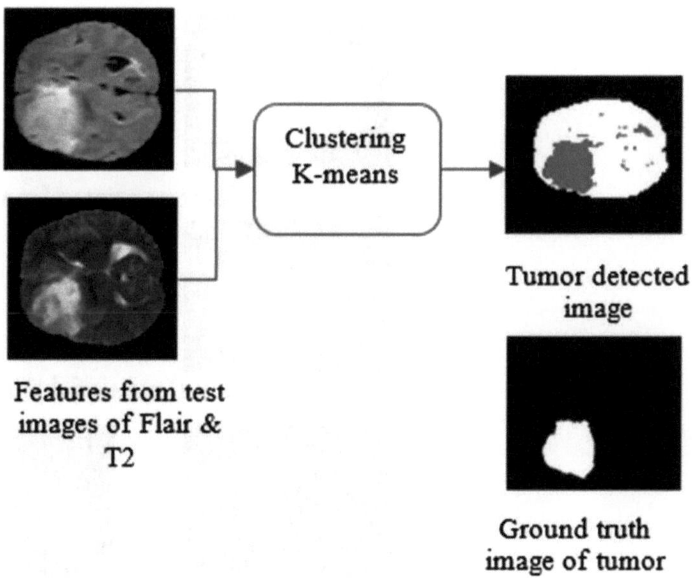

Fig. 6 Brain tumor detection using *K*-means

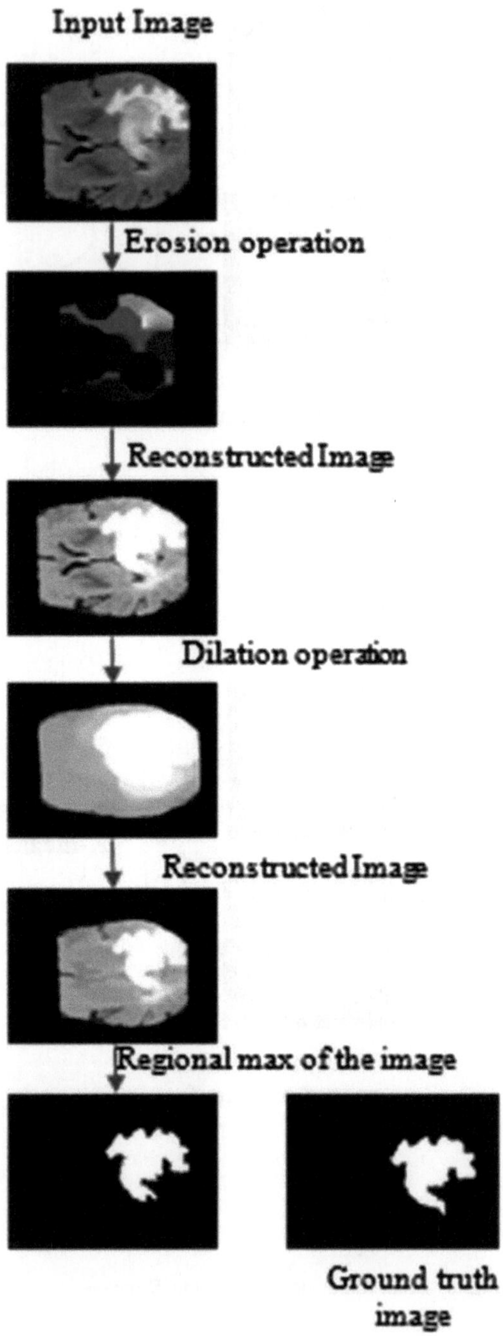

Fig. 7 Brain tumor detection using morphological operators

B. *Comparison of performance of different techniques*

Table 1 indicates the comparison of various techniques used for brain tumor detection and segmentation.

From Table 1, we can say that K-NN, *K*-means, and morphological operators are used for the purpose of tumor detection from MR images. As discussed earlier, the K-NN is a supervised learning technique as it uses a data which also contains the class label information, while other techniques such as *K*-means and morphological operators are used as unsupervised learning technique, because they do not have the class label information. So, these techniques are also used for brain tumor detection purpose.

C. *Comparison of results of different techniques*

We have performed our initial experiment with 100 images of BRATS dataset. We have used K-NN classifier as a supervised method, *K*-means as unsupervised methods and Morphological operators. Results that we have achieved are shown in Table 2.

Table 1 Comparison of different technique

Technique	Learning technique	Out come	Comments
K-NN	Supervised technique	Gives good results with good accuracy	Simple, classes do not have to be linearly separable and faster than SVM but sensitive to noisy data
K-means	Unsupervised Clustering	Accurately separates tumor and non-tumor regions	Minimize sum of the square distance between object and centroid. But difficult to predict K with limited no of the cluster
Morphological operator	Unsupervised technique	Separates foreground and background objects	Simple method for tumor detection. Not work well if foreground and background objects are not differentiated

Table 2 Results of various techniques used

Techniques	Features used	No of images	Results		
			Precision (%)	Sensitivity (%)	Dice (%)
K-NN classifier	Intensity-based features	100	79	68	71
K-means clustering	Intensity-based features	100	68	77	69
Morphological operator	NA	70	60	72	64

6 Conclusion

Accurate diagnosis of brain tumor is a major issue to the world. Using technology, we can give a small contribution that directly helps the experts or doctors for diagnosis of the tumor. Image processing concepts supervised and unsupervised learning techniques are very much helpful for the detection and segmentation of brain tumor region. In this phase, we have done an initial experiment for brain tumor detection which will be further explored with other supervised and unsupervised techniques to improve the results. As a future work, we can also segment the tumor region into subtumoral structure edema, necrotic, enhanced tumor, and non-enhanced tumor region.

References

1. Zhang, Su Ruan, Stephane Lebonvallet, Qingmin Liao and Yuemin Zhu. Kernel Feature Selection to Fuse Multi-spectral MRI Images for Brain Tumor Segmentation. Computer Vision and Image Understanding, 2011, 115(2):256–269.
2. Holland, Eric C. "Progenitor cells and glioma formation." Current opinion in neurology 14.6 (2001): 683–688.
3. Al-Tamimi, Mohammed Sabbih Hamoud, and Ghazali Sulong. "Tumor brain detection through MR images: a review of literature." Journal of Theoretical and Applied Information Technology 62.2 (2014): 387–403.
4. Menze, Bjoern H., et al. "The multimodal brain tumor image segmentation benchmark (BRATS)." IEEE Transactions on Medical Imaging 34.10 (2015): 1993–2024.
5. "Dataset" available at https://www.smir.ch/BRATS/Start2015. Accessed on: July, 17, 2016.
6. Sachdeva, Jainy, et al. "Segmentation, feature extraction, and multiclass brain tumor classification." Journal of digital imaging 26.6 (2013): 1141–1150.
7. Patil, Rajesh C., and A. S. Bhalchandra. "Brain tumour extraction from MRI images using MATLAB." International Journal of Electronics, Communication and Soft Computing Science & Engineering (IJECSCSE) 2.1 (2012): 1.
8. Cordier, Nicolas, et al. "Patch-based segmentation of brain tissues." MICCAI Challenge on Multimodal Brain Tumor Segmentation. IEEE, 2013. 6–17.
9. Chavan, Nikita V., B. D. Jadhav, and P. M. Patil. "Detection and classification of brain tumors." International Journal of Computer Applications 112.8 (2015).
10. Clarke, L. P., et al. "MRI: stability of three supervised segmentation techniques." Magnetic Resonance Imaging 11.1 (1993): 95–106.
11. Verma, Ritu, Sujeet Tiwari, and Naazish Rahim. "Unsupervised MRI Brain Tumor Detection Techniques with Morphological Operations." International Journal of Science, Engineering and Technology Research (IJSETR), Volume 4, Issue 11, November 2015.
12. Meenakshi, S. R., Arpitha B. Mahajanakatti, and Shivakumara Bheemanaik. "Morphological Image Processing Approach Using K-Means Clustering for Detection of Tumor in Brain." International Journal of Science and Research (IJSR). 2319–7064.
13. "morphological operatoe" online available at https://www.cs.auckland.ac.nz/courses/compsci773s1c/lectures/ImageProcessing-html/topic4.htm, Accessed on: November, 15, 2016.
14. Joseph, Rohini Paul, C. Senthil Singh, and M. Manikandan. "Brain tumor MRI image segmentation and detection in image processing." International Journal of Research in Engineering and Technology 3.1 (2014): 1–5.

EEG Signal Classification for Epileptogenic Zone and Seizure Zone

Hardika B. Gabani and Chirag N. Paunwala

Abstract ElectroEncephaloGgram (EEG) signals play an important role to identify epileptic disorders. Epilepsy is a neurological disorder that is an unexpected electrical disruption of the brain, because the activity of nerve cells in the brain becomes disrupted, causing people to experience "seizures." Nowa-day, researcher works and focuses on automatic analysis of EEG signals to classify epilepsy. The EEG signal recording system produces very long data. Thus, the classification of epileptic seizures requires a time-consuming process. This paper proposes a Support Vector Machine (SVM)-based automated seizure classification system using Approximation Entropy (ApEn). ApEn reduces patient data size without loss of information. ApEn is a statistical parameter that measures the amplitude value of an EEG signal current based on its previous amplitude value. In this paper, we measure sensitivity, specificity, and accuracy using SVM classifiers. The overall score as high as 98.62% can be achieved by using the proposed system to distinguish the epilepsy state (seizure class) from the normal state (non-seizure class) using the time domain method.

Keywords ElectroEncephaloGram (EEG) signal · Classification of epilepsy seizure · Approximate entropy (ApEn) · Support Vector Machine (SVM)

1 Introduction

Spikes of electrical activity in different brain regions are determined by the EEG signal, and we can determine the position and relative strength. Epilepsy is a disease that was taken using electricity unusual EEG activity. Fifty million people suffer

H. B. Gabani (✉) · C. N. Paunwala
Dr. R.K. Desai Marg, Opp. Mission Hospital,
Athwalines, Surat, Gujarat, India
e-mail: hardikagabani62@gmail.com

C. N. Paunwala
e-mail: chirag.paunwala@scet.ac.in

R. Kher et al. (eds.), *Proceedings of the International Conference on Intelligent Systems and Signal Processing*, Advances in Intelligent Systems and Computing 671,
https://doi.org/10.1007/978-981-10-6977-2_5

from epilepsy seizures [1] around the world. Factors that may cause epilepsy, brain injury, metabolic disorders, alcohol or drug abuse, brain tumors, and genetic disorders.

A moment in time, epilepsy is not possible to predict in most cases. Classification purposes, required for continuous recording of the EEG. In some cases, EEG recording requires a very large duration of time, perhaps until a week or two. Since the traditional method was boring and slow, an automatic epileptic seizure classification system was developed [2]. The proposed work is an automatic epileptic EEG classification system using SVM and feature extraction and reduction by using approximate entropy (ApEn).

This is shown in the image below, and we will mark the EEG inputs. ApEn technique [3] is used to mark features. Extracted features are then applied to the classifier to classify seizures or non-seizures data (Fig. 1).

Epilepsy since recording began in the mid-1970s, the programmer investigation and discover EEG Signal. Epilepsy and seizures EEG analysis of the placement, the current PC-based test and discuss two issues: Epilepsy seizure classification and EEG analysis. Many feature extraction techniques have been used for the classification of epilepsy seizure. SVM (Support Vector Machine) - based classification system for epilepsy seizure has been proposed by many researchers.

The Lyapunov exponent [4, 5] provides significant details about changes in EEG activity in turn facilitating early detection of epilepsy. The correlation dimension [6] is useful to measure correlation which quantifies complex neural activity of human brain. During epileptic seizure, the value of ApEn has been found to exhibit strong relationship with synchronous discharge of large groups of neurons. The features obtained from complexity analysis and spectral analysis of EEG signals has been effectively used for diagnosis of epilepsy [7]. Recently, the approximate entropy (ApEn) [3] - based methods have been developed for analyzing linear signals for classification of epileptic seizures in epilepsy seizure [8, 9]. The mean frequency parameter of IMFs has been proposed to discriminate well between seizure and seizure-free EEG signals. For classification between healthy and epileptic EEG signals, weighted frequency has been found to be some parameter [10]. Analysis of normal and epileptic seizure EEG signals by using area measured from the trace of

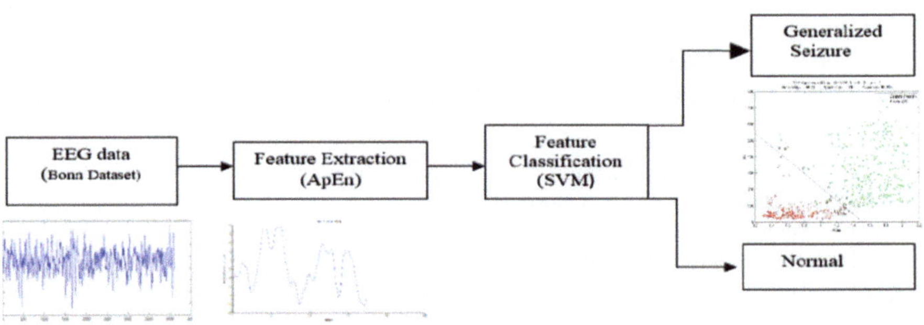

Fig. 1 Block diagram

analytical signal representation of intrinsic mode function (IMF) has been proposed in [11]. The area parameter and mean frequency of IMFs computed using Fourier–Bessel expansion are used for epileptic seizure classification in EEG signals [12]. Also, IMFs of EEG signals have been used for recognition of epileptic seizure [9].

In first experiment, all 100 time series of F and S are taken for training and testing. For frame size 173, entropy values are 690 for each time series, so if we take 100 time series, entropy values would be 69,000 for one class and it is double (138,000) by considering both seizure and non-seizure class. These procedures followed for all four features. Entropy values of both classes S and F for training and testing dataset for all frames are shown in Table 1 (Fig. 2).

For frame size $N = 2048$ and $m = 2$ and $r = 0.9$, gets optimum accuracy for ApEn. From that, we get highest accuracy 95.25% of the feature ApEn with SD for experiment and frame size $N = 2048$ and $m = 2$ and $r = 0.0$, gets optimum accuracy for ApEn. From that, we get highest accuracy 94.25% of the feature ApEn with mean for experiment.

Table 1 Number of entropy value for testing

Sr. no	Time series of F and S	Frame size	No. of entropy values for training	No. of entropy values for testing
1.	200	173	1,38,000	1,38,000
2.	200	256	96,000	96,000
3.	200	512	48,000	48,000
4.	200	1024	24,000	24,000
5.	200	2048	12,000	12,000

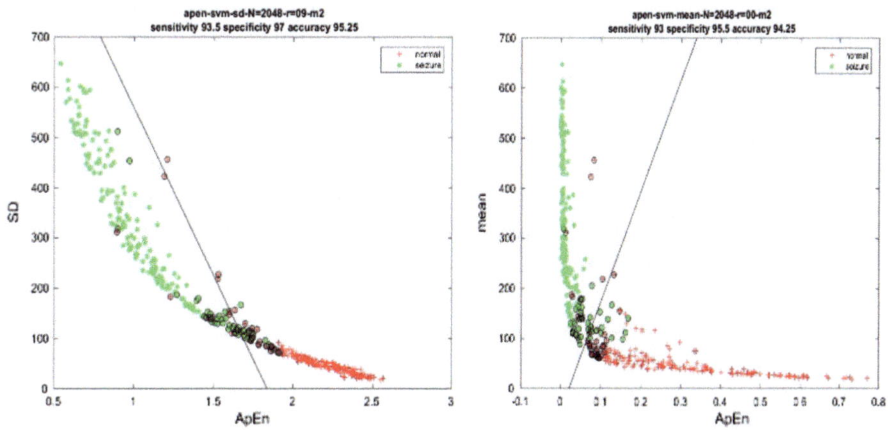

Fig. 2 ApEn for F and S file set for **a** $N = 2048$, $m = 2$, $r = 0.9$ with SD. **b** $N = 2048$, $m = 2$, $r = 0.0$ with mean

Table 2 Entropy value after 70% training

Sr. no	Frame size	Time series of F and S for training	Time series of F and S for testing	No. of entropy values for training
1.	173	140	200	96,600
2.	256	140	200	67,200
3.	512	140	200	33,600
4.	1024	140	200	16,800
5.	2048	140	200	8400

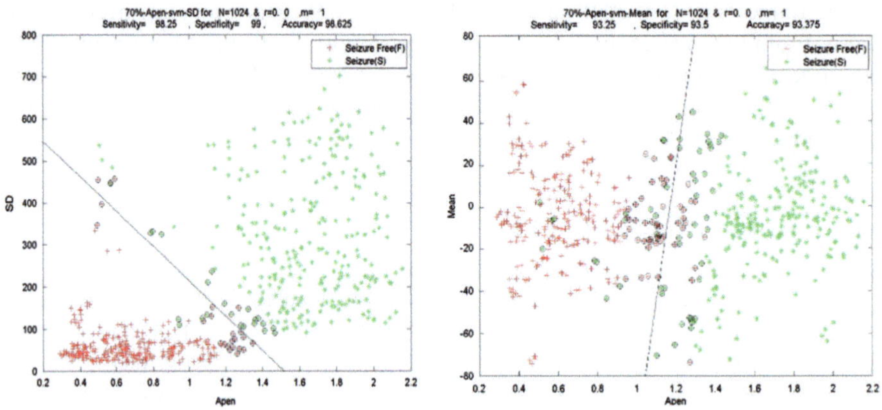

Fig. 3 ApEn for F and S file set after 70% training and testing **a** $N = 1024$, $m = 1$, $r = 0.0$ with SD. **b** $N = 1024$, $m = 1$, $r = 0.0$ with mean

For training purpose, all 70-time-series data for F and S and 100-time-series data are taken for testing. Entropy values of both classes S and F for training and testing dataset for all frames are shown in Table 2.

Figure 3 are for all optimum results of experiment feature dataset as shown in Tables 2. The figure shows the SVM classification for the seizure and normal class using radial basis kernel function, where seizure is denoted by * and normal by +. The line is describing linear classification of the dataset. The o describes wrongly classify data points of opposite class. Here, we measure the performance parameters like, standard deviation (SD) [13], accuracy [14], sensitivity [14], specificity [14], and mean [13].

2 Experimentation Results

In our work, we have extracted the features from the EEG signal and classification is done using SVM classifier [15] in two class, i.e., seizure-free and seizure patient data. ApEn values are measured in form of *m, r,* and *N*. The values of *m, r,* and *N* are as follows [3]:

1. Number of samples (*m*) = 1, 2, 3;
2. Normalization ratio (*r*) = 0–90% of SD of the data sequence in increments of 10%;
3. Frame size (*N*) = 173, 256, 512, 1024, and 2048.

Approximation entropy is extracted along with SD and mean. The randomness of EEG signal was extracted in the features, based on different size of frame (*N*), number of samples values (*m*), and normalized ratio (*k*). From the set of features, ApEn with SD and mean are used for classification using the SVM classifier.

We have used BONN dataset for EEG signals which is publicly available online and described by Andrzejak et al. [16]. The EEG dataset contains both seizures and non-seizures. The Bonn dataset consist of five subsets (Z, O, N, F, and S) each containing 100 single-channel EEG signals, each signal of 23.6 s in duration with the sampling rate of 173.61 Hz.

EEG recordings of five healthy volunteers with eyes open (Z) and closed (O) have been recorded on the surface, using standard electrode placement scheme. These two are recorded in seizure-free intervals from five patients in the epileptogenic zone (F-seizure free) and from the hippocampal formation of the opposite hemisphere of the brain (N-seizure free). The set S contained seizures signal which gives an ictal activity by using the same 128-channel amplifier system with an average common reference, and all EEG signals are recorded. In the proposed work, classification of the F (Seizure-free class) and S (Seizure class) is done by using approximate entropy (ApEn) feature extraction and reduction and SVM as classifier.

For all 100 EEG datasets, 70 datasets are used for training and the others are used for testing using SVM classifier. SVM classifier is used to classify unknown data properly. The highest accuracy is 98.625% for the feature set ApEn with SD for frame size $N = 1024$, sample value $m = 1$, and normalization ratio $r = 0.0$. In the proposed method, accuracy is achieved up to 98.625% for the feature set ApEn with SD. For training and testing purpose, we get different accuracy, sensitivity, and specificity as shown in Fig. 4.

As shown in Table 3, all the papers are worked on Bonn dataset and they achieved maximum accuracy 98.27%. In the proposed method, accuracy up to 98.625% is achieved for the feature set ApEn with SD.

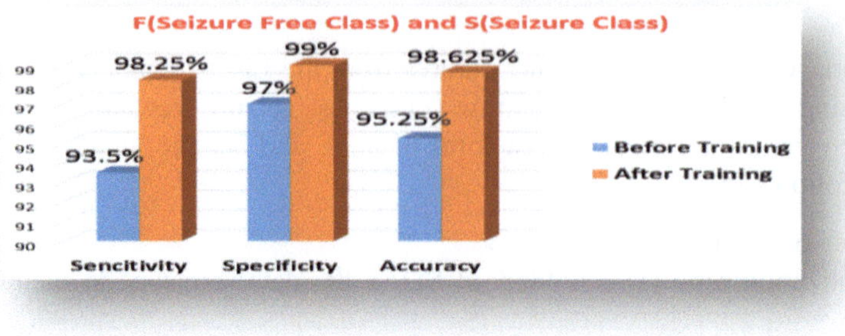

Fig. 4 Classification accuracy, sensitivity, and specificity at before training and after 50% training for F (seizure free) and S (seizure class)

Table 3 Comparison of methodology for same dataset

Methodology	Subset	Classification accuracy (%)
(KNN) classifier [17]	F and S	93
Clustering and support vector machine (SVM) classifier [18]	F and S	93.91
Elman neural network [19]	S and F	93.33
Multiclass SVM [20]	F and S	97.00
Thresholding [21]	F and S	96.00
ANN [22]	F and S	98.27

3 Conclusion

We have extracted the features from the EEG signal, and classification done using SVM classifier in to two class seizure and normal. Approximation entropy is extracted along with SD and mean. The randomness of EEG signal was extracted in the features, based on different size of frame (*N*), number of samples values (*m*), and normalized ratio (*r*). From the set of features, ApEn with SD and ApEn with mean were used for classification using the SVM classifier. The highest classification accuracy is 98.625% for F and S class. The main classes of the F and S that will be an classes of the seizure data and non-seizure data.

References

1. Klaus Lehnertz, Florian Mormann, Thomas Kreuz, Ralph G. Andrzejak, Christoph Rieke, Peter David, And Christian E. Elger, "Seizure prediction by nonlinear EEG analysis. IEE Eng Med BiolMag", article in IEEE engineering in medicine and biology magazine, Research gate January 2003.

2. N. Mc Grogan (1999). Neural network detection of epileptic seizures in the electroencephalogram, [Online]. Available: http://www.new.ox.ac.uk/~nmcgroga/work/transfer
3. Vairavan Srinivasan, "Approximate Entropy-Based Epileptic EEG Detection Using Artificial Neural Networks" IEEE Transactions On Information Technology In Biomedicine, Vol. 11, No. 3, May 2007.
4. N.F. Güler, E.D. Übeyli, I. Güler, Recurrent neural networks employing Lyapunov exponents for EEG signal classification, Expert Systems with Applications 29 (3) (2005) 506–514, October 5. Foster, I., Kesselman, C., Nick, J., Tuecke, S.: The Physiology of the Grid: an Open Grid Services Architecture for Distributed Systems Integration. Technical report, Global Grid Forum (2002).
5. E.D. Übeyli, Lyapunov exponents/probabilistic neural networks for analysis of EEG signals, Expert Systems with Applications 37 (2) (2010), pp. 985–992, March.
6. A. Accardo, M. Affinito, M. Carrozzi, F. Bouquet, Use of the fractal dimension for the analysis of electroencephalographic time series, Biological Cybernetics 77 (5) (1997), pp. 339–350, November 2.
7. S.F. Liang, H.C. Wang, W.L. Chang, Combination of EEG complexity and spectral analysis for epilepsy diagnosis and seizure detection, EURASIP Journal on Advances in Signal Processing (2010), vol. 2010, Article ID 853434.
8. R.B. Pachori, Discrimination between ictal and seizure-free EEG signals using empirical mode decomposition, Research Letters in Signal Processing 2008 (2008), Article ID 293056.
9. S. Li, et al., Feature extraction and recognition of ictal EEG using EMD and SVM, Computers in Biology and Medicine 43(7) (2013), pp. 807–816.
10. R.J. Oweis, E.W. Abdulhay, Seizure classification in EEG signals utilizing Hilbert–Huang transform, Bio Medical Engineering On Line 10 (2011) 38, December.
11. R.B. Pachori, V. Bajaj, Analysis of normal and epileptic seizure EEG signals using empirical mode decomposition, Computer Methods and Programs in Biomedicine 104 (3) (2011), pp. 373–381, December 7.
12. V. Bajaj, R.B. Pachori, EEG signal classification using empirical mode decomposition and support vector machine, in: Proceedings International Conference on Soft Computing for Problem Solving, AISC 131, 20–22 December, 2011, Roorkee, India, 2011, pp. 623–635.
13. Ram Bilas Pachori, Shivnarayan Patidar, "Epileptic seizure classification in EEG signals using second-order difference plot of intrinsic mode function," in Elsevier Transactions on Computer methods and programming in biometric(2014), vol. 9, pp. 494–502.
14. Varun Bajaj and Ram Bilas Pachori, "Classification of Seizure and Non seizure EEG Signals Using Empirical Mode Decomposition", IEEE transactions on information technology in biomedicine (2012), vol. 16, no. 6, November.
15. Nicoletta Nicolaou, Julius Georgiou, "Detection of epileptic electroencephalogram based on Permutation Entropy and Support Vector Machines," in Elsevier Transactions on Expert Systems with Applications (2012), vol. 8, pp. 202–209.
16. R.G. Andrzejak, et al., Indications of nonlinear deterministic and finite-dimensional structures in time series of brain electrical activity: dependence on recording region and brain state, Physical Review E (2001), vol. 64, Article ID 061907.
17. Guohun Zhu, Yan Li, Peng (Paul) Wen, "Epileptic seizure detection in EEGs signals using a fast weighted horizontal visibility algorithm," in Elsevier Transactions on Computer method in programming biometric(2014), vol. 12, pp. 64–75.
18. Siulya, Yan Li a, Peng (Paul) Wenb, "Clustering technique-based least square support vector machine for EEG signal classification," in Elsevier Transactions on Computer method in programming biometric(2011), vol. 15, pp. 358–372.
19. Giorgos Giannakakis, Vangelis Sakkalis, Matthew Pediaditis, And Manolis Tsiknakis "Methods For Seizure Detection And Prediction: An Overview" Springer Science+business Media New York 2014.
20. A.S. Muthanantha Murugavel And S. Ramakrishnan "Multi-class SVM For EEG Signal Classification Using Wavelet Based Approximate Entropy" Institute For Computer Sciences, Social Informatics And Telecommunications Engineering 2012.

21. U. Rajendra Acharya "Automatic Detection Of Epileptic Eeg Signals Using Higher Order Cumulant Features" International Journal Of Neural Systems, Vol. 21, No. 5 (2011).
22. Ling Guo, Daniel Rivero, Alejandro Pazos "Epileptic Seizure Detection Using Multiwavelet Transform Based Approximate Entropy And Artificial Neural Networks" Journal Of Neuroscience Methods 193 (2010) 156–163.

Design and Development of DAQ and Processing System for FMCW RADAR

Rucha Joshi, Amit Patel, Keyur Mahant and Hiren Mewada

Abstract Frequency-modulated continuous wave radar is generally used for calculating range and speed of any target. A research paper is intended to develop back-end electronics for frequency-modulated continuous wave (FMCW) radar with optimum usage of field-programmable gate array (FPGA)-based processing unit. Sampling of the received IQ (in-phase and quadrature-phase) signals is done at speed of 500 KSPS using 12-bit ADC128S102. Acquired data is further stored into internal memory of FPGA for post-processing. Zero-crossing algorithm is implemented on the received signal to obtain the beat frequency of the same. Moreover, stored data and processed data are transferred to the computer using SPI protocol. LabVIEW-based simulator is also developed to process the received data and retrieve the beat frequency from it.

Keywords Frequency-modulated continuous wave (FMCW) · Field-programmable gate array (FPGA) · Zero crossing

R. Joshi (✉)
Electronics and Communication Engineering, Charotar University of Science and Technology, Changa, India
e-mail: joshirucha27@gmail.com

A. Patel · K. Mahant · H. Mewada
Charusat Space Research and Technology Center, Charotar University of Science and Technology, Changa, India
e-mail: amitvpatel.ec@charusat.ac.in

K. Mahant
e-mail: keyurmahant.ec@charusat.ac.in

H. Mewada
e-mail: hirenmewada.ec@charusat.ac.in

R. Kher et al. (eds.), *Proceedings of the International Conference on Intelligent Systems and Signal Processing*, Advances in Intelligent Systems and Computing 671,
https://doi.org/10.1007/978-981-10-6977-2_6

1 Introduction

This paper describes design and development of the back-end electronics of FMCW radar using FPGA and LabVIEW-based simulator development. In RF signal processing, phase and frequency information is required to determine distance, for that FMCW radar is used in this approach, so basic principle of FMCW radar is distance measurement which is accomplished by calculating the differences in phase or frequency between the transmitted and received signal. Some signal processing algorithms are implemented to determine frequency and phase of signal. Fast Fourier transform (FFT), zero-crossing algorithm and Hilbert transformation are more likely used algorithms for phase and frequency estimation [1]. Here, zero-crossing algorithm is used as it is more accurate than other algorithms because of its high accuracy and low processing latency, and it is suitable for real-time applications. Furthermore, compatibility is increased as it gives 3 (past, current and previous)-point measurement of signal [9].

FPGA, microprocessors and DSP controllers are three main families of digital device technology which are generally used to design RF back-end electronics. FPGA is programmable logic device which allows parallel processing with high speed and memory resources on FPGA is higher compared to other processing devices [2, 3]. Thus, FPGA is utilized here. Moreover, the same FPGA is utilized to configure the frequency synthesizer and for transferring data to computer. Here, VHDL is chosen as software design language because it is standard language used for describing digital system. Xilinx is used for generating VHDL code.

2 System Block Configuration of Back-End Electronics

In this paper, we are focusing on FPGA implementation of back-end processing; back-end processing consists of I/Q demodulator, ADC and memory. In proposed system of back-end electronics, data is obtained from IQ demodulator [4]. Received IQ signals are digitized using 12-bit ADC128S102 [5] with the sampling rate of 500 KSPS. Furthermore, acquired data are stored into the 16-k (2048 $\times$ 8) block memory for the post-processing.

Here, instrument takes the data for 4096 μs, and it is stored into the block memory of FPGA. Further, the stored data is processed for calculating the beat frequency. FPGA takes 4096 μs to process the data and obtain the beat frequency. Once the data is processed, the data is transferred to computer using RS232 serial interface with the baud rate of 9600 (Fig. 1). FPGA takes 2.12 s to transfer the stored data to computer. In the computer, data is received using LabVIEW and stored into xls file which can be used for post-processing. In LabVIEW, zero-crossing algorithm is implemented to obtain beat frequency and verify the result of FPGA.

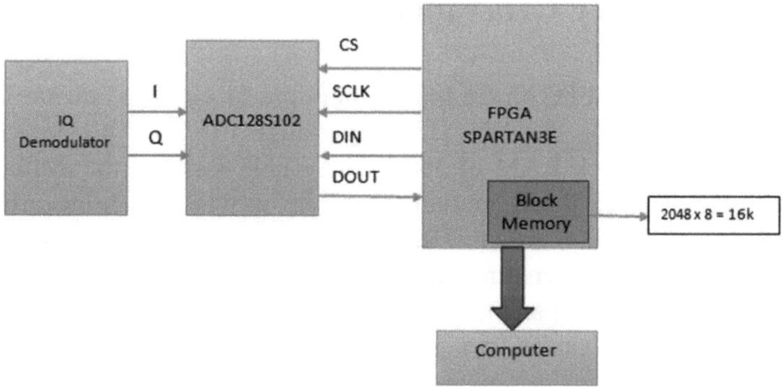

Fig. 1 Block diagram of back-end electronics

Here, ADC128s102 is selected on the basis of beat frequency. Spartan3E (XC3S500E) FPGA [6] board is used for configuration, and 16-k bits of block memory are used for data storage.

3 Calculation of Beat Frequency

$$F_b = T_R \times f'$$

where,

$$T_R = \frac{2R}{C}, f' = \frac{f_1 - f_2}{t}$$

T_R is time taken by signal to travel back from object, R is measurement of distance to the target, and C is speed of light in medium f_1 and f_2 are sweep frequency and t is sweep time [7].

R 1 m
C 3×10^8 m/s
t 20 ms
f_1 36 GHz
f_2 26 GHz

Here, the obtained value of the beat frequency as per the given specification is ~5 kHz, and as per the Nyquist criterion for sampling, "signal must be sampled at least twice as often as its highest frequency to enable correct reconstruction of the signal". As per that chosen analog-to-digital converter (ADC) is having the sampling rate of 500 KSPS–1 MSPS, which provides enough sampling rate to reconstruction the original signal from the taken samples.

4 Internal Memory Generation

All Spartan3 generation FPGAs are having multiple block RAM memories, which can be utilized for storing the data. Here, Spartan3E (XC3S500E) device is used, which contains total block RAM of 360-k bits. Table 1 shows the available block RAM device. Each block RAM contains 18-k bits, in which 16-k bits are allocated to data storage and 2-k bits are allocated to parity which is an optional.

Size of the generated internal memory is 16 k (2048 × 8) which is used to store digitized data of ADC. As per the project requirements and beat frequency specification, 16-k memory space is enough to store data.

5 Developed Hardware for Back-End Processing

The actual developed hardware for back-end processing of FMCW RADAR using FPGA is as shown in Fig. 2. Developed board is designed in such a way so it can be sandwiched on the available Papilio-One FPGA board.

Here, the green board is the developed board which is sandwiched on the available red Papilio-One FPGA board. Developed board contains 12-bit, single channel analog-to-digital converter with the conversion throughput rates of

Table 1 Available block memory of device

FPGA device	RAM columns	RAM blocks per column	Total RAM k bits
XC3S500E	2	10	360 k

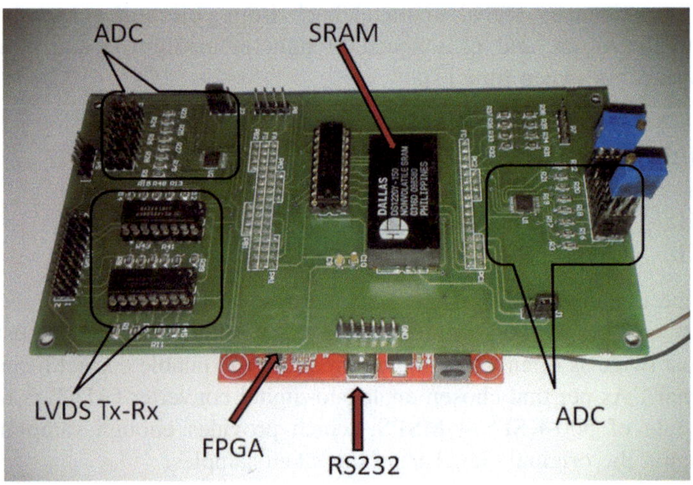

Fig. 2 Developed hardware for back-end processing

1.25 MSPS for processing data coming from IQ demodulator. Hardware also contains 12-bit, 8-channel analog-to-digital converter with the conversion throughput rates of 1 MSPS for monitoring the health parameter of whole system like temperature monitoring of DC–DC converters (ADP3336), VCO, LNA and SRAM. It has high-speed, low-power CMOS SRAM organized as 32,768 × 8 bits that operates on single 5 V power supply, which can be utilized in future for storing the data coming from the IQ demodulator. LVDS transmitter and receiver are used for transferring the data to control unit with high speed. The size of the developed board is about 121 mm × 62 mm [8].

6 Experimental Results

6.1 Simulation Results of Block Memory (2 k × 8)

Figure 3 shows utilized the available block memory of the XC3S500E, Spartan 3E FPGA. Available block memory into the FPGA is of 360 k. As per the project requirement we are suppose to take 2 k samples of the incoming data, which is sufficient to determine the beat frequency of the received signal. So, we have utilized 16-k (2048 × 8) bit of memory out of 360 k to store the acquired data. Result of the generated block memory is shown in Fig. 3. Here, in Fig. 3, data of incoming signal with the frequency of 5 kHz sampled with the sampling frequency of 500 KSPS is shown. Signal is having the input voltage of 2 V (p–p) and sample with 500 KSPS, so there is a change between two consecutive digital value stored into the block memory is of 32 in decimal. In future if the requirement arises to acquire the data for the longer time or to acquire the data with the higher sampling frequency, then SRAM or DRAM may be utilized to store the acquired data.

	0	1	2	3	4
0x0	000000000000	000000100000	000001000000	000001100000	00010000000
0x10	00100000000	001000100000	001001000000	001001100000	00101000000
0x20	01000000000	010000100000	010001000000	010001100000	01001000000
0x30	01100000000	011000100000	011001000000	011000100000	01100000000
0x40	01001000000	010001100000	010001000000	010000100000	01000000000
0x50	00101000000	001001100000	001001000000	001000100000	00100000000
0x60	00001000000	000001100000	000001000000	000000100000	00000000000
0x70	00011000000	000110100000	000111000000	000111100000	00100000000
0x80	00111000000	001110100000	001111000000	001111100000	01000000000
0x90	01011000000	010110100000	010111000000	010111100000	01100000000
0xA0	01010000000	010011100000	010011000000	010010100000	01001000000
0xB0	00110000000	001011100000	001011000000	001010100000	00101000000
0xC0	00010000000	000011100000	000011000000	000010100000	00001000000
0xD0	00010000000	000100100000	000101000000	000101100000	00011000000
0xE0	00110000000	001100100000	001101000000	001101100000	00111000000
0xF0	01010000000	010100100000	010101000000	010101100000	01011000000
0x100	01011000000	010101100000	010101000000	010100100000	01010000000

Fig. 3 Result of block memory

6.2 Results of ADC on MDO

Figure 4a, b shows the hardware results of digitized data obtained from ADC. Control signals required for interfacing ADC with FPGA are shown in Fig. 4a, b, which are CLK, Chip Select (CS), SCLK and DOUT. Here, CLK defines the operating frequency of the utilized FPGA, CS is used for the selection of the ADC chip, and SCLK defines the conversion rate of the ADC. Utilized ADC is of 12-bit, 8-channel; for the selection of the channel, we are supposed to provide the DIN to the ADC, and digitized output after conversion is received from the DOUT which is shown in Fig. 4a, b. DIN is provided on the 3–5th falling edge of the SCLK, and digitized data is received on 5–16th the falling edge of SCLK.

6.3 Simulation and Hardware Results of Zero-Crossing Detection

Figure 5a, b shows that data coming from I/Q demodulator is digitized using ADC128s102, acquired and processed using XC3S500E FPGA. Zero-crossing algorithm is implemented on acquired data from ADC in XC3S500E FPGA. Result

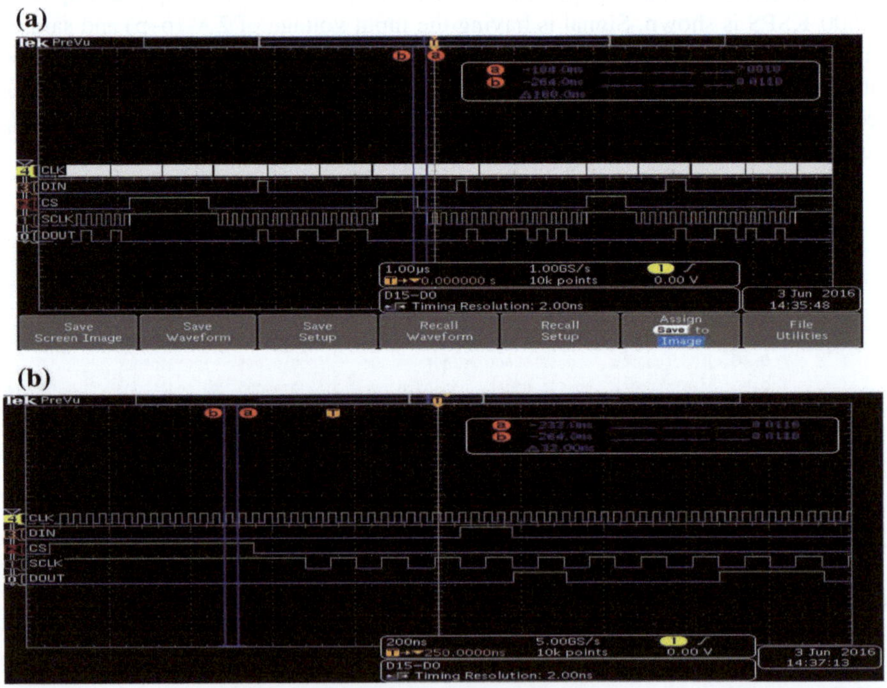

Fig. 4 **a** Result of data acquired from ADC and **b** Result of data acquired from ADC

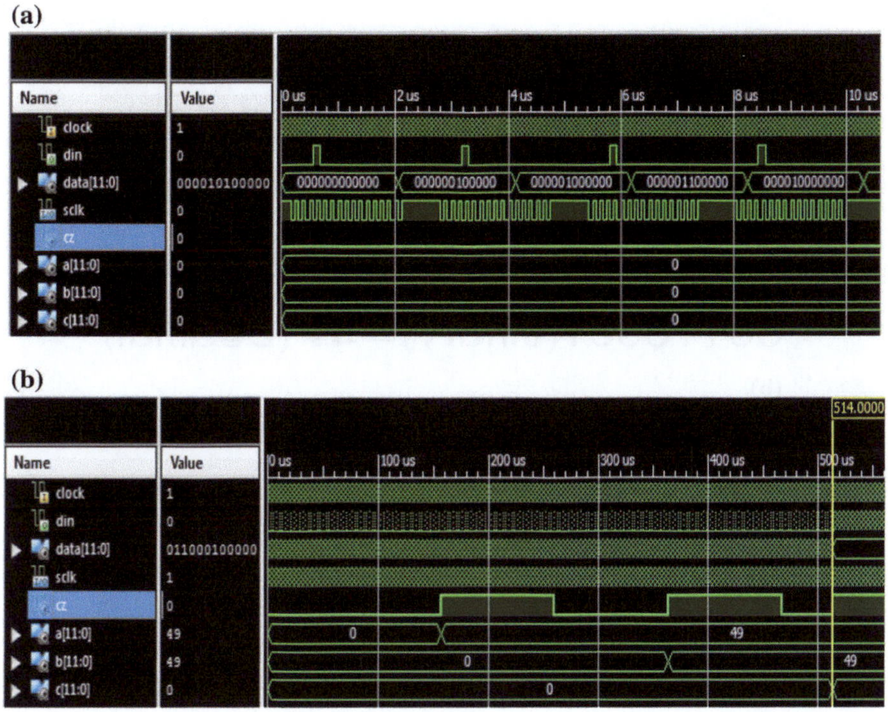

Fig. 5 **a** Simulation result of zero-crossing algorithm and **b** Received zero-cross output on simulation

of the zero crossing is stored into the variables a, b and c as shown in Fig. 5b. Count value of variables a, b and c is 49 which indicates that the beat frequency of the received signal is of ~ 5 kHz with sampling rate of 500 KSPS.

To determine the beat frequency, data acquired from the ADC128s102 needs to be processed. Zero-crossing algorithm has been implemented into the FPGA to determine the beat frequency. Result of the zero-crossing algorithm for the input frequency of 5 and 2.5 kHz with the input voltage of 2 V (p–p) is shown into Fig. 6a, b. Here, the acquisition is done with the sampling frequency of 500 KSPS (2 μs) for the input signal frequency of 5 and 2.5 kHz. As per the sampling rate and the input signal frequency successfully received count values between two zero-cross points are 49 and 99 respectively. By multiplying the count values with the sampling time, time period of the beat signal or the input signal has been successfully achieved which is shown in Fig. 6a, b.

(a)

00110001(Binary)= 49 (Decimal)

(b)

01100011(Binary) = 99(Decimal)

Fig. 6 **a** Result of zero-crossing algorithm for 5 kHz and **b** Result of zero-crossing algorithms for 2.5 kHz

7 Timing Diagram

Timing diagram of the data acquisition and transfer is shown in Fig. 7a, b. In Fig. 7a, it is shown that once the instrument received the trigger signal from the user to turn on the system, acquisition of the data is stared and acquired data is stored into the internal memory of FPGA. Once the process of data acquisition, processing and storing is completed, the data is transferred to the computer using serial interface, which is shown in Fig. 7b.

8 Results Obtained in LabVIEW Simulator

Once the data from the IQ demodulator is digitized by ADC, processed by FPGA and stored into the block memory of the utilized FPGA, it is transferred for the purpose of post-processing and verification of zero-crossing algorithm on the different platforms like LabVIEW and MATLAB. Here, the acquisition is done with

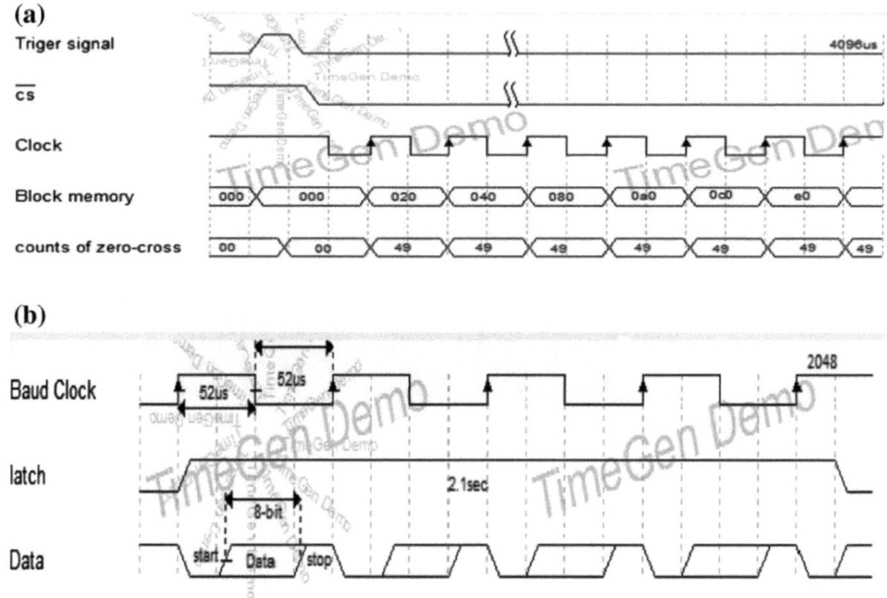

Fig. 7 **a** Timing diagram of data acquisition process and **b** Timing diagram of data transfer process

the sampling frequency of 37 kHz (27 μs), because here the acquisition and transfer process is done at the same time.

As per the sampling rate of 37 KSPS and the input frequency of 1, 2.5 and 5 kHz successfully receive count values are of 37, 14 and 7 respectively, which is shown in the Fig. 8a, b, c with the red highlighted callouts shapes. Detail of the sampling frequency and input signal frequency is depicted into Table 2.

9 FPGA Power and Resources Utilization

Tables 3 and 4 describe the FPGA resources and power utilization. Table 2 shows the details of FPGA resources utilization, number of slice FFs, LUTs and slices, like 16% of 4 input LUTs are utilized out of total available 4 input LUTs. Table 4 shows the power consumption of the FPGA, which is approximately 80 mW.

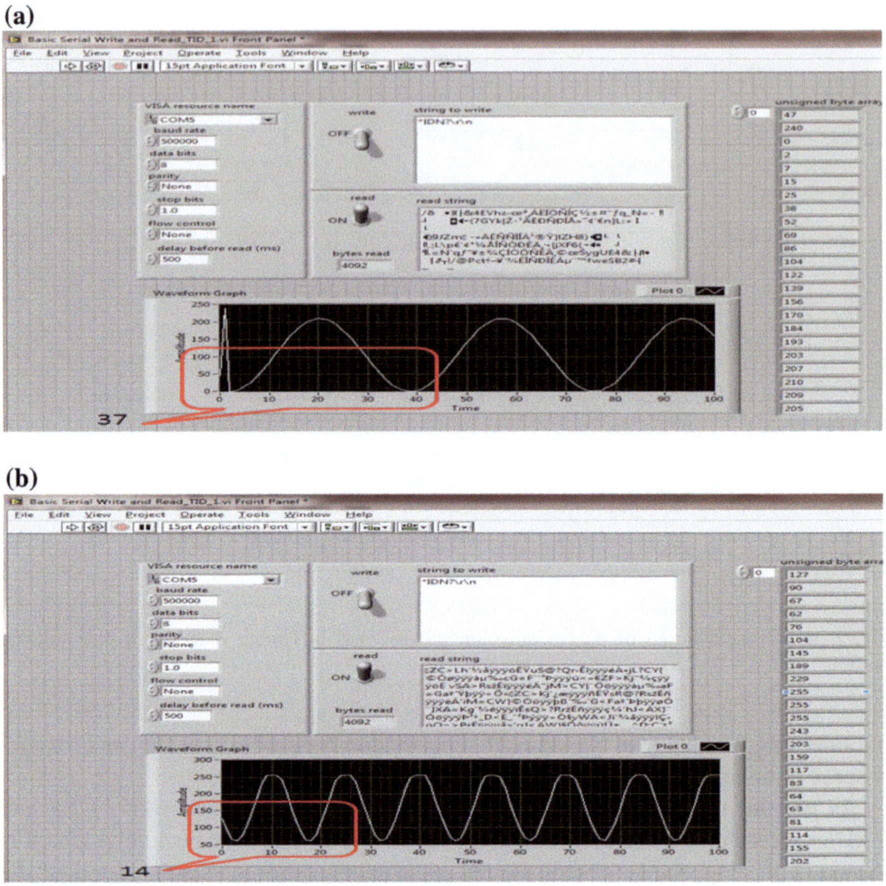

Fig. 8 **a** Result of transferred data on LabVIEW for input signal of 1 kHz and **b** Result of transferred data on LabVIEW for input signal of 2.5 kHz

Table 2 Count values as per the input signal frequency

Sr. no	Input signal frequency (kHz)	Input signal time period (ms)	Sampling time (μs)	Sampling frequency (kHz)	Count value
1	1	1	27	~37	37
2	2.5	0.4	27	~37	14

Table 3 Resources utilization summary

Logic utilization	Used	Available	Utilization (%)
Number of slice flip-flops	220	9312	2
Number of 4 input LUTs	1563	9312	16
Number of occupied slices	1753	9312	18
Number of bonded IOBs	28	66	41
Number of BUFGMUXs	3	24	12
Average fan-out of non-clock nets	8.17	–	–

Table 4 FPGA power consumption summary

Types of measurement	Power (W)
Quiescent	0.076
Dynamic	0.003
Total	0.080

10 Conclusion

FPGA-based back-end electronics can be designed for the acquisition, processing and transferring the data to the computers of RF-based receiver system. Here, we have utilized the same FPGA for other purposes such as configuration of frequency synthesizer, generating the tuning voltage of the free running VCO and controlling the SPDT switches. High-speed data processing can also be developed for data acquisition by using high-end FPGAs. Here, we have also implemented the zero-crossing algorithm to obtain phase and frequency information of beat signal, and distance information can be derived from the same.

Moreover, LabVIEW and MATLAB can be used for storing and processing the data. Obtained beat frequency from the FPGA is also verified into the developed MATLAB- and LabVIEW-based simulators. So, by using such software and processing unit like FPGA, a high-speed, highly reliable and powerful data acquisition and processing system can be designed.

11 Future Scope

Data acquisition can also be implemented using high-speed ADC of 80MSPS/100MSPS. SRAM/DRAM can be utilized for data storage. Also different signal processing algorithms can be utilized for obtaining the beat frequency of the reflected signal.

References

1. Yizheng Liao, "Phase and Frequency Estimation: High-Accuracy And Low-Complexity Techniques".
2. Vaughn Betz, Jonathan Rose, "Using Architectural "Families" to Increase FPGA Speed and Density".
3. Rodger H. Hosking "Use FPGA resources to boost radar system performance", RF Design Defense electronics.rfdesign.com/mag/510RFDSF2.pdf.
4. James J. Alter, Jeffrey O. Coleman, RADAR Handbook, Ch-25 Radar Digital Signal Processing, Naval Research Laboratory 1010505.
5. Texas Instruments, Analog to Digital Converters Available: www.ti.com/lit/ds/symlink/adc128s102.pdf.
6. XILINX. Spartan-3 Generation FPGA User Guide: Extended Spatan-3E, and Spartan-3 FPGA Families UG331,V1.8.available: http://www.xilinx.com/support/documentation/userGuides/ug331.pdf.
7. Yan Wu, J.P.M.G. Linnartz, "Detection Performance Improvement of FMCW Radar Using Frequency Shift". IEEE MAY 10–11, 2011.
8. Keyur K. Mahant, Amit V. Patel, Alpesh Vala, Riddhi Goswami "FPGA based temperature control and monitoring system for X-ray measurement instrument".
9. R.W.W all "simple methods for detecting zero crossing". IEEE, proceedings of the 29th annual conference of IEEE.

Side-Lobe Interference Minimization in MC-CDMA Based Cognitive Radio

Bhumika D. Kavaiya and Vinay M. Thumar

Abstract Cognitive radio system is a promising technique to pull off spectrum utilization requirements. Key feature is high data rate wireless communication being offered to the unlicensed user on the basis of request. MC-CDMA is one of the promising multi-carrier transmission methods to achieve this in cognitive radio (CR) system. It is quite possible that transmission of primary and secondary users may occur in an adjacent spectrum, interference is caused by secondary users to the primary user (PU) band. Fourier-based transmission of such systems is having many rewards, but there this method generates side lobes which can cause interference in PU band. This paper focuses on minimizing side-lobe interference caused by secondary users to the PU band. Power allocation-based methods are verified to minimize the interference. Interference is minimized with the use of water-filling-based power allocation method as compared to uniform power allocation method. This objective is accomplishment and explained in this paper.

Keywords Cognitive radio · MC-CDMA · Power allocation · Side-lobe interference · Water-Filling

1 Introduction

The era of wireless communication is demanding for high data rate applications to the wide range of users. This leads to deficiency of spectrum availability. To achieve high data rate over traditional single carrier transmission causes ISI due to multipath and may increase receiver complexity. Above reason leads to need of mechanism which can increase spectrum efficiency. This has initiated the idea of cognitive radio (CR), where unlicensed users are allowed to use a portion of spectra when licensed users are disabled, in a way that this temporary allocation should not

B. D. Kavaiya (✉) · V. M. Thumar
Department of Electronics and Communication Engineering,
DDU, Nadiad 387001, Gujarat, India
e-mail: kavaiya_b@yahoo.co.in

R. Kher et al. (eds.), *Proceedings of the International Conference on Intelligent Systems and Signal Processing*, Advances in Intelligent Systems and Computing 671,
https://doi.org/10.1007/978-981-10-6977-2_7

affect the primary transmission. Since like have been an increase among demand for greater information degree transmission, the structures are integrating the multi-carrier transmission techniques. The multi-carrier transmission approach employs the by Fourier transform mechanism. The entire bandwidth will remain cut up between numbers of tributary channels. As a result, a high information stream will be transformed among numbers of low rate channels, who are transmitted over distinct sub-channels. The splitting concerning high data rate channel of a range on lower information rate channels causes expand of symbol duration, whereas a decrease rate parallel sub-carriers reduce the relative amount of dispersion in period prompted via multipath delay spread. Various multi-carrier transmission systems are developed and used by many Telecom standards to fulfil requirements. MC-CDMA is one of the multi-carrier transmission techniques. MC-CDMA gives settlements of OFDM and CDMA. Earlier MC-CDMA was proposed by Linnartz in 1993 [1]. MC-CDMA has provided multi-carrier transmission with low complexity of receiver structure. In cognitive environment also MC-CDMA is providing suitability by fulfilling the criteria such as no hardware modification required to upgrade system, spectrum shaping and compatibility to adaptive modulation technique. It also provides advantage of CDMA such as diversity and spreading technique. MC-CDMA has variants which depend on spreading done in time domain or frequency domain. MC-CDMA uses frequency domain spreading, which has advantages of collecting energy in frequency domain at receiver side. Efficiency of this system also depends upon spreading code selection. Based upon the application in uplink or downlink, Walsh or PN codes are widely chosen for the system. In cognitive radio system, secondary user may get non-contiguous sub-carriers and thus the system transmission becomes non-contiguous. In this case, Walsh code lost its orthogonal property if number of sub-carriers are not in terms of 2^N, Where N indicates number of sub-carriers. The system performance gets poor in such environment. To overcome from this problem, Carrier Interferometry (CI) code is chosen. CI codes are introduced to mitigate the effect of losing orthogonal property in NC transmission. They are complex-valued code having any of the length. It also has capabilities to accommodate more number of users by providing extra phase shifting in code, no other modification required. In this work, CI code is chosen for NC-MC-CDMA system-based cognitive radio. Similar to NC-OFDM [2, 3] environment, NC-MC-CDMA also generates side lobes which may cause interference to nearer PU band [4, 5], which should be in limited form for any efficient CR system.

The paper is methodical as follows: Sect. 2 is about cognitive radio, Sect. 3 provides system model. Section 4 detailed the MC-CDMA system, Sect. 5 entitled power allocation and interference management. While Sect. 6 gives simulation results and Sect. 7 provides the conclusion.

2 Cognitive Radio

Cognitive radio is a smart Wi-Fi communication system so much is aware on its surrounding environment. It utilizes the methodology on understanding by building in conformity with learn beyond the environment yet inure its inner states according to statistical variations through two fundamental objectives, relatively reliable communication whenever yet any place needed and efficient utilization of the spectrum. The solution for enabling technology on dynamic spectrum gets right of entry to cognitive radio (CR) technology. CR provides the ability in accordance with portion the wireless channel including the unlicensed users of an opportunistic way. CRs are envisioned to stay in a position according to furnish the high bandwidth according to mobile users via heterogeneous wireless architectures or dynamic spectrum access techniques. The networked CRs also impose several challenges fit in imitation of the huge extent of reachable spectrum as well namely numerous QoS necessities about applications.

To share the spectrum with licensed customers except disturbing them, and associate the numerous quality of situation need of applications, every CR user into a CR Network have to determine the part concerning spectrum that is available—called spectrum sensing. To pick out the best accessible channel—referred to as spectrum decision. Coordinate get admission to this channel together with other users called spectrum distribution by means of following CR cycle shown in Fig. 1 [6].

3 System Model

Cognitive radio-based system model considering for this work is shown in Fig. 2. Single primary user and multiple secondary users are considered. All SUs transmissions are controlled by centralized control system. Channel condition between

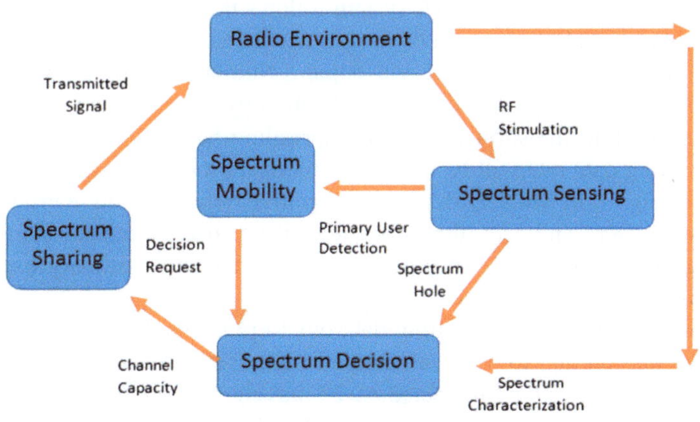

Fig. 1 Cognitive radio [6]

Fig. 2 System model

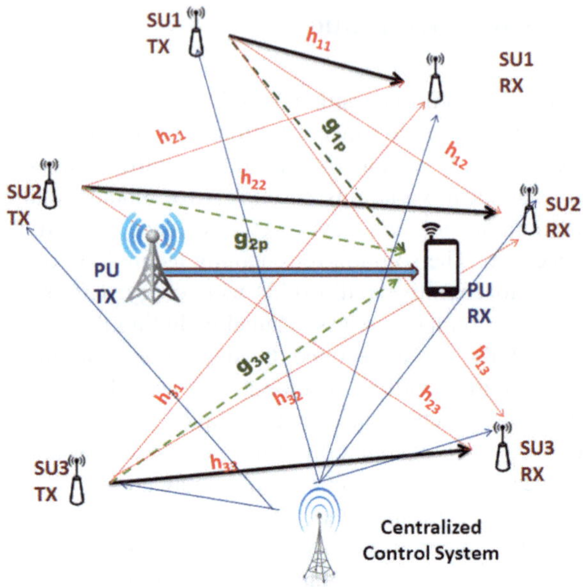

SU transmitter and SU receiver given by h_{kl}, intersecondary user interference considered by the channel gain of SU transmitter to near by SU receiver h_{kl}, Where l indicate path. Interference from SUs to PU gets affected via channel gain g_{kp}, where gain is from kth secondary user to pth PU band. Here, single PU with single band is considered.

4 Modelling of CI/NC-MC-CDMA

Implementation of NC-MC-CDMA system follows the NC-OFDM procedure. The difference is no serial to parallel conversion needed at input side and data bits are converted into number of copies via copier. Number of copier depends upon number of sub-carrier N_p of the system. Each copy is multiplied with spreading code and allocate to the sub-carrier. Thus spreading become in frequency domain and transmitting data rate will be same as input data bit [7]. Transmitter structure is as shown in Fig. 3. In MC-CDMA-based cognitive radio, few sub-carriers are used by licensed user and so that for secondary users transmission become non-contiguous transmission. In this case, sub-carrier under primary user band is considering as null sub-carriers.

CI codes are chosen for spreading data bits to preserve the orthogonal property in NC transmission and to achieve more efficient system response. Over a Walsh code whose code length is restricted to 2^N having bits +1 or −1, CI codes are complex orthogonal codes of any length of N. CI-based MC-CDMA provides the

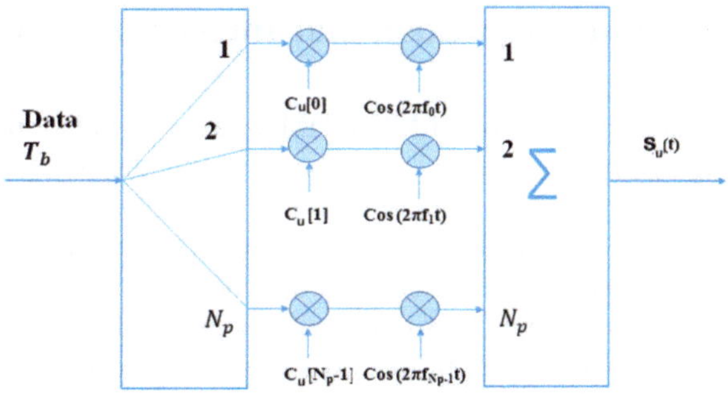

Fig. 3 MC-CDMA transmitter structure [8]

capability of supporting $U > N$ users using phase shift $\frac{\pi}{N}$ without expansion of bandwidth. The spreading sequence for CI code is generated by Eq. 1 [9].

$$c^u(t) = \sum_{i=0}^{N_p-1} \beta_i^u * e^{j2\pi\Delta fit} \quad (1)$$

where $\beta_i^u =$ is the uth user complex spreading sequence for $i = 0, 1,\ldots, N_P - 1$.

$$\beta_i^u = \beta_0^u, \beta_1^u, \ldots, \beta_{N_p-1}^u \quad (2)$$

Here, $\Delta f = 1/T_b$ where T_b is symbol duration.

For Hadamard code, β_i^u is −1, +1. For the case of CI codes, $\beta_i^{u\prime}$ are complex phases corresponding to Eq. 2 can be written as Eq. 3

$$\beta_i^u = 1,\ e^{j\Delta\theta u},\ e^{j2\Delta\theta u},\ e^{j3\Delta\theta u}, \ldots, e^{j(N_p-1)\Delta\theta u} \quad (3)$$

When $u = 1,\ 2, \ldots, N_p - 1$

$$\Delta\theta_u = \frac{2\pi}{N} u \quad (4)$$

And when $u = N_p, N_p + 1, \ldots, 2N_p$, it is defined by

$$\Delta\theta_u = \frac{2\pi}{N} u + \frac{\pi}{n} \quad (5)$$

In this way from Eqs. 4 and 5, $2N$ users are supported on codes of length N with minimal performance degradation as the number of users increase from N to $2N$. The transmitted signal for the all users for CI/MC-CDMA in [10, 11]. Effect of CI code on simulation is also verified and shown in [12]. Here, $D^U(t)$ is input data bits, u = Total number of users, A = Amplitude [9].

5 Power Allocation and Interference Management

Frequency response of MC-CDMA-based cognitive radio system is shown in below Fig. 4. It is observed that because of the Fourier-based transmission of the secondary system side-lobe power may cause interference to nearby primary user [13–15]. This work intended to minimize the side-lobe interference caused by SUs to PU. Here, effect is observed on whole PU band. PU band may use CDMA-based transmission or FDMA-based transmission. Interference effect will be same on all PU users as here only SUs to PU effect is considered.

Let total available bandwidth of secondary user is B_{su} Hz which is decided by the coherence bandwidth. B is the sub-carrier bandwidth for secondary user. N_{su} are the total number of sub-carrier allocated to secondary user. So the number of sub-carriers is given by $N_{su} = \frac{B_{su}}{B}$. Uniform bandwidth allocation done on each sub-carrier for all u users. For the given power budget and allocated bandwidth system overall throughput should be maximum, with the condition that the interference caused by SU users to PU band should be under the threshold defined by spectrum regulatory authority. The optimization problem for sub-carrier power allocation and PU interference management is as follows:

Problem Definition:

$$\text{Obj} = \max \quad C, \quad \forall u \tag{6}$$

Subject to,

$$I_m \leqslant I_{th} \tag{7}$$

$$p_{k,i} \geqslant 0 \tag{8}$$

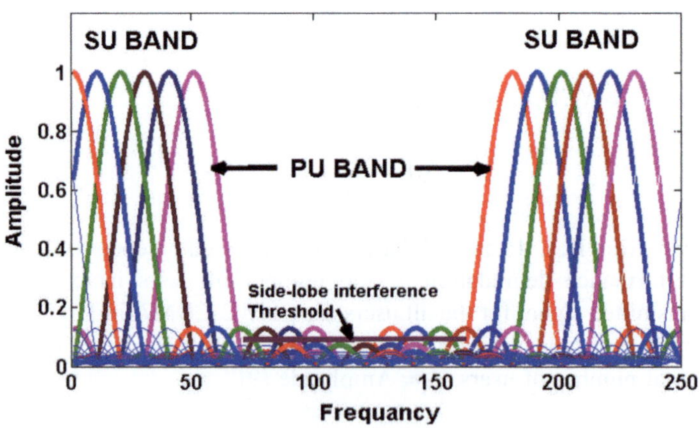

Fig. 4 Frequency response of NC-MC-CDMA

$$\sum_{k=1}^{u}\sum_{i=1}^{N_{su}} p_{k,i} \leqslant P_t \tag{9}$$

$$0 \leqslant B \leqslant \Delta f_{cu} \tag{10}$$

where,

$$C = \sum_{k=1}^{u}\sum_{i=1}^{N_{sc}} B\log_2\left(1 + \frac{p_{k,i} * h_{kl,i}}{\sigma^2}\right) \tag{11}$$

$$I_m = \sum_{k=1}^{u}\sum_{i=1}^{N_{sc}} p_{sc} * g_{k,p} \int_{\text{PUBand}} \operatorname{sinc}((f - f_{k,i}) * T_s)^2 \tag{12}$$

where B = uth user sub-carrier bandwidth, $p_{u,i}$ = uth user power on ith sub-carrier, $h_{kl,i}$ = kth user transceiver channel gain on ith sub-carrier, σ^2 = AWGN variance.

Equation 6 gives the objective that system should achieve maximum overall throughput for all the users with the constraints that interference to the PU band should be below threshold given by authority as in Eq. 7. Power allocated to all the secondary users should satisfy the constraint given in Eqs. 8 and 9. Sub-carrier bandwidth should be lesser than coherence bandwidth as in Eq. 10. Overall capacity of the system is given as Eq. 11. Interference to the PU band by each secondary user is calculated by Eq. 12.

To fulfil the objective of obtaining maximum throughput with the constraint that side-lobe interference should be below the threshold interference, algorithms are developed using two different power allocation methods. Uniform power allocation-based on–off method is modified using the water-filling power allocation method to minimize the side-lobe interference. Algorithm is stated here shows that after allocating resources to the system, power allocation is done. Interference is calculated for allocated power and compared with threshold interference (decided by Telecom regulatory authority). If the interference is above the threshold level, highest interference causing sub-carrier is elected and power level to that sub-carrier is making null. Thus, on–off method is applied here to reduce the side-lobe interference level. Uniform power allocation does not give satisfied performance; hence, water-filling power allocation is applied to achieve the objective of this work. Algorithm is as shown below.

6 Simulation Results

In this section, calculation of side-lobe interference from SUs to PU and management of interference below threshold using different power allocation methods are carried out. Let system having total allocated bandwidth 6 MHz out of which 1 MHz uses by PU and 5 MHz used by SU users. Here, we are considering only single PU in system but for the presence of more PU scenario is remain same. Here, we are considering total 64 sub-carrier allocated to both of them by centralized cognitive system. Sub-carrier allocated to the PU will be null sub-carrier for secondary user. Threshold interference arbitrary chosen for simulation is I_{th} = 8e−11 W and input power of single user is 2 W. Side-lobe interference calculation from SUs to PU is carried out by considering PU as a single band.

Algorithm 1 Modified Interference Management algorithm

procedure ALLOCATION OF RESOURCES
  Total Available Sub carrier Bandwidth $=B$
  Total Available Power for Each User $=P_t$
  for each user $i \in u$ **do**
    let i^{th} user Bandwidth$=B_{su}$ |C $\in$ max C_t
    let $N_{si} = \frac{2B_{su}}{B-1}$
    for each Subcarrier $j \in$ **do** N_{si}
      let Channel gain $h_{i,j}$,$\forall$ N_{si}
      P_{ij}=$\frac{P_t}{N_{si}}$
    end for
  end for
end procedure
procedure CALCULATION FOR INTERFERENCE TO PU
  While $(P_{ij} \geq 0)$
  Compute I_{ij},$\forall$ N_{si},$\forall$ u
  I_{tot}=$\sum_{i=1}^{u} I_{ij}$
  Calculate Throughput ,$\forall$ N_{si},$\forall$ u |C $\in$ max C_t
  Use Water filling Algorithm for $\sum P_{ij} \leq P_t$
  Calculate Throughput ,$\forall$ N_{si},$\forall$ u |C $\in$ max C_t
  Compare Throughput with previous result
end procedure
procedure INTERFERENCE MINIMIZATION
  for each user $i \in$ **do** u
    for each Subcarrier $j \in N_{si}$ **do**
      Do Get max I_{ij}
      $h_{i,j}$=0
      Compute I_{ij},$\forall$ N_{si},$\forall$ u
      I_{tot}=$\sum_{i=1}^{u} I_{ij}$
      While $I_{tot} > I_{th}$
      Calculate Throughput ,$\forall$ N_{si},$\forall$ u |C $\in$ max C_t
using power allocation by uniform power allocation
      Calculate Throughput ,$\forall$ N_{si},$\forall$ u |C $\in$ max C_t
using power allocation by water filling method
      Compare Results
    end for
  end for
end procedure

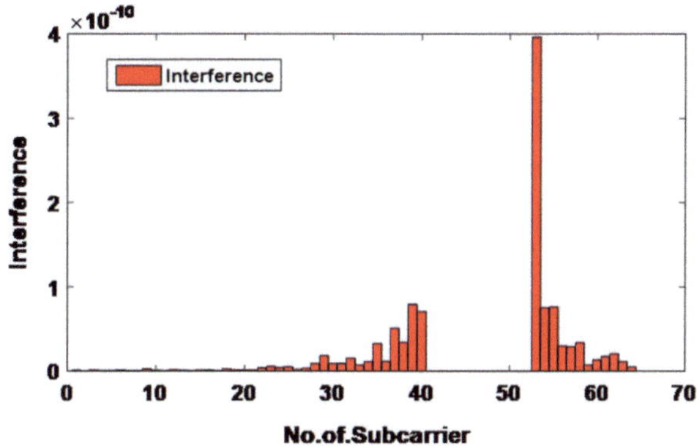

Fig. 5 Interference from SU to PU

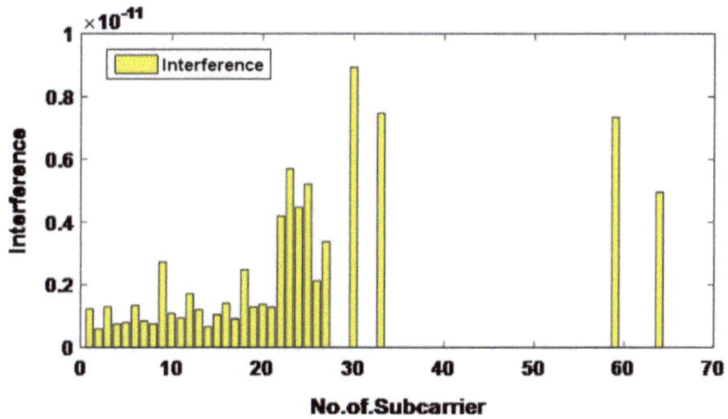

Fig. 6 Interference after minimization to PU band

Figure 5 shows the interference caused by each sub-carrier of SU to PU band. We get total interference is 1.11e−9 W to PU band which is above the threshold interference. We need to reduce this interference below threshold value. Algorithm based upon uniform power allocation is applied to minimize interference below the threshold value.

Figure 6 shows the interference caused by each sub-carrier after minimizing total interference. Result gets 7.8e−11 W interference to PU which is below threshold. To minimize this algorithm has null the 21 sub-carrier of secondary user which causing maximum interference to PU band.

Throughput performance for single user before and after interference minimization is shown in Fig. 7, where throughput is degraded as number of sub-carriers to be null after minimizing the interference.

Another simulation carried out for such system using multi-users, as more than single user carried by MC-CDMA system interference causing to PU is also increase. For four users, each of having 2 W power on total 64 sub-carrier where PU having 1 MHz and SU 5 MHz and by keeping interference threshold I_{th} = 1e−10 W. Each user power equals distribution on 64 sub-carrier that is 2/64 W. Before minimization total interference to PU is carried out 3.4e−9 W to PU. Each sub-carrier interference is

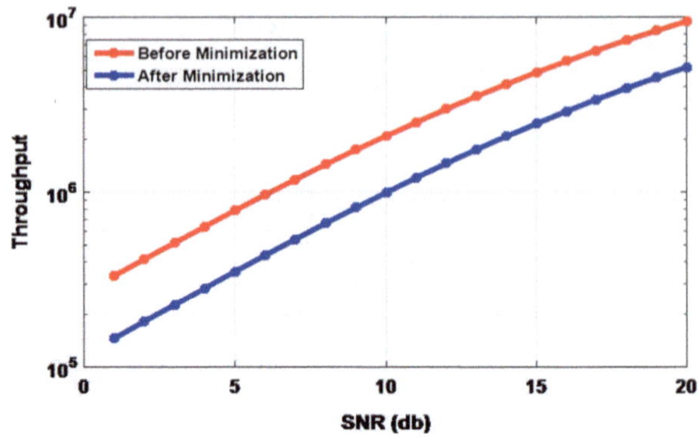

Fig. 7 Throughput performance for single user

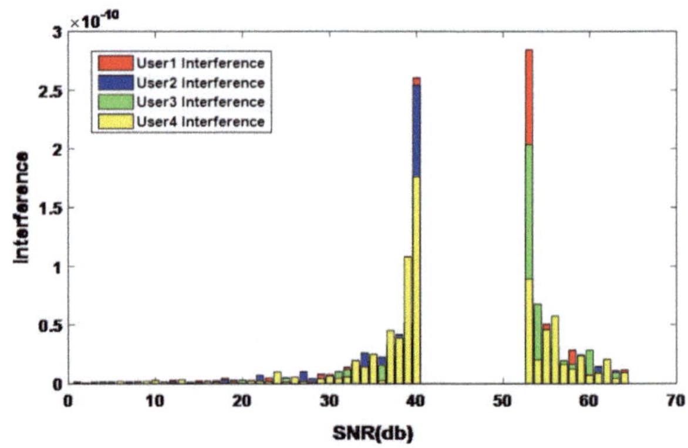

Fig. 8 Interference by four secondary users to PU band

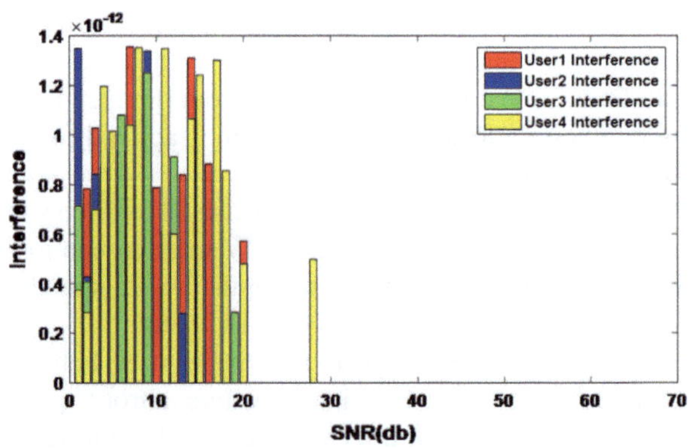

Fig. 9 Interference by four users after minimization

showing in Fig. 8. After applying algorithm, interference reduces to PU is 9.5e−11 W which is obtained by null 160 SU sub-carrier which causing highest interference to PU shown in Fig. 9.

Throughput calculation for both scenarios is shown in Fig. 10.

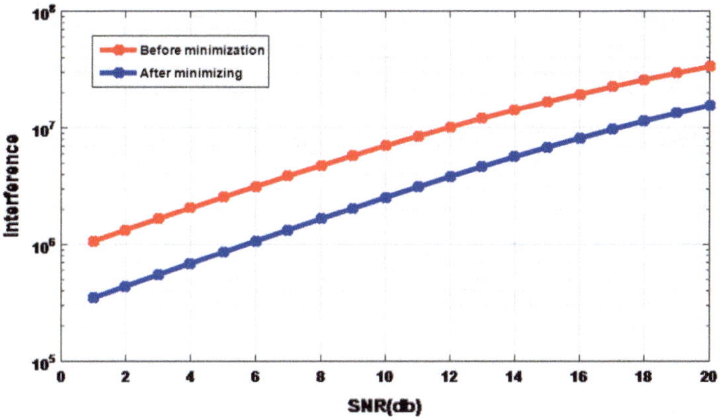

Fig. 10 Throughput performance of four users

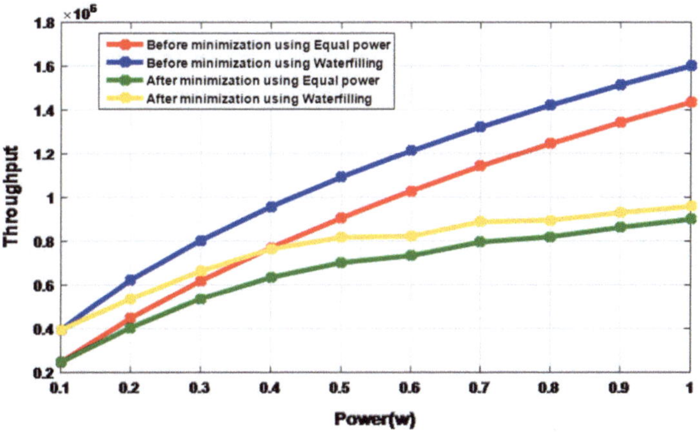

Fig. 11 Throughput versus power performance with modified algorithm

6.1 *Interference Management and System Performance*

Based upon algorithm using water-filling power allocation method as discussed in previous section interference management is performed. For the single user having input power range of 0.1–1 W and 64 sub-carrier system algorithm is applied to reduce the side-lobe interference causing from SU to PU. Here interference threshold is kept at 1e−11 W. Here, system performance analyzed based upon throughput of the system. Throughput of the system before management and after management using developed algorithm is shown in Fig. 11.

Out of 64 sub-carrier, 20 sub-carriers became null to reduce interference level below threshold. After the management again power allocation done and throughput is calculated. In Fig. 11, if we compare result with previous algorithm

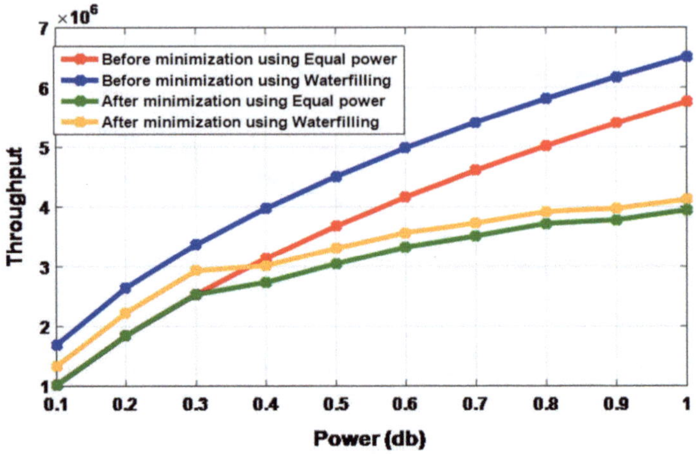

Fig. 12 Throughput versus power performance for multi-user MC-CDMA

here throughput is (shown in yellow) increased. It proves that the new mechanism is inferior compare to conventional on–off method. For the multi-user NC-MC-CDMA, above algorithm is also verified. For the four user and 64 sub-carrier, each user having input power range 0.1–1 W. Interference threshold is assumed 5e−10 W.

Figure 12 shows the result for throughput of four-user system, here also it is verified that throughput of system is increased compare to conventional on–off method.

7 Conclusion

From the system model and interference calculation, it is shown that SU side lobe creates interference in PU band. The aim of this paper is to enhance the performance of SU transmission such that SU creates interference below threshold to the PU band. Interference calculation is carried out for Secondary users to PU band over a range of input power. When SU's interference to the PU band is above threshold, different power allocation-based algorithms can be applied to manage this interference below threshold. On–off method-based algorithm is applied with uniform power allocation; it is observed that it minimizes interference but at the same time throughput is degraded. To combat this limitation, on–off algorithm is used with water-filling-based power allocation and throughput is improved. Results are verified for single user as well as multiple users.

References

1. Y. Nathan, J.-P. M. Linnartz, and G. Fettweis, "Multi-carrier cdma in indoor wireless radio networks," *IEICE Transactions on Communications*, vol. 77, no. 7, pp. 900–904, 1994.
2. V. Thumar, T. Nadkar, U. B. Desai, and S. Merchant, "Distributed resource allocation for cognitive radio networks: Sub-carrier power and bandwidth sizing," in *Vehicular Technology Conference (VTC Fall), 2013 IEEE 78th.* 1em plus 0.5em minus 0.4em IEEE, 2013, pp. 1–6.
3. G. T. S. M. Taskeen Nadkar, Vinay Thumar and U. Desai, "Adaptive guard interval and power allocation for ofdm based cognitive radio," *ICTACT Journal On Communication Technology: Special Issue on Wireless Networks and application*, vol. 2, 2011.
4. S. Zhao, "Power allocation in ofdm-based cognitive radio systems," *School of Graduate Studies at Ryerson University*, 2013.
5. Z. Yuan and A. M. Wyglinski, "On sidelobe suppression for multicarrier-based transmission in dynamic spectrum access networks," *IEEE Transactions on Vehicular Technology*, vol. 59, no. 4, pp. 1998–2006, 2010.
6. S. Haykin, "Cognitive radio: brain-empowered wireless communications," *IEEE Journal on Selected Areas in Communications*, vol. 23, no. 2, pp. 201–220, Feb 2005.
7. S. Hara and R. Prasad, "Overview of multicarrier cdma," *IEEE communications Magazine*, vol. 35, no. 12, pp. 126–133, 1997.
8. L. Hanzo, L. Yang, E. Kuan, and K. Yen, "Single and multi-carrier cdma: Multi-user detection, space-time spreading, synchronisation and standards," 2003.
9. B. Natarajan, C. R. Nassar, S. Shattil, M. Michelini, and Z. Wu, "High-performance mc-cdma via carrier interferometry codes," *IEEE Transactions on Vehicular Technology*, vol. 50, no. 6, pp. 1344–1353, 2001.
10. B. Xie, E. Like, M. Temple, and Z. Wu, "Performance of downlink mc-cdma and ci/mc-cdma systems in the presence of narrowband interference," in *Proceedings of the 2009 International Conference on Wireless Communications and Mobile Computing: Connecting the World Wirelessly.* 1em plus 0.5em minus 0.4em ACM, 2009, pp. 1121–1125.
11. C. R. Nassar, B. Natarajan, and S. Shattil, "Introduction of carrier interference to spread spectrum multiple access," in *Wireless Communications and Systems, 2000. 1999 Emerging Technologies Symposium.* 1em plus 0.5em minus 0.4em IEEE, 1999, pp. 4–1.
12. B. D. Kavaiya and V. M. Thumar, "Suitability of carrier interferometry code for nc-mc-cdma based cognitive radio," *International Journal of Current Engineering and Scientific Reaserch (IJCESR)*, vol. 3, no. 9, pp. 123–129, oct 2016.
13. M. Rajabzadeh and H. Khoshbin, "A novel multicarrier cdma transmission scheme for cognitive radios with sidelobe suppression," *International Journal of Communication Systems*, vol. 26, no. 11, pp. 1485–1499, 2013.
14. L. D. H. Sampaio, T. Abrão, B. A. Angélico, M. F. Lima, M. L. Proença, and P. J. E. Jeszensky, "Hybrid heuristic-water filling game theory approach in mc-cdma resource allocation," *Applied Soft Computing*, vol. 12, no. 7, pp. 1902–1912, 2012.
15. J. Zhu and Y. Bar-Ness, "Power allocation algorithm in mc-cdma," in *Communications, 2002. ICC 2002. IEEE International Conference on*, vol. 2. 1em plus 0.5em minus 0.4em IEEE, 2002, pp. 931–935.

References

Evaluation of RTT-Based Techniques for Network State Classification

Dipali Shah, Purvang Dalal, Mohanchur Sarkar and Sejal Dalal

Abstract RTT is the only parameter available with TCP sender for estimating network capacity. Accordingly, legacy TCP revises its sending rate for utilizing maximum of the network capacity, without overloading. Since in wireless network, RTT is weakly correlated to the congestion, any estimate using only the recent RTT sample may lead to performance penalty. This paper presents a simulation-based study of the two RTT-dependent mechanisms, targeting fair network state classification by taking RTT statistics into account. It is demonstrated that the performance of TCP scheme equipped with one technique can be improved further by tweaking using the other technique.

Keywords Round-trip time · Wireless network · Network state classification
Available bandwidth · Congestion window

1 Introduction

Transmission control protocol (TCP) [1], the de facto transport protocol over Internet, is primarily designed and turned for wired network with an initial objective to utilize maximum of available network bandwidth without overloading the network [2]. TCP believes packet loss and delay variations as the network congestion indicators and accordingly regulates its sending rate to lessen the congestion [3]. For that it revises one of its state variables congestion window (cwnd) either on

D. Shah · P. Dalal (✉)
Department of Electronic and Communication Engineering,
Dharmsinh Desai University, Nadiad, India
e-mail: pur_dalal.ec@ddu.ac.in

M. Sarkar
Space Applications Centre (ISRO), Ahmedabad, India

S. Dalal
S. V. Patel Institute of Technology, Vasad, India

R. Kher et al. (eds.), *Proceedings of the International Conference on Intelligent Systems and Signal Processing*, Advances in Intelligent Systems and Computing 671,
https://doi.org/10.1007/978-981-10-6977-2_8

detection of a packet loss or on arrival of a positive acknowledgment (TCP-*ack*) from the receiver [4].

The advancement in wireless technology has extended Internet services to end users equipped with wireless devices. Because more than 90% of Internet traffic (including the most popular applications like Web (HTTP), file transfer (FTP), and e-mail (SMTP), etc.) uses TCP as transport layer protocol for reliable data transfer [5], it is apparent that the wireless Internet segment must also use TCP for extending similar services. In wireless domain, the quality of the radio link can fluctuate greatly in time because of channel fading, interference, shared channel access, and mobility [5]. Consequently, a TCP flow may encounter packet losses and delay variations, despite there is no congestion. In the absence of mechanisms for accurate and in time discrimination, TCP's response to most of the packet losses and delay variations is found inappropriate [3]. Consequently, TCP fails in utilizing the available network bandwidth gracefully [6].

In order to mitigate TCP performance issues arising mostly due to its unfair reaction to the non-congestion losses and delays, extensive research has been carried out in last two decades [5–7]. The link layer techniques attempt to shield TCP from wireless transmission losses [8]. However, link recovery is not guaranteed and has adverse impact on TCP's RTT-dependent flow control [3]. Several attempts are made to revise TCP congestion control parameters while attempting loss recovery. The revision is in accordance with the input from explicit or implicit mechanisms for loss differentiations [9]. In addition to the deployment issues in the Internet, TCP proposals with explicit mechanisms are incongruous particularly when loss discrimination is merely indicative and information about current network state is either ignored or not available. On other side, TCP schemes with implicit mechanisms are much appreciated due to minimum changes leading to ease of deployment in Internet. However, the efficiency of such TCP schemes is not guaranteed in a shared wireless environment due to the limited ability of a TCP sender in observing the network resource together with the fast-changing network bandwidth availability [10]. Consequently, several techniques have been proposed in literature for enhancing their performance by taking advantage of round-trip time (RTT) statistics for network state classification.

In this work, two promising techniques, proposed in recent past, are elucidated and examined through simulations. The first technique, presented in [11], exploits RTT statistics over a period of data transfer and proposes a model for network state classification. The second technique is presented in [12], wherein the TCP sender is modified (i.e., Adaptive TCP) to estimate the degree of congestion by observing RTT evolution since inception of data transfer. The estimate is utilized further to prevent undue reduction in TCP sending rate, in favorable network condition. The results highlight the scope for further performance improvement of Adaptive TCP by exploiting RTT variations observed over a limited period instead of entire data transfer phase. Rest of the paper is organized as follows: Sect. 2 discusses the basic TCP mechanisms related to flow control and loss recovery. In Sect. 3, two RTT-based techniques stated earlier are briefed. Section 4 discusses on performance evaluation of both techniques over an IEEE 802.11-based wireless network. Section 5 concludes this paper.

2 TCP Mechanisms

TCP's efficiency for network bandwidth utilization is attributed to its basic functionality such as flow control and loss recovery. While these mechanisms have a well-presented definition of their own, in practice, they are closely coupled with each other.

2.1 Flow Control

In the beginning of the TCP connection (or after a prolonged disconnection), the sender is unaware of available network bandwidth. In this situation, it executes the slow-start phase for progressive probing and quick utilization of available bandwidth. In this phase, sender sets the cwnd size to initial window size (1 or 2 maximum segment size (MSS) [13]). Then after, cwnd increases by one MSS for each arrival of a TCP-ack. In normal conditions, all sent packets are acknowledged in RTT interval and hence in slow-start phase, growth of cwnd is exponentially over RTT. This happens until either a packet loss event occurs or cwnd reaches the slow-start threshold (ssthresh) [1]. Once cwnd reached the ssthresh, TCP enters the congestion avoidance phase. In this phase, as its name implies, it tries to avoid network congestion by restricting the growth of cwnd with an objective to utilize additional bandwidth gradually, i.e., without overloading the network. In this phase, growth of cwnd is linear over RTT, i.e., cwnd is increased by one segment for each RTT. This happens until a loss event occurs. Thus, in both of the above-mentioned phase, the growth in the size of cwnd is greatly dependent on RTT, estimated by the TCP sender on end-to-end basis. When delay is large, then RTT value increases; so TCP sender takes more time to utilize maximum of the available network bandwidth [6].

2.2 Loss Detection and Recovery

A packet loss is assumed if the sender does not receive an acknowledgment for the packet within a time-out interval called RTO, which is dynamically computed as the estimated smoothed RTT plus four times its mean deviation [14]. TCP achieves reliable data delivery by retransmitting the lost packet. Upon a time-out, the sender first revises ssthresh to the half of the current cwnd size (minimum two MSS [1]) and then after initiates the slow-start mechanism with the cwnd size of initial window. This procedure is also called the additive increase and multiplicative decrease (AIMD) algorithm [15]. In this way, packet loss detection using time-out and the growth of cwnd on arrival of each TCP-ack are greatly dependent on RTT. When the RTO value is large, waiting for a time-out event to trigger retransmission is considered inefficient [16]. Due to this, fast retransmit was proposed to provide a timely detection of a lost packet. It triggers when the sender receives three duplicate acknowledgments (DupAcks) from the receiver. Upon receiving these DupAcks, TCP infers packet loss and retransmits the unacknowledged packet with smallest sequence number. Fast recovery is executed after fast retransmit [16]. The sender exits the fast recovery when a new TCP-ack is received and deflates the cwnd; that is, the cwnd is set to half of the current ssthresh plus the number of DupAcks [1].

The said fast recovery mechanism can recover from for a single packet loss per RTT but cannot avoid time-out action with multiple packet losses per RTT [15]. In order to recover from multiple packet losses from a window of data, the TCP mechanism is enhanced further, wherein the sender exploits the partial acknowledgment for recovering from multiple packet losses from a single window of data. However, for recovering from multiple packet losses, it requires multiple RTTs [17]. Thus, like its flow control mechanism, TCP's loss detection and recovery mechanisms are also RTT-dependent.

3 Proposals with RTT-Based Classification

Since TCP throughput is limited by the augmentation in cwnd per RTT [18], the efficiency claimed by TCP is essentially depending upon its accuracy for RTT estimation, in addition to the size of cwnd [19]. In wireless networks, the delay variations are in general weakly correlated with the network congestion and hence any network state discrimination taking instantaneous value of RTT into account leads to poor TCP performance. Therefore, the TCP sender would not regulate the data flow according to the actual state of network, especially related to congestion. This demands for robust technique for fair network state classification. In the following subsection, two RTT-based proposals for the same are explained.

3.1 Model for RTT-Based Network State Classification [11]

The model proposed in [11] operates on the long-term and short-term RTT evolutions for network state classification. Since the sampling rate of RTT is non-uniform and mostly at lower rate particularly in wireless networks, this model considers mean of all the RTT values pertaining to the same window as a representative value of RTT for classification. This model classifies network into four states: '00' (no congestion, no error), '01' (no congestion, high error), '10' (high congestion, no error), and '11' (high congestion, high error). The proposed model uses following RTT statistics.

Normalized Rise of the mean RTT

The total duration of a TCP connection is divided into a series of windows. For a given window (say *i*), there will be a series of RTT values obtained corresponding to all packets transmitted in that window. In this context, $\mathrm{RTT}_{i,j}$ is the RTT measurement for the *j*th packet for the *i*th window. Accordingly, mean RTT for the *i*th window (MW*i*) is derived using Eq. (1).

$$\mathrm{MW}_i = \frac{\sum_{j=0}^{\mathrm{cwnd}_i} \mathrm{RTT}_{i,j}}{\mathrm{cwnd}_i} \tag{1}$$

Here, cwnd_i denotes the value of cwnd for the *i*th window. In order to estimate the short-term condition of the network load, the proposed model computes normalized rise of the mean RTT. The normalized rise of the mean RTT for last k windows ($\mathrm{MR_k}$) is obtained using Eq. (2),

$$\mathrm{MR}_k = \frac{M_k - \mathrm{IRTT}}{\mathrm{IRTT}} \tag{2}$$

where IRTT denotes the minimum RTT observed and M_k denotes the window average of the mean values MW*i* for the last *k* windows. M_k is computed using Eq. (3).

$$M_k = \frac{\sum_{j=k-i}^{i} \mathrm{MW}_i}{k} \tag{3}$$

Standard Deviation of the mean RTT

In order to signify the extent to which the RTT values have been dispersed from the ideal condition, the ratio of the standard deviation to the ideal RTT for last k windows (R_k) is obtained using Eq. (4).

$$R_k = \frac{\sigma_k}{\mathrm{IRTT}} \tag{4}$$

where, the standard deviation of the mean RTT values, MW*j* for the last *k* windows (σ_k) is derived using Eq. (5).

$$\sigma_k = \sqrt{\frac{\sum_{j=k-i}^{i} \left(\mathrm{MW}_j - M_k\right)^2}{k}} \tag{5}$$

In order to classify operating state of the network, the above statistical parameters from the measured RTT are required to be observed for short or long number of windows (*k*). The high congestion state of the network exhibits smooth RTT and hence produces higher value for normalized rise of mean RTT. On the other side, in

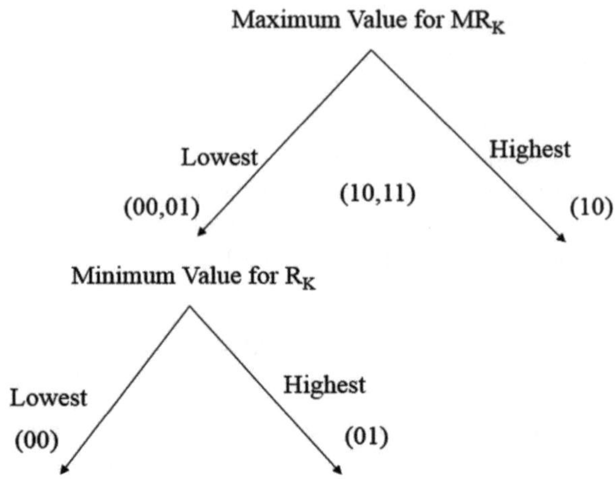

Fig. 1 Model for network state classification

the network with less congestion and less amount of wireless losses, exhibits very small amount of deviation for mean RTT. The proposed model for network state classification using above-mentioned RTT statistics is shown in Fig. 1.

3.2 *Adaptive TCP [12]*

Adaptive TCP sender attempts to estimate degree of congestion by exploiting long-term evolution of RTT and its statistics. For that on arrival of each TCP-ack, sender computes the value of congestion estimator (α) using Eq. (6).

$$\alpha = 1 - \frac{\mathrm{RTT}_n - \mathrm{RTT}_{\min}}{\mathrm{RTT}_{\max} - \mathrm{RTT}_{\min}} \tag{6}$$

The value of α varies from 0 to 1 based on the estimated value of RTT. This mechanism not only facilitates sender for knowing the state of network (i.e., congestion or non-congestion) but also assists TCP sender to quantify the degree of congestion. This will guide sender about rising or decreasing trends of congestion in the network. Adaptive TCP sender utilizes the above value of α and adjusts the size of cwnd after nth loss recovery using Eq. (7).

$$W_{\mathrm{LA}(n)} = \alpha \times \mathrm{cwnd}(n) + (1 - \alpha) \times W_{\mathrm{LN}(N)} \tag{7}$$

Here, cwnd(n) represents the maximum size of cwnd at sender prior to the nth loss detection. $W_{LN(N)}$ is the size of cwnd, which a native TCP sender will have after nth loss recovery action. Equation (7) itself concludes that the Adaptive TCP sender never advances the size of cwnd after loss recovery beyond the size seen just before loss detection. In this context, Adaptive TCP protects conservative approach of TCP Newreno, which is essential for maintaining fairness.

4 Performance Evaluation

Performance evaluation for both of the techniques has been carried out through simulations under ns-2. Firstly, the TCP codes in ns-2 are modified to implement the stated behavior of TCP and also to obtain stated RTT statistics at the sender. The simulation scenario used for obtaining all analysis is as shown in Fig. 2. During simulations, wireless errors were introduced using a well-known two-state Markov error model in the unit of frame error rate (FER). The results were obtained for 200 s of TCP data transfer between sender(s) on wireless segment and receiver (s) on wired segment. The backbone is configured for 10 Mbps of data rate and 45 ms of one-way delay. The queue size at intermediate nodes is kept higher than bandwidth-delay product over end-to-end path. Unless specified, all other wired links have 100 Mbps data rate and 1 ms of one-way delay. The wireless transmission follows IEEE 802.11b specifications.

The simulations were carried out under four different network states as described below.

State—'00'

In this scenario, a single TCP/FTP flow between $W1$ and $S1$ was introduced. The wireless FER was set to 0%. It must be noted that the maximum TCP throughput from each IEEE 802.11 AP is limited to 66% of its maximum data rate support (≈5.5 Mbps). Therefore, with single flow, the total data delivered over backbone is less than its capacity and it will not cause congestion packet loss.

State—'01'

This scenario was almost identical to that described above for state—00. In addition, the wireless link errors were introduced using FER in the range of 0–10%.

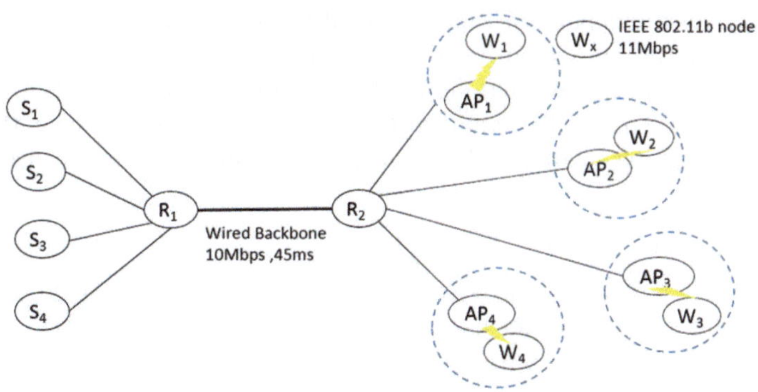

Fig. 2 Performance scenario

State—'10'

In this scenario, four identical TCP/FTP flows (each through dedicated AP) were introduced between wireless senders (*W*1 to *W*4) and receivers (*S*1 to *S*4) in wired domain. The wireless FER is set to 0%. Thus, the traffic TCP delivered over backbone increases with increase in number of TCP flows (each flow from different AP). This causes congestion packet loss over backbone.

State—'11'

This scenario was almost identical to that described above for state—10. Additionally, the wireless link errors were introduced using FER in the range of 0–10%. This reassembles the network characteristics, identical to the terrestrial environment for Internet access.

4.1 Analysis for RTT-Based Model

With an objective to validate RTT-based model proposed in [12], the simulations were carried out in four different states of network environments. The statistics were obtained based on evolution of RTT over a short period and long period. For short-term analysis, RTT statistics are obtained considering the previous window count (k) as 10 or 20; whereas, k is increased above 20 (say 40) for long-term analysis.

Classification using MR_k

Table 1 gives the maximum value obtained for the normalized rise of the mean RTT. The values are presented for different values of k as mentioned along with.

The obtained values resemble the model proposed for network state classification. The highest value of MR_k is obtained for network state '10'. The proposed state classification with $k = 10$ and $k = 20$ is shown in Fig. 3.

It was also observed that the difference between maximum and minimum of MR_k reduces with increase in k (i.e., number of previous window into consideration).

Classification using R_k

Table 2 gives the maximum value obtained for the normalized rise of the mean RTT. The values are presented for different values of k as mentioned along with.

The obtained values resemble the model proposed for network state classification (refer Fig. 4). The smallest value of R_k is obtained for network state '00'. It was also observed that the smallest value of R_k reduces with increase in k (i.e., number

Table 1 Analysis for maximum MR_k

Network state	00	01	10	11
MR10 (k = 10)	0.177	0.152	0.720	0.641
MR20 (k = 20)	0.177	0.152	0.211	0.211
MR40 (k = 40)	0.177	0.177	0.211	0.211

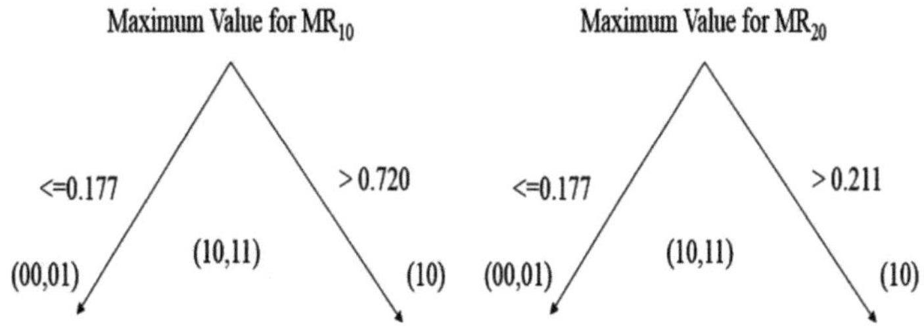

Fig. 3 Network state classification using MR_k

Table 2 Analysis for maximum R_k

Network state	00	01	10	11
R10 (k = 10)	0.300	0.428	0.616	0.610
R20 (k = 20)	0.217	0.322	0.477	0.532
R40 (k = 40)	0.174	0.279	0.462	0.516

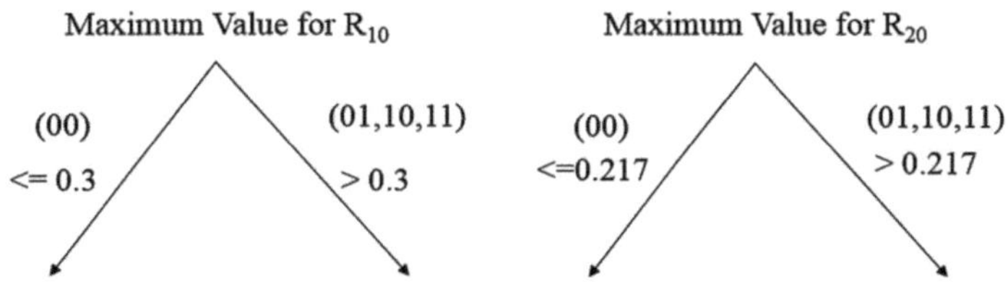

Fig. 4 Network state classification using R_k

of previous window into consideration), which is obvious as the standard deviation reduces with increase in the number of samples.

Figure 5 shows the approach for the network state classification by combining both of the RTT statistics obtained.

The results obtained through simulations direct us for development of algorithm that continuously updates MR_k and R_k and same can be utilized further for classifying the instantaneous network state, while resuming transmissions after loss recovery.

4.2 Analysis for Adaptive TCP

The performance of Adaptive TCP is compared with Newreno through simulations under all different states described earlier. Figure 6 shows the comparison in terms of highest TCP sequence number of the sent packet in a network with state '00'. The analysis is presented for the entire range of FER. It is observed that with

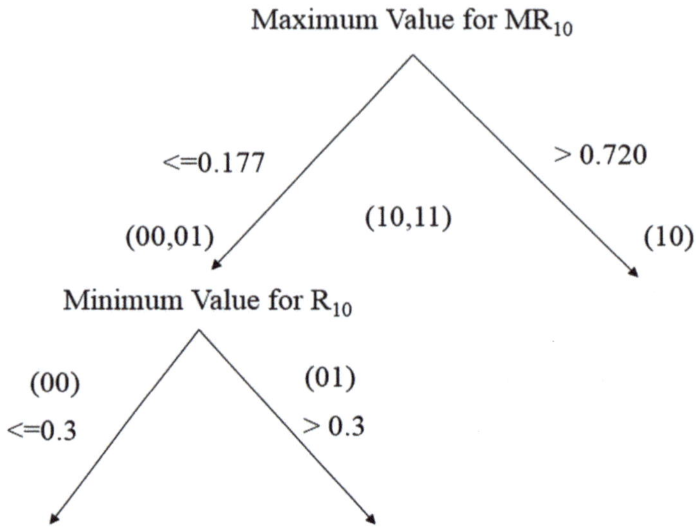

Fig. 5 Network state classification using MR_{10} and R_{10}

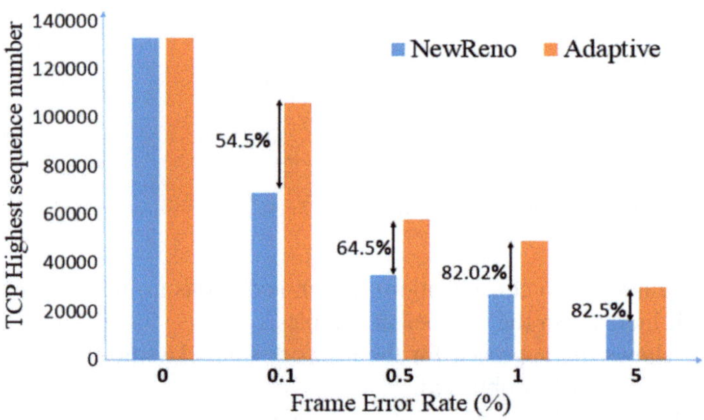

Fig. 6 Comparison of TCP sequence numbers (1 TCP flow)

increase in FER, the percentage improvement demonstrated by Adaptive TCP over Newreno has also increased.

Figure 7 shows the comparison in terms of total TCP packets sent in a network with state '10'. Here, the number of concurrent TCP flows varies from one to four. Though the results were obtained for entire range of FER, the analysis is presented for the FER of 1%. It is observed that with increase in number of competing traffic over backbone (i.e., degree of congestion), the percentage improvement demonstrated by Adaptive TCP reduces. However, the performance of Adaptive TCP is

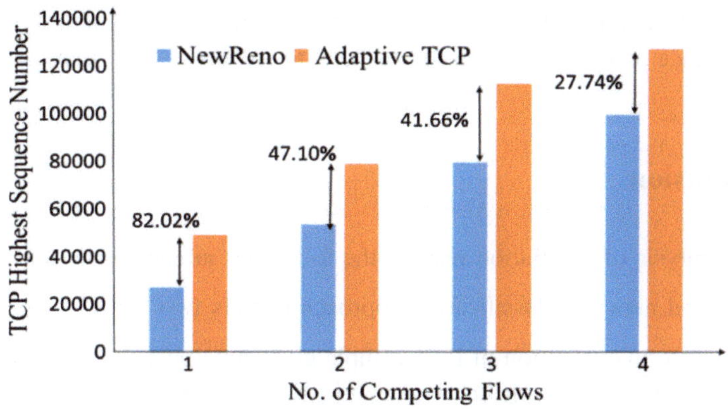

Fig. 7 Comparison of TCP sequence numbers (FER = 1%)

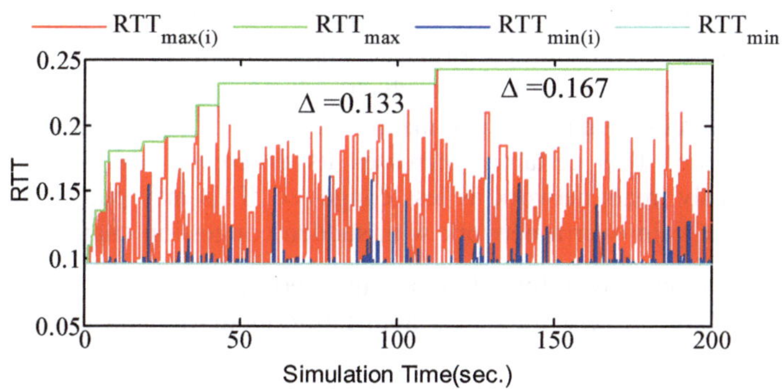

Fig. 8 RTT statistics over simulation period

never seen inferior to that observed with Newreno. Both of the above results validate the claims made by authors in [12].

The interesting outcome of simulation results is shown in Fig. 8. In Fig. 8, maximum (RTTmax) and minimum (RTTmin) values of RTT observed during the data transfer are plotted. It was noticed that with the progress in the data transfer, their difference Δ (i.e., RTTmax−RTTmin) increases. This in turn wrench value of α toward 1 (i.e., most of the time network is treated as non-congested). Such kind of behavior may have adverse impact on Adaptive TCP's performance, particularly with abrupt delay variation that causes sudden congestion.

Fig. 8 also shows the minimum and maximum values of RTT, recorded during uninterrupted evolution of cwnd. To explain, after each loss recovery, TCP initializes RTTmin and RTTmax using the first RTT estimate made. These values are referred as RTTmin(i) and RTTmax(i), respectively. It is seen from the figure that, the progress in the data transfer does not result in the consistent increase in the difference

RTTmax(i)−RTTmin(i). By utilizing this difference in computation of α (ref Eq. 6), the performance penalty (as suspected earlier) to Adaptive TCP, may be reduced.

5 Conclusion

From the analysis of simulation results, the following inferences are drawn.

(1) RTT-based network classification approach reveals two facts.

   (a) Long-term evaluation of RTT sample gives much better estimate regarding negligible congestion in the network. However, short-term evaluation can be useful for identifying strong congestion in the network.
   (b) Hierarchical model using MR_i and R_i statistics can be utilized for confident network state classification related to congestion in the network.

(2) Adaptive TCP proposal uses RTT deviation for estimating degree of network congestion to demonstrate performance improvement. However, performance improvement is high in a network which is less congested.
(3) The algorithm uses minimum and maximum values of RTT throughout the data transfer phase. This certainly increases its difference and hence algorithm unconsciously treats network as less congested. This may put up limit on the performance gain over a network with sudden congestion change.

The results show possibility of improving performance of Adaptive TCP by combining RTT-based network state classification model for fair estimate of congestion in the network.

References

1. Allman, M., Paxson, V. and Blanton, E. (2009) TCP Congestion Control. RFC5681. http://dx.doi.org/10.17487/rfc5681.
2. Lar S-U, Liao X. An initiative for a classified bibliography on TCP/IP congestion control. *Journal of Network and Computer Applications* 2012; **36**(1):126–133. DOI:10.1016/j.jnca.2012.04.003.
3. Dalal, Purvang, et al. "Refining TCP's RTT dependent mechanism by utilizing link retransmission delay measurement in Wireless LAN." *International Journal of Communication Systems* (2015).
4. Henderson T, Floyd S, Gurtov A. *The Newreno modification to TCP's fast recovery algorithm*, 2012. RFC6582.
5. Al-Jubari, Ammar Mohammed, et al. "TCP performance in multi-hop wireless ad hoc networks: challenges and solution." *EURASIP Journal on Wireless Communications and Networking* 2011.1 (2011): 1.
6. Ros, David, and Michael Welzl. "Less-than-best-effort service: a survey of end-to-end approaches." *IEEE Communications Surveys & Tutorials* 15.2 (2013): 898–908.

7. Cheng, Rung-Shiang, and Hui-Tang Lin. "A cross-layer design for TCP end-to-end performance improvement in multi-hop wireless networks." *Computer Communications* 31.14 (2008): 3145–3152.
8. Sun, F., Li, V.O.K. and Liew, S.C. (2004) Design of SNACK Mechanism for Wireless TCP with New Snoop. *IEEE Wireless Communications and Networking Conference*, **5**, 1046–1051.
9. Dalal, Purvang, et al. "Link Layer Correction Techniques and Impact on TCP's Performance in IEEE 802.11 Wireless Networks." *Communications and Network* 2014 (2014).
10. La, Richard J., Jean Walrand, and Venkatachalam Anantharam. *Issues in TCP vegas*. Electronics Research Laboratory, College of Engineering, University of California, 1999.
11. Sarkar, Mohanchur, K. K. Shukla, and K. S. Dasgupta. "Network State Classification based on the Statistical Properties of RTT for an Adaptive Multi State Proactive Transport Protocol for Satellite based Networks." *International Journal of Computer Networks and Communication (IJCNC)* 2.6 (2010): 155–174.
12. Dalal, Purvang, et al. "Adaptive TCP: A Sender Side Mechanism with Dynamic Adjustment of Congestion Control Parameters for Performance Improvement in WLAN." *International Journal of Communications, Network and System Sciences* 8.5 (2015): 130.
13. M. Allman, S. Floyd, C. Partridge. Increasing TCP's Initial Window, 1998. RFC 2414.
14. Paxson V, Sargent M, Allman M, Chu J. Computing TCP's retransmission timer, 2011. RFC6298.
15. Henderson T, Floyd S, Gurtov A. *The Newreno modification to TCP's fast recovery algorithm*, 2012. RFC6582.
16. S. Floyd, T. Henderson The NewReno Modification to TCP's Fast Recovery Algorithm. 1999. RFC2582.
17. Lin, Dong, and H. T. Kung. "TCP fast recovery strategies: Analysis and improvements." *INFOCOM'98. Seventeenth Annual Joint Conference of the IEEE Computer and Communications Societies. Proceedings. IEEE.*
18. Sarolahti P, Kojo M Forward RTO-recovery(FRTO): an algorithm for detecting spurious retransmission timeouts with TCP and the stream control transmission protocol (SCTP), 2005. RFC4138.
19. Lim CH, Jang JW. Robust end-to-end loss differentiation scheme for TCP over wired/wireless networks. *IET Communication* 2008; **2**(2):284–291.

CMOS Image Sensor Parameter Configuration Using LabVIEW

Dhvani R. Patel, Brijesh N. Shah and Karan Jasani

Abstract Imaging identifiers (detectors) are the main components of optical equipment and space missions that have sensors placed on board committed to Earth perception (high-determination imaging, environment spectroscopy, etc.), solar system investigation (small-scale cameras, direction for self-governing vehicle, etc.) and universe perception (space telescope central planes, controlling sensors, etc.). UART–SPI interface is used for configuring the different parameters of image sensor which includes inserting the training pattern to sync LVDS (low-voltage differential signalling) receivers, setting the offset at the input of the ADC, inserting a pattern to generate a test pattern, enabling sequencer for image capture, image type selection settings like monochrome/grey image or colour image and many more. The image sensor has a number of LVDS high-speed outputs that transfer image data over longer distances. A special training mode enables the receiving system to synchronize the incoming data stream when switching to master, slave or triggered mode. A programmable offset and gain amplifier for each channel also is integrated by the image sensor. It requires only one master clock for operation up to 500 fps. The sensor is available in a monochrome version or Bayer (RGB)-patterned colour filter array.

Keywords CMOS image sensor · FPGA · SPI · UART · LabVIEW

D. R. Patel (✉)
Electronics and Communication Engineering, Charotar University of Science and Technology, Changa, India
e-mail: pateldhvani22@gmail.com

B. N. Shah · K. Jasani
Charusat Space Research and Technology Center, Charotar University of Science and Technology, Changa, India
e-mail: brijeshshah.ec@charusat.ac.in

K. Jasani
e-mail: karanjasani.ec@charusat.ac.in

R. Kher et al. (eds.), *Proceedings of the International Conference on Intelligent Systems and Signal Processing*, Advances in Intelligent Systems and Computing 671,
https://doi.org/10.1007/978-981-10-6977-2_9

1 Introduction

Market for image sensor was dominated by charge- coupled device (CCD). Effortlessness of on-chip integration, higher framing rate, less power utilization and lower fabricating costs pushed CMOS APS (Active Pixel) innovation to make up with CCDs [1] . For various scientific and technological applications, one of the most powerful sensing devices is CMOS image sensor [2].

Communication is an important factor which affects the performance of the system [3]. UART is a gadget that permits exchanging of information in non-concurrent and in a sequential manner from a serial port of PC [4]. UART makes the serial communication possible between two devices. UART contains parallel-to-serial converter for the information transmitted from PC and serial-to-parallel converter for the information coming in through the serial line. Likewise, UART has a support for briefly putting away information for rapid transmission [3]. Exchange of information between the FPGA and secondary devices in full-duplex concurrent mode is allowed by Serial Peripheral Interface (SPI). There are two modes of operation: master and slave [5]. SPI allows master device to initiate communication with slave device.

In this paper, the optimal configuration of various parameters of the CMOS image sensor is done using the UART–SPI interface. UART is used for the communication between FPGA and PC. SPI is used for the communication between FPGA and CMOS image sensor. CMOS image sensor is configured using SPI uploads. Once the CMOS image sensor parameters are configured, the CMOS image sensor data is read through SPI.

2 CMOS Image Sensor (CIS) Architecture

Figure 1 demonstrates the design of the CMOS image sensor. The sensor comprises a pixel array, data block, analog front end, and LVDS transmitter and receiver.

Separate modules for SPI, clock division and sequencer are also integrated. This architecture enables programmable addressing. The starting point of address can be uploaded by SPI. In this system, CMOS image sensor parameter configurations such as inserting the training pattern to sync LVDS receivers, setting the offset at the input of ADC, inserting a pattern to generate a test pattern, enabling a sequencer for image capture, image type selection settings like monochrome/grey image or colour image and many more things are done using SPI.

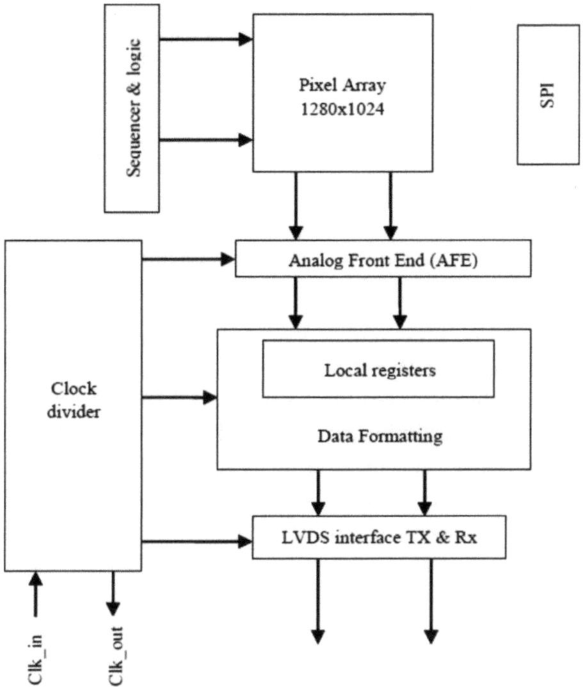

Fig. 1 Block diagram of CMOS image sensor

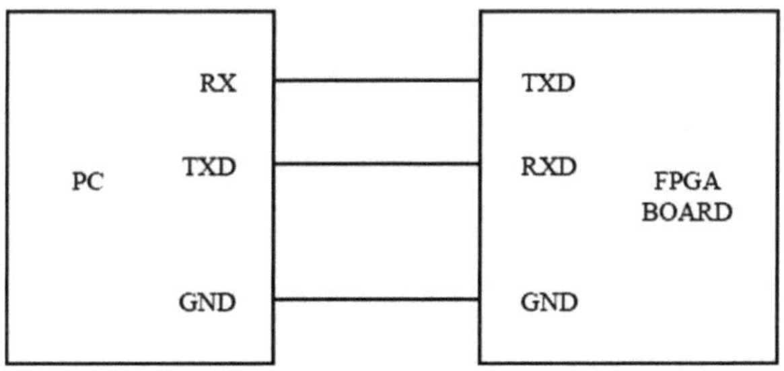

Fig. 2 Serial communication

3 UART (Universal Asynchronous Receiver/Transmitter)

In this system, UART core functions are implemented using VHDL and integrated them into FPGA. With help of this integration, we achieved stable, compact and reliable data transmission [6]. In this system, UART is implemented for the communication between labVIEW PC and FPGA. For establishing full-duplex data communication between two devices, UART needs two lines for transmission (TXD) and reception (RXD) (Fig. 2).

TXD on transmit side is the yield of Universal Asynchronous Receiver/Transmitter; RXD on receiver side is the contribution of Universal Asynchronous Receiver/Transmitter. “Start bit” is included by UART at the start of every word

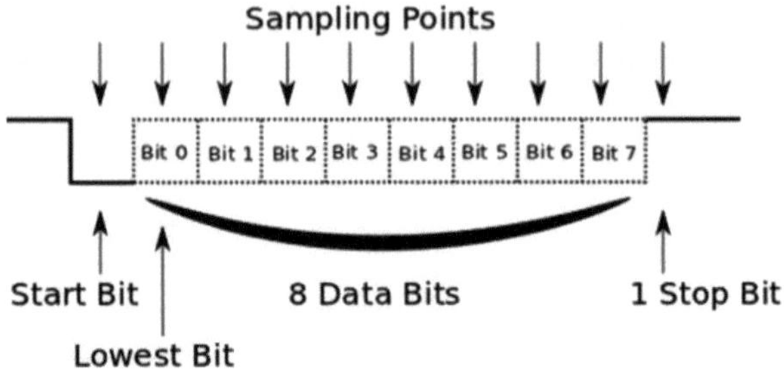

Fig. 3 UART frame structure

that will be transmitted. Start bit is utilized to recognize the recipient that now the information will be transmitted. LSB of the individual information bits is transmitted after the start bit. Even or odd parity bit may add by the sender (transmitter) after the entire information word has been sent. To check an error, parity bit is used. After parity bit, one stop bit is sent by sender (transmitter) [6].

UART incorporates sender (transmitter) and recipient. At sender (transmitter) side, parallel-to-serial change is performed, and on the recipient side, serial-to-parallel transformation is performed. Signal of the UART waits for the center and begins at the centre of sampling tick when the incoming signal changed to “0” (start bit). When the center of the sampling tick reaches 7, the middle position of the start bit will be reached by the incoming signal as shown in Fig. 3. It clears the center and starts again. When the center of the sampling tick reaches 15, the center of the first information bit will be reached by the signal. Restart at the centre and repeat the above step $N-1$ times to receive the rest of the information bits; if parity bit is utilized, then repeat it one more time [7].

Figure 4 shows the front panel for serial communication with FPGA. To start the serial communication with FPGA using front panel we need to set the different parameters, for example, baud rate, serial port, no. of stop bits, no. of information bits, no. of stop bits, parity, and timeout of the serial port of the PC for correspondence. After setting all such parameters, communication is made possible between PC and FPGA. Figure 5 shows the transmission and reception of character “*A*”, transmitting “*A*” to FPGA and receiving same character on PC at a baud rate of 9600 bps.

4 Serial Peripheral Interface

Communication with a slave device can be initiated by master device which is allowed by synchronous protocol such as Serial Peripheral Interface (SPI). Transfer of data is performed from master to slave devices and vice versa [7]. Rapid data transmission is allowed by SPI between two or more devices, and implementation of SPI is also very easy. Master device provides the clock signal to maintain the synchronization. Changing and reading of data are under the control of the clock

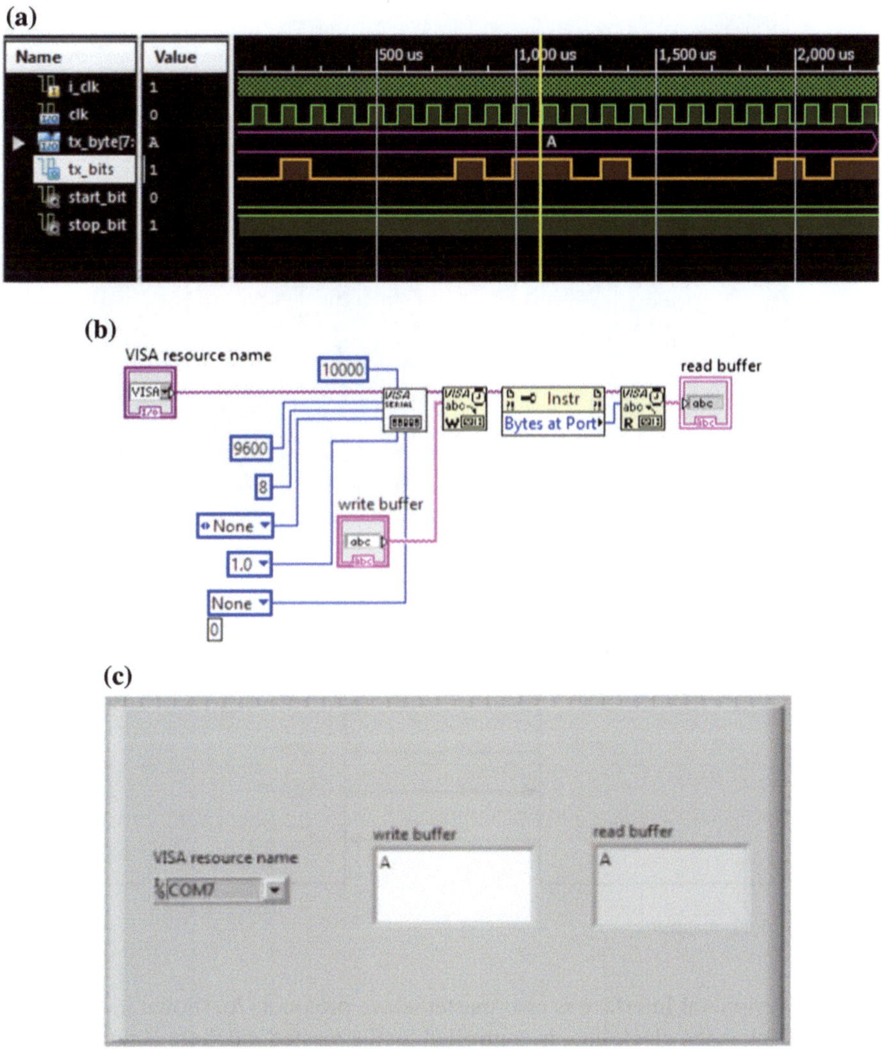

Fig. 4 **a** Software simulation result, **b** block diagram and **c** front panel in LabVIEW

signal. The communication of the data between two devices is done with reference to the clock signal, so SPI is known as synchronous protocol [8]:

- SCLK (serial clock)—generated by the master;
- MOSI (master out slave in)—used to transfer data to the slave device;
- MISO (master in slave out)—used to transfer data out of the slave device;
- SS (slave select)—for enabling device operation.

Fig. 5 Implementation on hardware

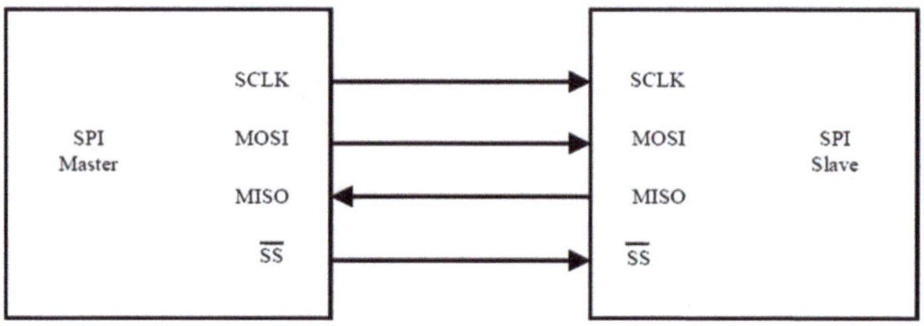

Fig. 6 SPI block diagram

Serial Peripheral Interface is also master–slave protocol. As shown in Fig. 6, the SCLK, which is the clock line, is controlled by the master and there is no transfer of data performed until and unless the clock is manipulated. Multiple slave devices can also be allowed with individual slave select line. All slaves are controlled by master.

The information on the data "in" line is:

- Command bit, indicating write (1) or read (0) access;
- 7-bit address;
- 8-bit data word.

Data is always written on the bus on the falling edge of the clock and will be sampled on the rising edge of the clock. As shown in Figs. 7 and 8, the system clock must be active to keep SPI uploads stored on the chip. The SPI clock speed must be slower by a factor of 30 when compared to system clock. The "out" line is

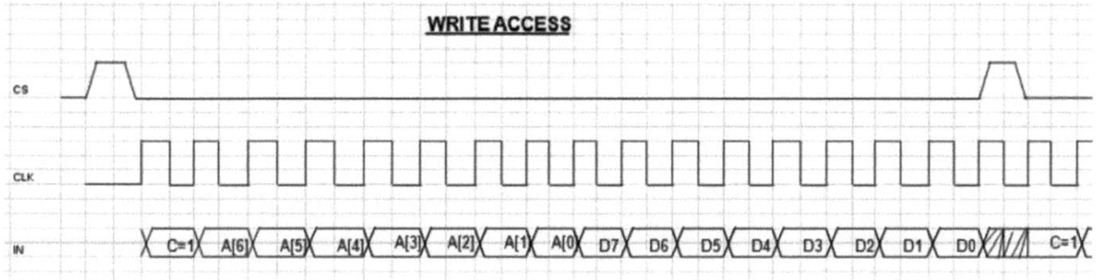

Fig. 7 Write access ($C = 1$)

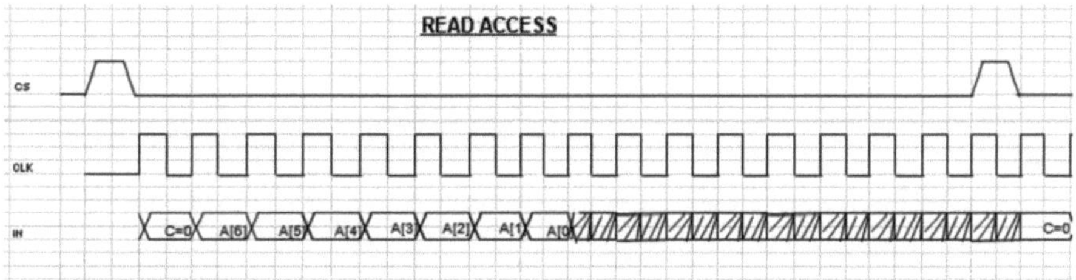

Fig. 8 Read access ($C = 0$)

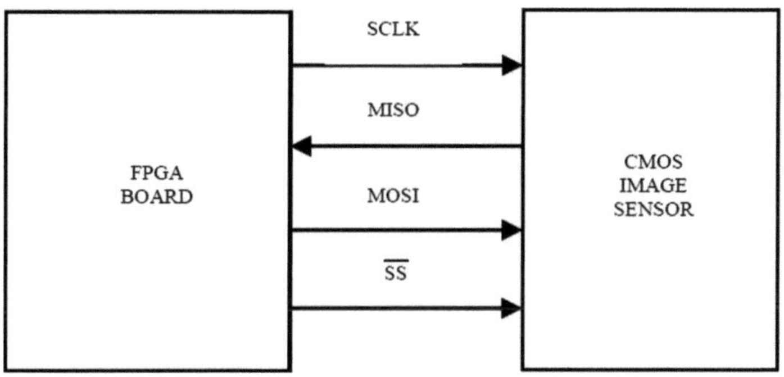

Fig. 9 CMOS image sensor configuration

held high *Z*. The data for the address *A* is transferred from the shift register to the active register bank (i.e. sampled) on a rising edge of cs_n. Only the register block with address *A* can write its data on the "out" bus. The data on "in" is ignored.

Figure 9 shows the signals that are used in SPI for the CMOS image sensor configuration. This configuration sets the values for the CMOS image sensor parameters. These parameters include inserting the training pattern to sync LVDS receivers, setting the offset at the input of ADC, inserting a pattern to generate a test pattern, enabling a sequencer for image capture, image type selection settings like monochrome/grey image or colour image setting and many more (Figs. 10 and 11).

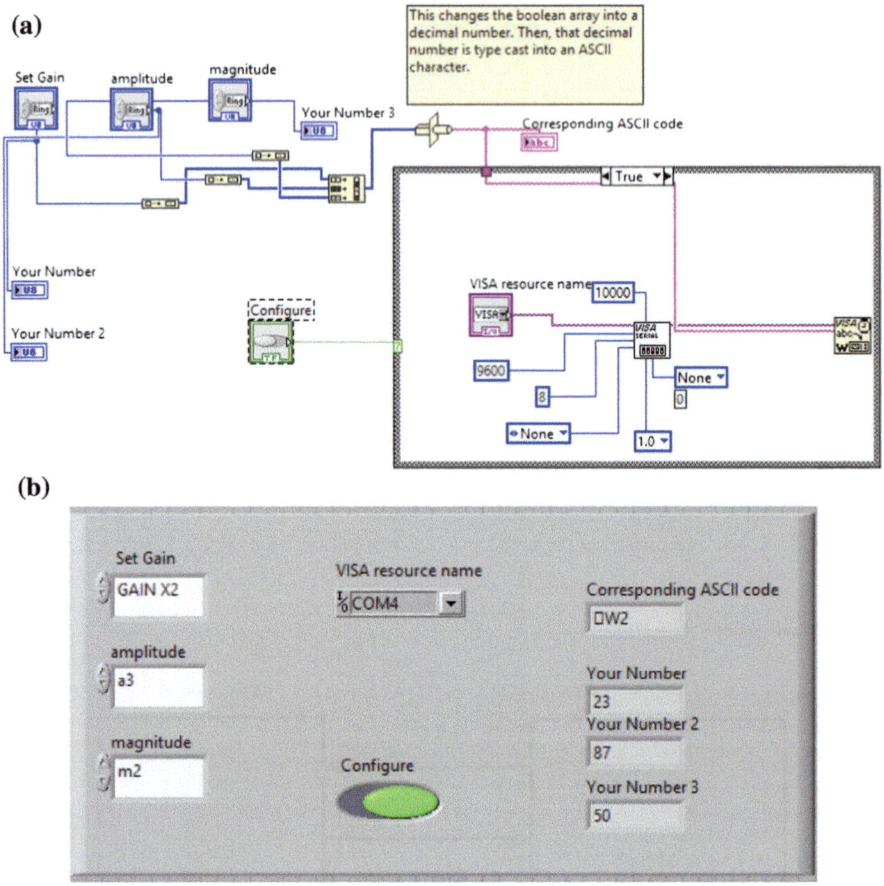

Fig. 10 **a** Block diagram and **b** front panel in LabVIEW

5 System Design

Figure 12 shows the full system design for the configuration of CMOS image sensor from LabVIEW through FPGA using UART (for the communication between LabVIEW and FPGA) and SPI (for the communication between FPGA and CMOS image sensors) interfacing. With the help of UART, PC can now communicate with FPGA. Data from FPGA can be received on PC, and it can also receive data from PC. If we configure the CMOS image sensor directly from FPGA, there is need to make changes in the code every time when the data changes. After saving the changes, load it into FPGA. But if we give/set the values of the different parameters (gain, amplitude, magnitude, etc.) for the CMOS image sensor configuration from the LabVIEW using the VI shown in Fig. 10b, there is no need to make any changes in the VHDL code, and you only need to select data that are

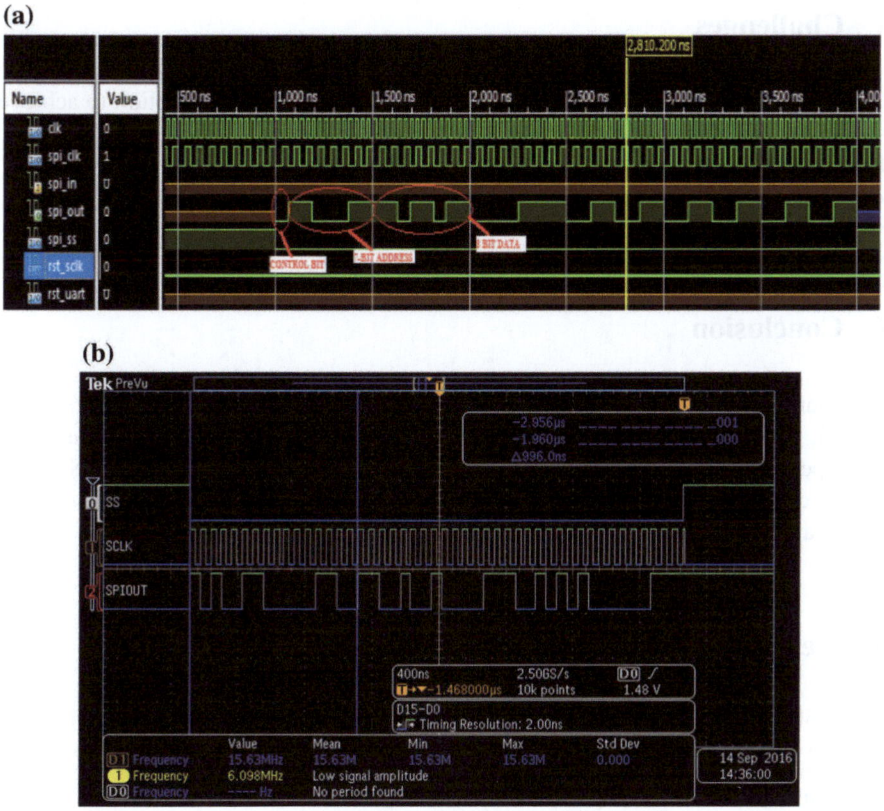

Fig. 11 **a** Software simulation result and **b** result on MDO

Fig. 12 Complete system using UART-SPI interface

listed in LabVIEW VI, then select the COM port to send the selected data values to the CMOS image sensor, at last press the configure button as shown in Fig. 10b and then run the VI. So, by using this way we can give any data values and no need to generate and load VHDL file again. All this can be done through UART–SPI interfacing.

6 Challenges

(a) Deciding various values for image sensor registers for configuration to achieve best possible image;
(b) Synchronization problems during UART–SPI interfacing;
(c) Working with different clock domains.

7 Conclusion

Communication using UART is less speedy, but there are no time constraints in the system, so UART is used. Also, it is cheaper than PCI/PXI card that is used for high-speed data transfer. UART–SPI interface is implemented for the CMOS image sensor configuration. Using UART–SPI interface, we can easily configure the sensor and no need to change the code again and again.

References

1. Orit Skorka and Dileepan Joseph, "CMOS digital pixel sensors: technology and applications", SPIE, 2014.
2. Takshi Tokuda, Toshihiko Noda, Kiyotaka Sasagawa, Jun Ohta, "Optical and Electric Multifunctional CMOS Image Sensors for On-Chip Biosensing Applications", 2011.
3. Vibhu Chinmay, Shubham Sachdeva, "A Review Paper on Design and Simulation of UART for Serial Communication", IJIRT, 2014.
4. R. K. Agrawal and V. R. Mishra, "The design of high speed UART", 2013 IEEE Conference on Information & Communication Technologies, JeJu Island, 2013, pp. 388–390.
5. Prof. Jai Karan singh, Prof. Mukesh Tiwari, Vishal Sharama, "Implementation of SPI-Slave on FPGA", IJAET, 2012.
6. Y. Y. Fang and X. J. Chen, "Design and Simulation of UART Serial Communication Module Based on VHDL", 2011 3rd International Workshop on Intelligent Systems and Applications, Wuhan, 2011, pp. 1–4.
7. T. Praveen Blessington, B. Bhanu Murthy, G. V. Ganesh and T. S. R.Prasad, "Optimal implementation of UART-SPI Interface in SoC", 2012 International Conference on Devices, Circuits and Systems (ICDCS), Coimbatore, 2012, pp. 673–677.
8. M. Poorani and R. Kurunjimalar, "Design implementation of UART and SPI in single FGPA", 2016 10th International Conference on Intelligent Systems and Control (ISCO), Coimbatore, 2016, pp. 1–5.

Bridge Software Between Zigbee and CAN Protocols for Distributed Embedded System

Hiteshkumar J. Lad and Vibhutikumar G. Joshi

Abstract An industrial plant production quality can be improved by applying robotic arms and automated system. Distributed embedded system shares information and cooperates to control automatic operations in real-time environment. Automated distributed embedded system network technology, communication media topology (wired or wireless or heterogeneous), and protocol selection are highly depended on the application environment. The challenging part in system development is to provide compatibility and transaction of data between two different network protocols used in heterogeneous distributed embedded system. In this paper, heterogeneous distributed embedded system (DES) is proposed using controller area network (wired network) and Zigbee (wireless) protocol. Zigbee and CAN network bridge application software is implemented on Freescale MCS9s08DZ60 microcontroller. Software establishes bridge between Zigbee and CAN protocols for data transaction and protocol compatibility. Such bridge software establishes communication between two physically separated CAN networks via wireless Zigbee-based network.

Keywords Controller area network (CAN) · Zigbee · Distributed embedded system (DES) · Automation

1 Introduction

In industrial plant, product cost and quality can be improved by using advanced technology-based automatic system. Such advanced system monitors different parameters from the distributed sensors and controls particular actuators and

---

H. J. Lad (✉)
Department of Information and Communication Technology, Surat, India
e-mail: hiteshlad.msc@gmail.com

V. G. Joshi
Department of Physics, Veer Narmad South Gujarat University, Surat, India
e-mail: joshi_vibhuti@yahoo.co.in

R. Kher et al. (eds.), *Proceedings of the International Conference on Intelligent Systems and Signal Processing*, Advances in Intelligent Systems and Computing 671,
https://doi.org/10.1007/978-981-10-6977-2_10

physical activity in the field. The automatic system operation efficiency can be improved by coupling the sensing, computing, actuating, and communication activities in real-time controlled manner [1]. Normally in the plant, heavy machinery section and mechanical machines are controlled through electronic circuits and its internal communication with field bus (wired)-based DES. Industrial plants have machinery installed in wide area and therefore to reduce wire harness, complexity and cost, short range wireless communication is used between small machinery sections [2]. Industrial real field applications need to establish heterogeneous network using wired and wireless distributed embedded system which collaboratively performs actions and fulfills application requirements. The heterogeneous networked DES connected with Internet and run time processes information access and control can be possible from remote location.

The percentage of production cost of electronic circuits (including software) increased from 19% in 2004 to 40% in 2010, and it will eventually reach 50% in 2020 [3]. The software development and modification approach can be better option to reduce development time, complexity, and cost. In this paper, concept of bridge software is used to establish interconnections between CAN protocol of field bus and Zigbee protocol of wireless network and also to establish heterogeneous network. Two physically separated CAN networks are connected through Zigbee wireless network which will work as single heterogeneous network.

2 Introduction of Protocols

2.1 Controller Area Network (CAN)

CAN uses message broadcasting method for message transmission in which single CAN node transmits a message on two-wired field bus and all other connected nodes receive message. Multi-master policy of CAN allows that multiple nodes can be a master and at the same time it can start communication but only one node can do transmission on a serial field bus at a time. Figure 1 shows data frame structure of CAN 2.0A field bus protocol. CAN 2.0A data frame's First field is Start of Frame (SOF), which is of one bit field is used to synchronize CAN bus connected nodes. Next '11'-bit arbitration field is considered as frame identifier field, and it is useful to decide frame transmission priority during arbitration process. Next six-bit control field contained information about frame type and payload size. Data field length is about 8 byte, which contains payload of the message. Followed by cyclic redundancy check (CRC), 15-bit field helps to decide received message content is same as transmitted message or not. Next field is for acknowledgment field and End of Frame (EOF) field [4]. CAN protocol uses non-return-to-zero coding for information transmission [5, 6]. The CAN field bus can transmit data maximum up to 1 Mbps data rate.

Fig. 1 CAN 2.0A data frame format

Start of frame	Arbit ration field	Control field	Data field	CRC field	ACK field	End of frame

Fig. 2 Zigbee data frame format

Frame control	Sequence number	Address information	Payload Data	Frame Check Sequence

2.2 *Zigbee Protocol*

Figure 2 shows Zigbee protocol (IEEE 802.15.4 standard specification) data frame. Three license-free ISM frequency bands: 2.4–2.4835 GHz, 868–870, and 902–928 MHz can be used to establish Zigbee protocol-based wireless communication [7].

Zigbee nodes follows Carrier Sense Multiple Access with Collision Avoidance (CSMA—CA) policy to access communication channel. Zigbee communication supports at three different maximum data transfer rates (1) 250 kbps at 2.4 GHz, (2) 40 kbps data rate at 915 MHz, and (3) 20 kbps data rate at 868 MHz [8]. Zigbee data frame's 'frame control field' carries information regarding frame type, delivery mode, addressing mode, security, and acknowledge service information. The frame is numbered to ensure a track that all frames are received.

The data frame payload field is carrying data. A frame-check sequence helps to check received frame at receiver is errorless or not. This frame structure improves reliability in difficult conditions [8].

3 System Architecture

As shown in Fig. 3, CAN nodes contain 8051 core microcontroller and sensor device. Once CAN node receives sensor data, it transmits on CAN bus. If transmitted message is only for local CAN network, then CAN gateway is inactive. If transmitted message needs to transfer to other CAN network also, then CAN gateway becomes active and transfers received data frame content of the message to bridge software. Bridge software transfers CAN data frame arbitration field, control field and payload content to Zigbee payload field. Finally, using Zigbee transmitter node data are transmitted wirelessly to the Zigbee receiver module and computer-connected Zigbee node. At receiver network, microcontroller extracts Zigbee data frame payload information in the variable. Receiver bridge software sets arbitration field, control field, and data payload field at CAN gateway and

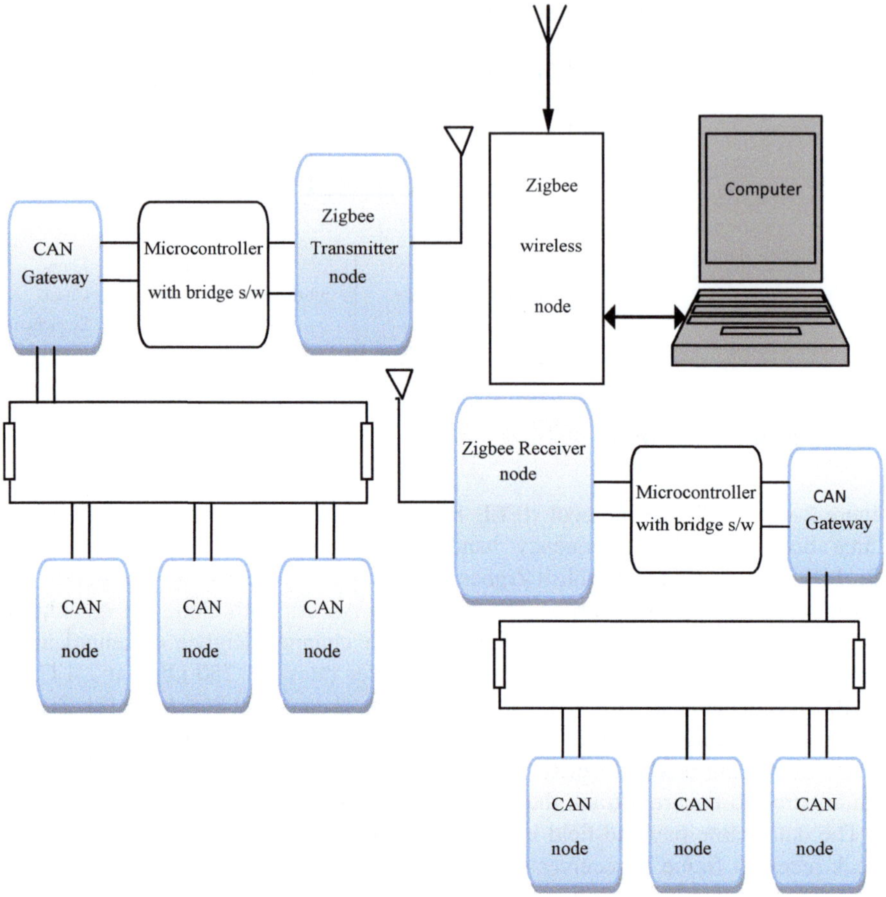

Fig. 3 Heterogeneous network-based system architecture

finally transmits it on CAN network. Zigbee-connected computer maintains data log, and it can perform remote control actions also as per system requirements.

4 Experimental Work

Free scale code-worrier software has debug facility with simulator and emulator, which provides great support. In the experiment two physically separated CAN based wired DES are established and communication between these was established through Zigbee nodes. All CAN modules contain microcontroller MCS9sDZ60 and CAN trans-receiver module TJA1040. One of CAN-based DES functions as measurement network, and its one CAN module senses temperature data using on chip available temperature sensor of MCS9sDZ60 microcontroller and sends it on CAN network. Second CAN module functions as CAN network gateway which loaded with CAN to Zigbee protocol bridge program software. This bridge program transforms content of arbitration field, control field, and payload of CAN data frame to Zigbee data frame payload field as shown in flowchart (Fig. 4). Maximum 12 bytes of Zigbee payload is require to store information about CAN data frame header and 8-byte payload. Two CAN-connected nodes make one sensor cluster of CAN network.

Another CAN-based DES performs as actuator network, and its one CAN module functions as CAN network gateway which is loaded with Zigbee to CAN protocol bridge program software. This module microcontroller extracts information from Zigbee data frame and transfers it to CAN data frame fields and transmits it on CAN network as shown in flowchart (Fig. 5). All other CAN-connected node/nodes functions as actuator control node whereas LED functions as an actuator. In implemented system, two microcontroller nodes make actuator cluster of CAN network. Third Zigbee module connected with computer and communication activities like signal strength and packet loss can be observed through XCTU software. CAN module and Zigbee module communication is established using UART communication at 9600 bps baud rate.

5 Conclusion

Bridge software for CAN and Zigbee protocols is successfully implemented with the help of Freescale 8051 core microcontroller MCS9S08DZ60. In this system, totally two CAN network clusters: (1) sensor cluster and (2) actuator cluster are established. Cluster node internal communication has been successfully established using CAN protocol-based communication at the distance of 100 m with 125 kB/s data rate. Intra-cluster communication has been also successfully established using Zigbee protocol-based communication up to the distance of 80 m. Each cluster CAN network gateway is implemented using Zigbee-CAN bridge software successfully and established wireless communication between two-wired CAN network-connected clusters. Zigbee module transmits frame signal strength measured at various distances in open field with maintaining line of sight as shown in Table 1. Zigbee received frame's received signal strength indicator (RSSI) is reduced with the increasing of distance. As shown in Table 1, received frame RSSI value after 80 m distance frame indicates frame information power loss is increased. RSSI values of frame for 80 m and more distance indicate wireless communication might generate information loss and increase number of same frame retransmissions on network.

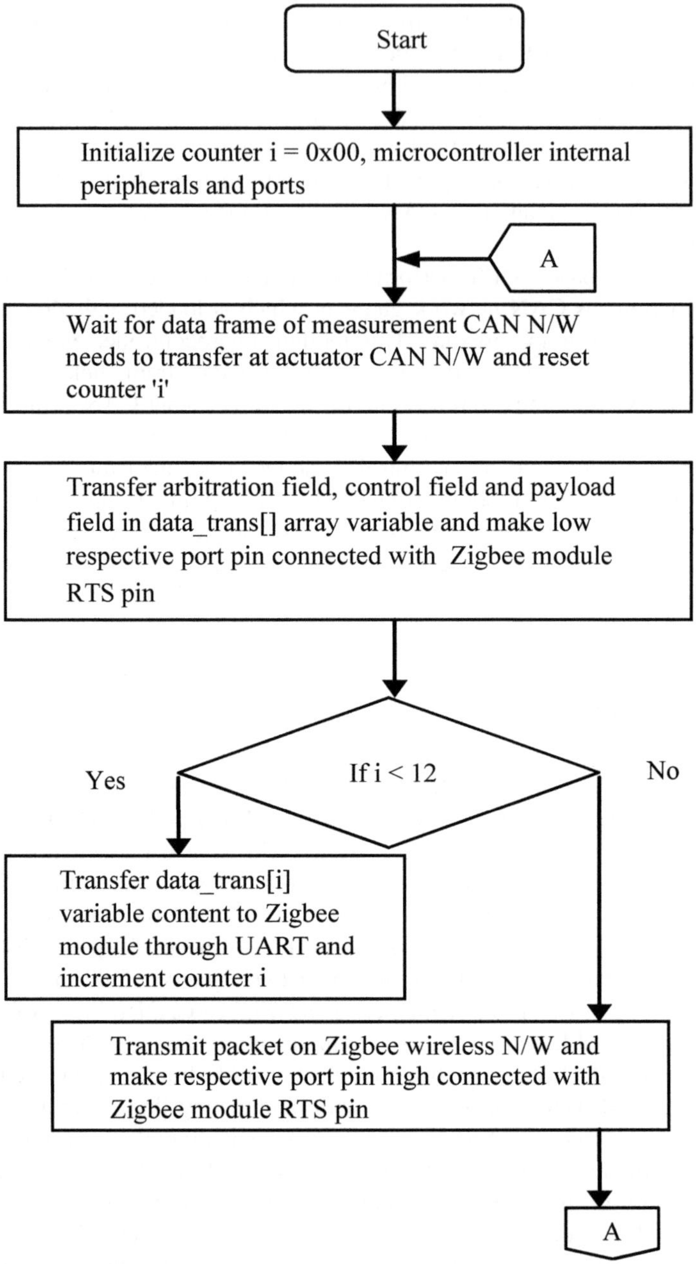

Fig. 4 Flowchart of bridge program at measurement

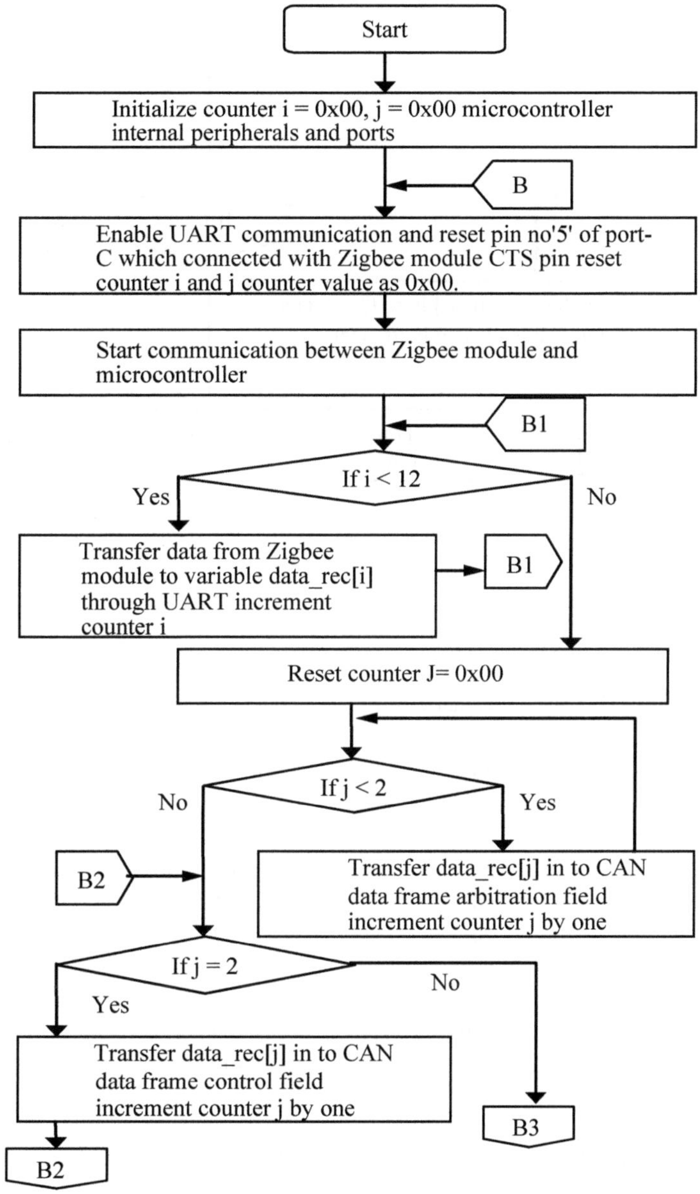

Fig. 5 Flowchart of bridge program at actuator network

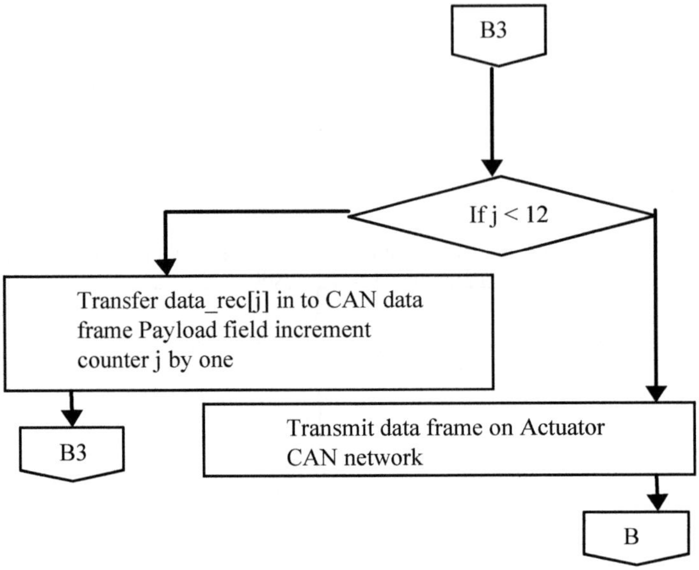

Fig. 5 (continued)

Table 1 Measurements of received signal strength indicator (RSSI) at different distances

Distance (m)	Percentage of RSSI (dbm)
10	−36
20	−53
30	−58
40	−64
50	−67
60	−71
70	−77
80	−85
90	−88
100	−90
110	−92

The bridge-based internetworked DES can establish electronic circuit network where wired network installation is not possible. In future, heterogeneous network utilities can be developed by establishing the bridge protocol between advanced protocols of the wired and wireless distributed embedded system for industrial or open field applications.

References

1. Bo Shen, Xingshe Zhou, Ru Wang, " BER Analysis for Controller Area Network Impaired by the Impulse Noise in Cyber-physical Systems", IEEE International Conference on Computer and Information Technology, 978-1-4799-6239-6/14, pp. 425–429, 2014.
2. Xiaohong Ren, Chenghua Fu, Tianwen Wang, Shuxiang Jia, "CAN Bus Network Design based on Bluetooth Technology", International Conference on Electrical and Control Engineering, 978-0-7695-4031-3/10, pp. 560–564, 2010.
3. Armin Wasicek, Oliver Höftberger, Martin Elshuber, Haris Isakovic, Andreas Fleck, "Virtual CAN Lines in an Integrated MPSoC Architecture", IEEE 17th International Symposium on Object/Component-Oriented Real-Time Distributed Computing, 1555-0885/14, pp. 158–165, 2014.
4. Thomas Nolte, Hans Hansson, and Christer Norstrom, "Probabilistic Worst-Case Response-Time Analysis for the Controller Area Network", Proceedings of the 9th IEEE Real-Time and Embedded Technology and Applications Symposium IEEE, 1080-1812/03,2003.
5. Mouaaz Nahas, "Developing a Novel Shared-Clock Scheduling Protocol for Highly-Predictable Distributed Real-Time Embedded Systems", American Journal of Intelligent Systems, DOI:10.5923/j.ajis.20120205.06, pp. 118–128, 2012.
6. Mouaaz Nahas, Michael J. Pont, Michael Short, "Reducing message length variations in resource-constrained embedded systems implemented using the Controller Area Network (CAN) protocol", Journal of Systems Architecture 55, Elsevier, 1383-7621. pp. 344–354, 2009.
7. P. Susmitha, V. Bhavya Reddy, Maninder Kaur, "Low Power Zigbee Technology in Wireless Mesh Networks", International Journal of Advanced Research in Electrical, Electronics and Instrumentation Engineering, Vol. 2, Issue 2, ISSN(Online): 2278–8875, February 2013.
8. Dr. S.S. Riaz Ahamed, "The role of Zigbee Technology in future Data Communication", Journal of Theoretical and Applied Information Technology, JATIT, www.jatit.org, 2009.

A New Global Shutter 8T CIS Architecture with CDS Operation

Kishankumar Suthar and Rajesh Thakker

Abstract CMOS image sensor (CIS) is heart of neoteric imaging system. Global shutter image sensor is more popular. In global shutter image sensor, each pixel has inbuilt storage capacity, by which it is possible to capture whole image at a time. Hence, it used for high-quality and motion blur-free imaging. Correlated double sampling (CDS) mechanism is introduced to reduce noise, and also it enables high-speed imaging. To incorporate CDS mechanism, at least two storage elements are required to obtain difference of reset signal and light signal. In this paper, new 8T global shutter pixel (GSP) based on 3T active pixel sensor (APS) is proposed. This 8T CIS architecture incorporates two storage elements for CDS operation. Theoretical analysis and simulation results based on spice model are reported in this paper. Simulation of proposed architecture is carried out in 0.35 μm technology using virtuoso analog design environment.

Keywords Global shutter pixel · CMOS image sensor (CIS) · In-pixel CDS
Low-noise image sensor · Low-power CIS

1 Introduction

A sensor is nothing but one type of transducer which converts energy from one form to another. Image sensor converts photon energy of light into electrical form, typically, in charge or voltage domain, depending upon the type of image sensor [1]. Photosensitive components like photodiode (PD) can perform this task. A charge-coupled device (CCD)-based structure can store photo-generated electrons and is capable to transfer it toward the column amplifier, but it is not compatible with standard CMOS process [2]. A pinned photodiode (PPD)-based structure is used in which transfer gate is implemented in pixel to transfer electrons

K. Suthar (✉) · R. Thakker
Vishwakarma Government Engineering College, Chandkheda, Ahmedabad 382424, India
e-mail: sutharkishan12@gmail.com

R. Kher et al. (eds.), *Proceedings of the International Conference on Intelligent Systems and Signal Processing*, Advances in Intelligent Systems and Computing 671,
https://doi.org/10.1007/978-981-10-6977-2_11

from the photodiode to the floating diffusion node [3]. It needs complicated process during fabrication.

The camera shutter allows light to pass through for a determined period of time. It can be mechanical or electronic mechanism. Method of light exposure can be global, block, or rolling. In rolling shutter, the image rows are sequentially reset and readout, while in a global shutter camera, all pixels in a frame capture whole image at a single time. Note that readout operation in both cases is same. Global shutter image sensors are essential where entire image is to be captured at a same time, which is required to eliminate motion distortions. Such applications are high-speed and machine-vision imaging. Global shutter imagers offer several advantages such as improved SNR, lower power consumption, and high-speed readout [2]. The global shutter imagers have number of challenges to be addressed. It should have high dynamic range (DR), good fill factor (FF), low-noise, high-speed readout, etc. Further, CDS is necessary for low-noise imaging. CDS operation can be performed by external circuitry or can be performed in pixel itself. In-pixel CDS operation reduces time consumption, hence, increases frame rate and easy to implement. But, it increases area consumption and complexity of pixel architecture, and hence, restricts the FF.

This paper describes the architecture and the operation of a new 8T global shutter pixel image sensor with in-pixel CDS. The proposed CDS architecture eliminates some of the shortcomings of the recent global shutter designs [4–10] such as the capacitive division between the reset and signal samples. The sensor can be used for traditional global shutter application and other high-speed imaging applications such as machine-vision imaging.

2 Literature Review

Solid-state sensors were introduced in 1960. Passive image sensor was introduced first which contains photodiode and one switch transistor only [1]. Active pixel sensor (APS) was introduced in 1968 to overcome limitations such as large smear, k_BTC noise, column fixed pattern noise (FPN) of early passive image sensor. 3T APS is basic active pixel sensor with lowest complexity and good dynamic range [11, 12]. 3T APS has trade-off between the full-well capacity and the conversion gain. It also suffers from k_BTC noise. To overcome these limitations, 4T and 5T structures were proposed [13–17]. They are used in modern imagers as pioneer blocks. Recent pixel architectures are proposed to incorporate in-pixel CDS and global shutter compatibility [4–10].

Capacitive transimpedance amplifier (CTIA) is also used for in-pixel CDS or delta difference sampling (DDS) operation [18, 19]. CTIA CIS has good voltage swings with 44 dB SNR, 70 frames/s frame rate and having improved sensitivity. CTIA has good linearity and sensitivity but needs larger area and high power; hence, it is used where area and power are not design constraints. Global shutter pixel architectures [6–10] have low power consumption and can be operated at fast

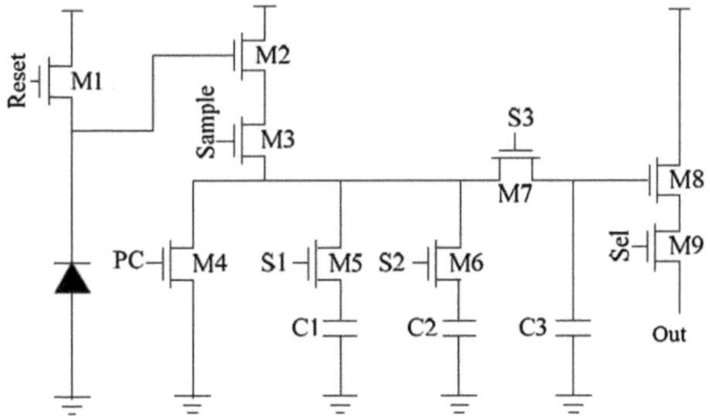

Fig. 1 Schematic of 9T CIS [7]

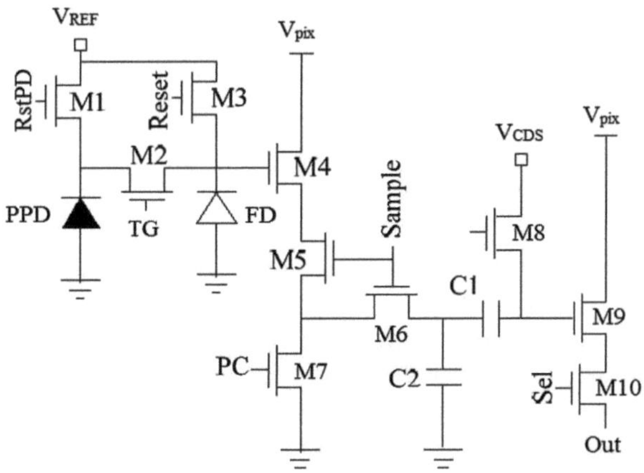

Fig. 2 Schematic of 10T CIS [10]

frame rate. To incorporate in-pixel CDS, 8T CIS was introduced. After that, 9T CIS was introduced in [7]. 9T CIS has 800 mV 'voltage swing' with less than 1% linearity error. Figure 1 shows 9T CIS pixel architecture. Implementation of new method for CDS operation was proposed in 10T CIS [10]. Figure 2 shows 10T CIS pixel architecture. 10T CIS has approximately 1 V voltage swing, read out noise ratio of 3400:1, maximum image lag of 10 ke^{-}, and 71 dB dynamic range (DR).

3 Proposed Architecture

The design of the pixel architecture can be divided into three blocks: (1) the photodiode, (2) 3T active pixel, and (3) CDS mechanism. The pixel architecture proposed in this work is shown in Fig. 3.

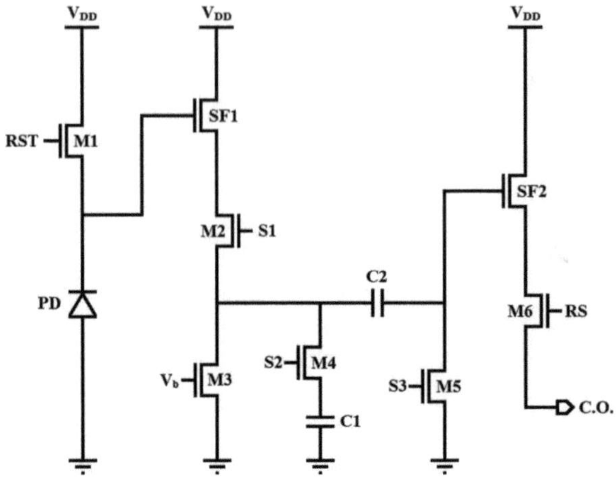

Fig. 3 Schematic of 8T CIS. RST—Reset, SF—Source follower, C—Capacitor, M—Transistor, RS—Row select, V_b—Bias voltage, CO—Column out, PD—Photodiode, V_{DD}—Supply voltage, S—Switch pulse

3.1 Photodiode

PD is a PN junction photodiode, and it is only photosensitive component in the circuit. With standard CMOS fabrication technique, n-well/p-sub photodiodes can be made. This topology gives advantages like low-doped n-well which results into wider depletion region and hence increases collection efficiency of the junction. Here, n-wells are created by diffusion with junction tends to be deeper than n+/p-sub junctions which further increases collection efficiency, and also wider depletion region leads to a smaller capacitance. It improves charge-to-voltage conversion ratio. Note that all modern CISs are based on the pinned photodiode (PPD) due to its ability to perform efficient CDS and capability to eliminate reset noise present in traditional 3T APS. Also the dark current in PPD is typically very low [10]. We can also incorporate PPD instead of PD.

3.2 3T Active Pixel

Transistors M1, SF1, M2, and PD together form a traditional 3T APS, in which M1 works as a reset switch and clears the charge stored in PD. Electron-hole pair is generated due to light exposure (sometimes called photo-generated carrier) and changes potential of PN junction. It is reflected between terminals of PD. After amplification, this potential can be sampled outside. Due to high input resistance, source follower is best choice to perform this task. Threshold voltage of source follower should be as low as possible (ideally zero), and also it should be less noisy.

M2 works as select switch for this basic 3T APS. Voltage across PD will be reflected to column readout circuitry or CDS mechanism, only when select switch, i.e., M2, is in 'ON' state. One load transistor M3 is incorporated with 3T APS to maintain required small standing current I_{SF1} for the first-stage source follower SF1. M3 is also used for two different purposes as explained later in this paper.

3.3 CDS Mechanism

Transistors M4–M5 and storage elements C1–C2 together form CDS mechanism, where storage elements, i.e., capacitors, are used for dual purpose to sample the reset and light signals and perform arithmetic operation between them. M4 works as a switch and decides when to sample reset signal to capacitor C1. Same way transistor M5 works as a switch and decides when to sample light signal to capacitor C2. Different combinations of state of transistors M2, M3, M4, and M5 are used for different purposes, which is more clearly described in next section. Transistors SF2 and M6 together form second-stage source follower which works same as combination of SF1 and M2 works. Output of this second-stage source follower is not just voltage level but it contains CDS-processed image information on which necessary data processing can be performed directly.

4 Analysis and Operation

For proper pixel capture operation, four processes should be followed: (1) reset, (2) integration, (3) sampling light information, and (4) row readout. Pixel capture operation can be controlled by giving required pulse sequences to pixel.

4.1 Timing Control

The timing sequence of various control signals of proposed 8T CIS circuit (Fig. 3) is shown in Fig. 4, and required voltage level is given in Table 1. Process to be followed and related signal voltage are given in respective column. Figure 5 shows timing scheme for entire frame operation. Detailed explanation with equations is provided in the following text.

(A) For reset operation, M1 should be in 'ON' condition. At same time, we can sample voltage of PD to capacitor C1. Note that at this time, transistor M3 works as load. Column 'A' represents required signal voltages to be given for these two tasks.

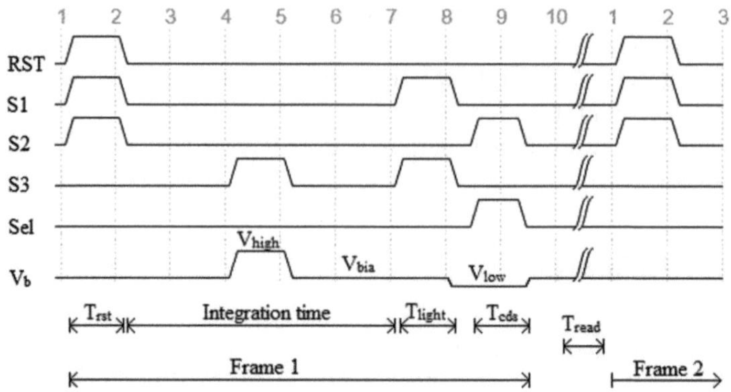

Fig. 4 Control signal timing

Table 1 Voltage level for different control signals of 8T CIS shown in Fig. 3

Signal	A	B	C	D
RST	V_{high}	V_{low}	V_{low}	V_{low}
V_b	V_{bias}	V_{high}	V_{bias}	V_{low}
S1	V_{high}	V_{low}	V_{high}	V_{low}
S2	V_{high}	V_{low}	V_{low}	V_{high}
S3	V_{low}	V_{high}	V_{high}	V_{low}
RS	V_{low}	V_{low}	V_{low}	V_{high}

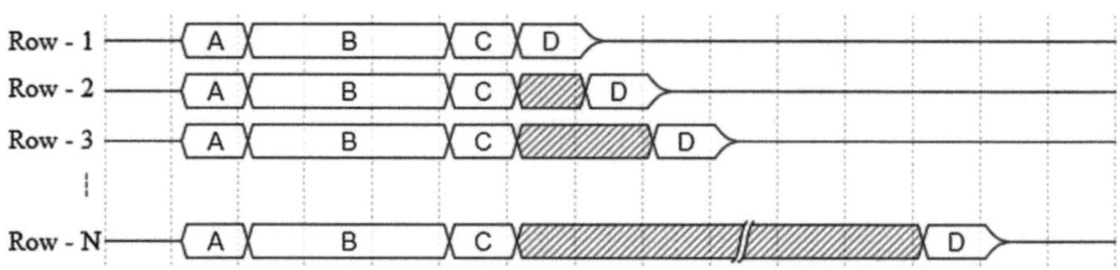

Fig. 5 Frame operation timing scheme

(B) After that, some time should be allocated as exposure time. In between, capacitor C2 can discharge through M3 and M5. In this process, M3 is in 'ON' state along with M5. Discharging time is not necessary to be long as exposure time because capacitor can discharge in very short time compare to exposure time.

(C) After exposure, voltage across PD is sampled across capacitor C2. This voltage is denoted by V_t. M2 and M5 should be in 'ON' condition.

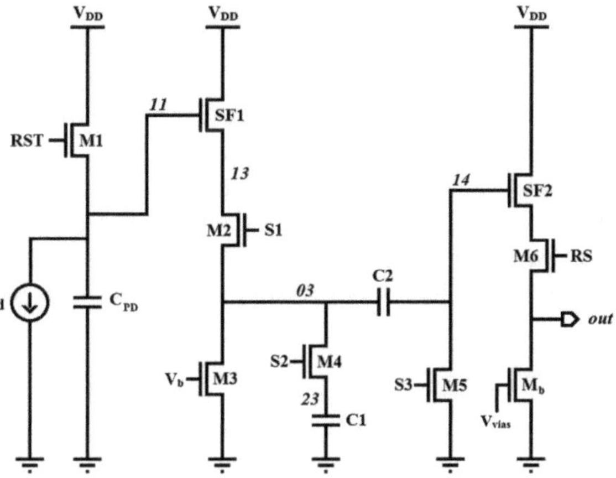

Fig. 6 Simulation scheme

(D) During this time, CDS and row readout operations can be performed simultaneously. M6 and M4 should be in 'ON' condition. Here, M3 should be in cutoff condition to prevent unnecessary discharge which causes information loss.

4.2 CIS Circuit Operation

Traditional 3T structure and its working are reported in many literatures [1–3, 11–13]. For accumulation mode, PD can be considered as parallel combination of capacitor and discharging current source [2, 3]. During reset operation, the voltage across PD will be

$$V_{\text{reset}} = V_{\text{DD}} - V_{\text{th}} \tag{1}$$

where V_{th} is threshold voltage of transistor M1. Reset voltage (V_{reset}) is sampled across capacitor C1 through transistors M2 and M4. Here, transistor M3 should be biased in such a way that maximum potential drops across M3. Reset sample (voltage across C1) is $V_{\text{reset}} - V_{\text{th}}$. Here, V_{th} is threshold voltage of transistor SF1. Next, M1 is turned off, and the PD is electrically floated. The photo-generated carriers accumulate in the PD junction capacitance as light is incident on it. The potential of the PD varies due to accumulated charges. The voltage of the PD decreases depending upon intensity of input light. Meanwhile, capacitor C2 is discharged through transistors M5 and M3, by which charges of previous frame are removed and the capacitor is ready for next frame.

Note that capacitor C1 is not needed to be discharged as reset voltage will be approximately same, i.e., variance of reset voltages is very low. After an accumulation time, the transistor M2 is turned on, and the signal in the PD is readout and sampled across storage element C2 through transistors M2 and M5. As in reset signal sampling, M3 is in biased condition.

The voltage sampled across C_2 corresponding to light intensity is of $V_t - V_{th}$. Here, V_{th} is threshold voltage of transistor SF1, and V_t is voltage of photodiode after accumulation time. After sampling, transistors M2, M3, and M5 are turned off; M4 and M6 are turned on, by which CDS operation is performed. Voltage on gate of SF2 is

$$V_{g,SF2} = (V_{reset} - V_{th}) - (V_t - V_{th}) \tag{2}$$

$$V_{g,SF2} = V_{reset} - V_t \tag{3}$$

Therefore, voltage at output node is,

$$V_{out} = V_{g,SF2} - V_{th} \tag{4}$$

Here, V_{th} is threshold voltage of second-stage source follower SF2. Threshold voltages of M1, SF1, and SF2 greatly affect voltage swing of this architecture. We may use hard reset on M1 and special purpose transistor with zero threshold voltage, to increase voltage swing or voltage overhead.

5 Discussion/Advantage

The proposed pixel architecture in this paper offers several advantages over previous designs. The bias current in the first-stage source follower is controlled by adding the switch M2 which greatly reduces power consumption.

Proposed architecture is better compare to existing 8T and 9T designs. This circuit does not use charge sharing between the reset and the signal samples stored in the in-pixel capacitors. CDS signal amplitude is not attenuated because there is no capacitive division between the two samples. It also increases the device sensitivity under identical conditions. Compare to 10T CIS, no additional voltage source is required for CDS operation.

Final output is containing difference of V_{reset} and V_t, which are the voltages before and after accumulation time, respectively. Thus, irrespective of reset voltage, we get exact light information, i.e., $V_{reset} - V_t$. It will help us to reduce noise. Assuming noise component as V_{noise},

$$V_{C1} = V_{reset} - V_{th,SF1} + V_{noise} \tag{5}$$

$$V_{C2} = V_t - V_{th,SF1} + V_{noise} \tag{6}$$

$$V_{out} = (V_{C1} - V_{C2}) - V_{th,SF2} \tag{7}$$

$$V_{out} = \left(V_{reset} - V_{th,SF1} + V_{noise}\right) - \left(V_t - V_{th,SF1} + V_{noise}\right) - V_{th,SF2} \tag{8}$$

$$V_{out} = (V_{reset} - V_t) - V_{th,SF2} \tag{9}$$

It can be seen that no noise component is present in final output. If threshold voltage of SF2 is brought to zero voltage, output voltage will be $V_{reset} - V_{light}$. The fill factor of the proposed sensor is approximately 27% because of the two storage capacitors and eight transistors per pixel.

6 Simulation

To verify the performance of proposed pixel architecture, a spice simulation is performed. Capacitor with parallel current source can be used instead of a photodiode [12]. Alternate method is given in [7] in which PD is replaced by voltage source. In simulation circuitry, a transistor M_b is incorporated to maintain required output current I_{out}. Here, one frame is assumed to be of 1 us time duration, and result of two frames is shown in Fig. 7. It is assumed that in first frame, light is incident in PD while in second frame, no light is present. To realize this assumption, discharging current source is set to 40 nA for first frame and 1 pA for second dark light frame. Control signals are applied as shown in Fig. 4.

To understand results of simulations, two basic characteristics of photodiode are to be understood; photocurrent of PD linearly increases with input light intensity, and voltage of PD decreases linearly when it operates in 'accumulation' mode [2]. It should be noted that the accumulation of electrons is interpreted as the process of discharge in the charged capacitor by generated photocurrent. Figure 8 shows relationship between photocurrent and equivalent CDS output. Output voltage is measured for different photocurrent I_{pd} in light frame. Voltage swing is difference between maximum output voltage and minimum output voltage. Maximum output swing of PD is limited in such a way to that linearity error of output voltage remains less than 1%. Here, output swing is found to be 1.2. Dynamic range (DR) is defined as the ratio of the largest non-saturating photocurrent to the minimum detectable photocurrent [20].

$$\mathrm{DR} = 20\log(\mathrm{i_{max}}/\mathrm{i_{min}}) \tag{2}$$

Power consumed per frame can be obtained by integrating current drawn from voltage source within frame time and multiplying it with supply voltage. It can vary according to intensity of light incident on PD. Standard deviation of light samples can be referred as random noise. Here, random noise is calculated of eight light

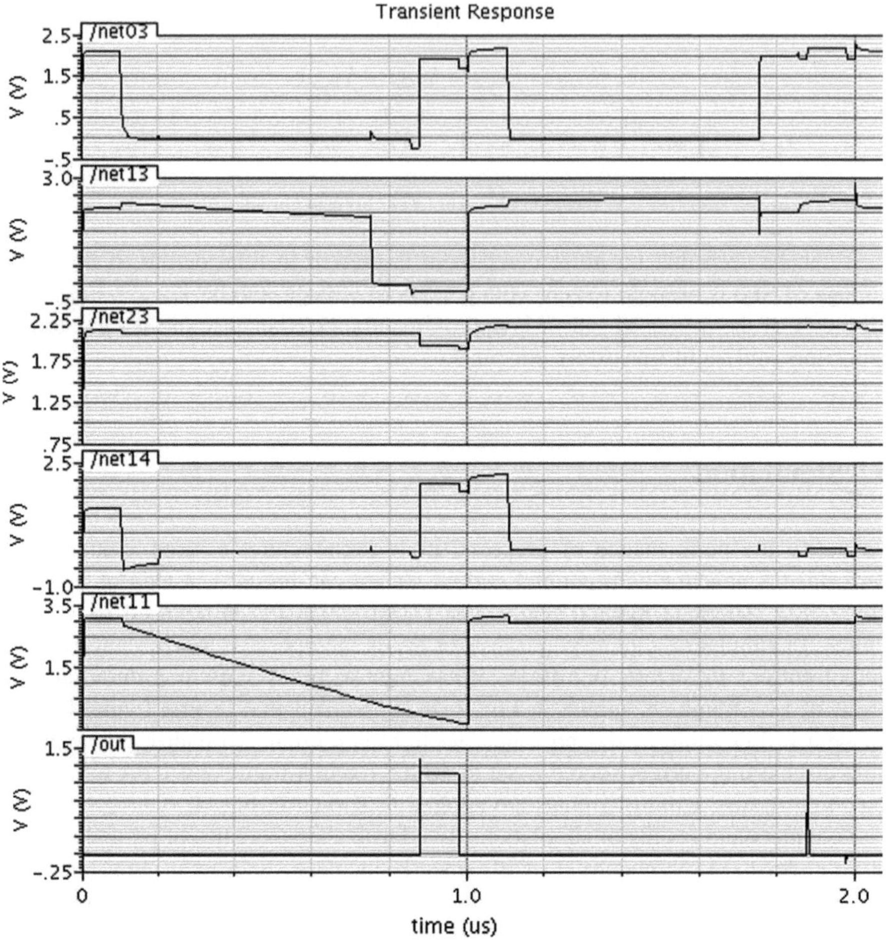

Fig. 7 Waveforms at different nodes in the simulation scheme of the proposed circuit shown in Fig. 6 obtained using simulation

samples. Simulation results of proposed architecture with 9T and 10T pixels are given in Table 2. It is carried out in 0.35-μm CMOS technology.

7 Conclusion

We have presented an 8T global shutter pixel with CDS mechanism. It can be used for traditional global shutter applications such as high-speed and hyper-spectral imaging. The proposed architecture offers several advantages over existing designs.

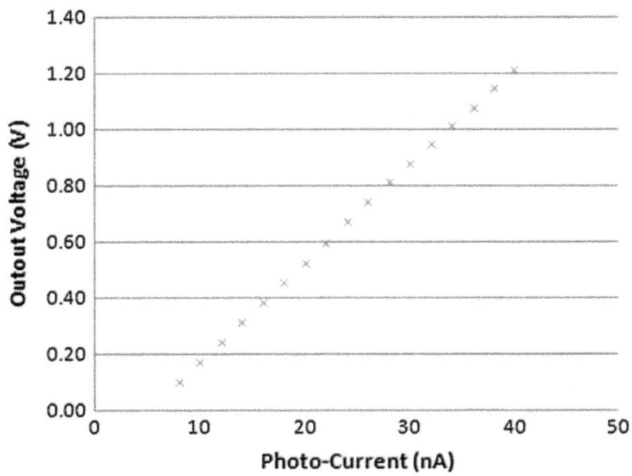

Fig. 8 Linearity of output voltage with photocurrent

Table 2 Parameter comparison

Parameter	9T CIS	10T CIS	This work (8T CIS)
Voltage swing	0.7 V	1.1 V	1.2 V
DR	66 dB	71 dB	72 dB
Power (per frame)	4.2 nW	5.2 nW	4.8 nW
Random noise	1.5 mV	1.1 mV	0.5 mV
SNR	53.3 dB	60 dB	67 dB
Supply voltage	3.3 V	3.2, 2.6, 2.8 V	3.3 V
Estimated pixel size (um^2)	15 × 15	14 × 14	13.5 × 13.5
FF	21%	25%	27%

Overall KTC noise is found to be reduced. Voltage swing is approximately 1.2 V, which can be further improved by using source follower transistor with low threshold voltage. Dynamic range is found to be 72 dB.

References

1. Abbas El Gamal and Helmy Eltoukhy, "CMOS Image Sensor," *IEEE Circuits & Devises Magazine*, Vol. 21, No. 3, May 2005.
2. Jun Ohta, "Smart CMOS Image Sensors and Applications," Book, CRC Press, 2007.
3. Eric R. Fossum and Donald B. Hondongwa, "A Review of the Pinned Photodiode for CCD and CMOS Image Sensors," *IEEE Journal of the Electron Devices Society*, Vol. 2, No. 3, May 2014.

4. Xhakoni, A. and Gielen, G., “A 132-dB Dynamic-Range Global-Shutter Stacked Architecture for High-Performance Imagers,” *IEEE Trans. Circuits Syst. II Exp. Briefs 61(6),* 398–402, 2014.
5. Masaaki S, “An 83 dB-Dynamic-Range Single-Exposure Global-Shutter CMOS Image Sensor with In-Pixel Dual Storage”, *Solid-State Circuits Conference Digest Technical Paper (ISSCC),* Feb, 2012.
6. X.Y. Wang, “A 2.2 M CMOS Image Sensor for High Speed Machine Vision Applications”, proc. *Society of Photographic Instrumentation Engineers (SPIE),* Vol. 7536, San Jose, Jan. 2010.
7. Yang Liu, Cheng Ma, Quan Zhou, and Xinyang Wang, “A New 9T Global Shutter Pixel with CDS Technique,” Proc. of *Society of Photographic Instrumentation Engineers (SPIE),* Vol. 9522 952210-1.
8. G. Meynants, “Global shutter imagers for industrial applications,” Proc. *Society of Photographic Instrumentation Engineers (SPIE),* Vol. 9141, 914108, 2014.
9. Wu, X. and Meynants, G., “High speed global shutter image sensors for professional applications,” Proc. *Society of Photographic Instrumentation Engineers (SPIE),* Vol. 9522, 95220 N, 2015.
10. K. Stefanov, B. Dryera, D. Halla, A. Hollanda, J. Pratlongb, M. Fryerb, and Andrew Pikeb, “A Global Shutter CMOS Image Sensor for Hyperspectral Imaging,” Proc. of *Society of Photographic Instrumentation Engineers (SPIE),* Vol. 9602 96020 K-1, 2015.
11. E. Fossum, “CMOS Image Sensor: Electronics Camera-On-A-Chip,” *IEEE Transaction on Electron Devices*, Vol. 44, No. 10, October 1997.
12. T. Reiner, B. Mishori, T. Leitner, A. Horovitz, Y. Vainbaum, M. Hakim, A. Lahav, S. Shapira, and A. Fenigstein, “CMOS Image Sensor 3T Nwell Photodiode Pixel Spice Model,” *Proceedings 23rd IEEE Convention*, Electrical and Electronics Enginner in Israel, 2004.
13. S. Mendis, S. Memeny, and E. Fossum, “CMOS Active Pixe Image Sensor,” *IEEE Transaction on Electron Devices*, Vol. 41, No. 3, March 1994.
14. V. Lalucaa, V. Goiffon, P. Magnan, C. Virmontosis, Guy Rolland, and S. Petit, “Single Event Effects in 4T Pinnend photodiode Image Sensor,” *IEEE Transaction on Nuclear Science*, Vol. 60, No. 6, December 2013.
15. Zhiyuan Gao, Jiangtao Xu, Yiming Zhou, and Kaiming Nie, “Analysis and Modeling of the Light-Dependent Full Well Capacity of the 4-T Pixel in CMOS Image Sensors,” *IEEE Sensors Journal*, Vol. 16, No. 8, April 15, 2016.
16. E. Martin, T. Nuns, J. P. David, O. Gilard, J. Vaillant, P. Fereyre, V. Prevost, and M. Boutillier. “Gamma and Proton-Induced Dark Current Degradation of 5T CMOS Pinned Photodiode 0.18 um CMOS Image Sensors,” *IEEE Transactions on Nuclear Science*, Vol. 61, No. 1, February 2014.
17. Liqiang Han, Suying Yao, and Albert J. P. Theuwissen, “A Charge Transfer Model for CMOS Image Sensors,” *IEEE Transactions On Electron Devices*, Vol. 63, No. 1, January 2016.
18. M. Rafat, M Elmahlawy, A. Zaki and M. Hamid, “Low-Power Low-Noise CTIA Readout Integrated Circuit Design for Thermal Imaging Applications,” *Communications on Applied Electronics (CAE),* Vol. 2, No. 8, September 2015.
19. Kartikeya Murari, Ralph Etienne-Cummings, Nitish V. Thakor, and Gert Cauwenberghs, “A CMOS In-Pixel CTIA High-Sensitivity Fluorescence Imager,” *IEEE Transactions on Biomedical Circuits And Systems*, Vol. 5, No. 5, October 2011.
20. C. Jun, R. Feng, Y. Hui, and X. Meihual, “A CMOS Image Sensor with Self-Reset Circuit in Active Pixel.” *Image and Signal Processing, 2009. CISP’09. 2nd International Congress on.* IEEE, 2009.

Time Series Regression of Weather Parameters Over the Last Century for Cotton Crop

Riddhi Patel and Geetali Saha

Abstract Agriculture is the largest provider of livelihood, not only to rural India in terms of produce but also to the urban sector in many fruitful ways varying from yielding harvest to processing to transporting to exporting these crop products amongst various commodity markets. Cotton, sugar, wheat and rice are the major Indian cash crops that are exported to other countries. The productivity of these crops is mainly affected by change in climatic conditions. However, at times extreme variations in climatic conditions such as temperature, vapour pressure, crop evapotranspiration can cause losses. The objective behind this research is to figure out upcoming weather parameter values for the crop of cotton. Different time series methods, such as moving average, single exponential methods and neural network-based methods are used to predict weather parameters and relate it to the productivity of major cotton-producing states of India such as Gujarat and Maharashtra covering around different districts. Data is taken for over a century (1901–2002), and based on that, forecast is determined. To check the efficiency of the different models, mean absolute percentage error (MAPE), mean square error (MSE) and root mean square error (RMSE) are evaluated and tabulated.

Keywords Crop evapotranspiration · MA · SES · Neural network
Regression analysis

1 Introduction

India is mainly an agricultural country. Agriculture continues to play a dominant part in the overall economic scenario as it produces many essential cash crops. Cash crops such as cotton, coffee, sugarcane, rice, wheat, onion, mangoes play a very

R. Patel (✉) · G. Saha
G. H. Patel College of Engineering and Technology, Vallabh Vidyanagar, India
e-mail: patelriddhi96@yahoo.com

G. Saha
e-mail: geetalisaha@gcet.ac.in

R. Kher et al. (eds.), *Proceedings of the International Conference on Intelligent Systems and Signal Processing*, Advances in Intelligent Systems and Computing 671,
https://doi.org/10.1007/978-981-10-6977-2_12

significant role in Indian economy thereby contributing to the foreign exchange. Amongst all cash crops, cotton is the major cash crop of India. Some of the cotton-producing states of India are Gujarat, Maharashtra, Rajasthan, Punjab, Haryana, Andhra Pradesh, etc. However, Gujarat and Maharashtra are the major cotton-producing states of India (approximately 46%). Next to Gujarat, Maharashtra is the second state in terms of cotton production.

Cotton is one of the major crops of Gujarat state. Gujarat contributes more than 30% of the total cotton production in the country as cotton grows in more than 30 lac hectares of the land. Due to the black soil and rainfall of 80–100 cm—parameters that are required for cotton production—Gujarat is most advantageous to the country. The major cotton-producing districts of Gujarat are Rajkot, Bharuch, Vadodara, Panchmahal, Mehsana, Ahmedabad and Surendranagar [1]. A comparison of cotton productivity over 16 years from 1999 to 2015 is compiled for reference.

Maharashtra is the next major contributor of cotton production in India. The lava soil of this state makes it more suitable for cotton production. Three main regions are Khandesh, Vidarbha and Marathwada which involve different districts such as Akola, Nagpur, Wardha, Dhule, Amravati [2]. A comparison of cotton production in Maharashtra for 8 years is quoted for reference [3].

Prediction of weather parameters such as temperature, vapour pressure and crop evapotranspiration plays a very dominant role for cultivation of cotton, and Also it would be a big benefit to the farmers if such parameters could be estimated beforehand. Temperature in the range of 21°–30 °C, vapour pressure in between 15 and 30 hPa and crop evapotranspiration 2.54–7.11 mm/day are the basic requirements for cotton production. There are mainly four types of cotton: *Gossypium arboreum*, *Gossypium herbaceum*, *Gossypium hirsutum* and *Gossypium barbadense* produced in Gujarat and Maharashtra; the first two known as Asiatic cotton, and the last two are known as New World cotton. Features of all types of cotton are explained in detail in [4].

A dissertation report [5] is a very neatly compiled book by Ratnadip and Agarwal. They have covered many prediction methods right from basic mathematical moving average methods to soft computing techniques such as artificial neural network and support vector machine in sufficient details. To calculate the accuracy amongst different models fitted to time series, many error parameters are discussed.

In [6], there is a detailed survey of different techniques which are used for prediction of different datasets. Also, based on their experimentation work, they have identified techniques that give fruitful results to particular sets of data. It contains a detailed explanation of traditional-, stochastic-, soft computing-based and fuzzy-based forecasting techniques along with their merits and demerits over each other. For comparing accuracy of different prediction techniques, University of Alabama's dataset and TAIEX-Taiwan Stock Exchange data are used based on which error measures are tabulated.

Moving average is the most popular and easy technique of prediction. In [7], variety of uses of moving average is well explained, and also the merits and demerits of different types of moving average methods are discussed. There are some unique types of moving average.

In [8], required parameters of weather for different crops are introduced. The behaviour of climatic conditions is beneficial to the farmers for proper cultivation of crops. Weather parameters such as average temperature, average rainfall, vapour pressure and cloud cover around 101 years are used to predict the next year weather situations. Prediction is done using different algorithms of TSA methods, and accuracy of each model is analysed by calculating RMSE, MAPE and MSE.

Evapotranspiration plays a very important role in estimating water cycle and moisture content of the atmosphere in the nearby vicinity. It has started becoming a large contributor in predicting crops and planning of resources. Three regression models, namely multivariate fractional polynomial, robust regression and Bayesian regression, are used to estimate the evapotranspiration on monthly basis for three cities of Iran [9]. Evapotranspiration is a complex and nonlinear element. So multivariate fractional polynomial is best tool for nonlinear problems as per their experimental results.

In [10], Penman–Monteith and artificial neural network methods are compared for the prediction of monthly reference evapotranspiration in sub-humid region of Dehradun. For NN modelling, different training algorithm and varied number of neurons are properly utilized to get the most suitable structure of ANN. The ANN with Levenberg–Marquardt with nine numbers of neurons in single hidden layer gives best prediction result with respect to their database.

The [11] provides the prediction of one day-ahead temperature in Kermanshah city of Iran. About 10 years of data from 1996 to 2006 was trained and tested using Multi-Layer Perceptron. It has been proven that with the help of Levenberg–Marquardt algorithm, MLP is the best model of prediction and has least error.

With reference to the database from Indiawaterportal [12], a comparison of basic time series analysis methods is executed using MATLAB tool, version (R2015b). In this paper, we used meteorological data such as average temperature, vapour pressure and reference crop evapotranspiration of 101 years from 1901 to 2001 and predict the same for the year 2002.

The present paper is organized as follows: Sect. 2 gives a detailed explanation of the various time series models for prediction. Section 3 shows the error measures which are used to check the accuracy of predictive algorithm. Section 4 provides the information about the data used for proposed work. Section 5 is about the result and discussion obtained using the time series methods. Section 6 gives the conclusion and suggestion for future work.

2 Time Series Models

A time series is essentially a set of collection of data points for specific time interval. A time series is an array of observations of variable measured at successive time periods. The measurements are captured at any regular time interval such as every hour, every day, month or year. The pattern of data is an important factor in understanding how the time series has behaved over the past. The past data values play an important role for forecasting the nature of data values in the future. There are many application areas where time series are well being utilized. The aim of time series analysis is to describe and summarize time series data and make an accurate forecast. Here, in-site data is fed and simulated using software for data analysis.

2.1 *Moving Average Model*

A moving average is a statistical method to analyse the data points by creating series of averages of different subsets of the full dataset. It is used to smooth out short-term fluctuations and highlight longer-term trends or cycles. It is an average of the previous n data points in time series data.

Each value in the time series data is equally weighted, so there are no weighting factors applied to any of the data value. The moving average method formula is given below:

$$P_{t+1} = \frac{S_t + S_{t-1} + \cdots + S_{t-n+1}}{n} \tag{1}$$

where

P_{t+1} Forecast for the next period
S_t Actual value for current period
n period over which averaging is done.

This method is easily applicable as it is a simple one but it is relevant to short-term forecasting. Also, the selection of the period of moving average is a difficult chore. For different MA interval, forecasting error may differ.

2.2 *Single Exponential Smoothing (SES)*

It is also called as simple exponential smoothing. It is mostly preferable for simple data without trend and seasonality, but the mean of the time series is slightly changing over time. SES model provides larger weights to the most nearest data, and the weights decrease exponentially as the data becomes more distant. The formula for SES method is as below:

$$G_{t+1} = \alpha * S_t + (1 - \alpha) * G_t \tag{2}$$

where

G_{t+1} next forecast value of the variable;
G_t present forecast value of the variable;
S_t present actual value of the variable;
α smoothing constant between 0 and 1.

Trend is a long-term increase or decrease in the time series data. This is the type of tendency which continues to exist for a very long period over the years.

Seasonality is defined as periodic fluctuation of time series data within certain limits. This is short-term instant in time series with repetitive variations in weather within the one year.

If the time series data has two components, trend and seasonality, there are some other methods such as double exponential smoothing (DES) and triple exponential smoothing (TES) which also give better forecasting results.

Also there are some other techniques which are useful for prediction of future data. Neural networks are widely used in forecasting.

2.3 *Neural Networks*

Neural networks consist of many nodes. Each node has an activation function. Networks are basically information processing system. In this system, elements called neurons which process the information. The signals are transmitted by means of connection links. Every link possesses an associated weight, which is multiplied along with the incoming signal of net. The net is trained by applying activations to the net input.

Neural network uses mathematical algorithms to learn the relationship and patterns from a given dataset. These activation functions determine the actual output from the network. By choosing a combination of these functions, the network can be trained, validated and tested. Once trained, this net is given a set of input and output for validation, and after validation, the net is used for testing (Fig. 1).

Different types of activation functions used for the proposed algorithm are given below.

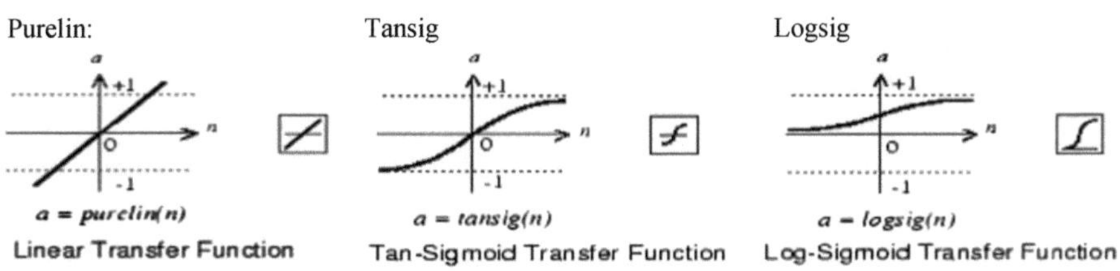

Fig. 1 Different activation functions

3 Error Measures

After prediction, the accuracy of the models is tested by means of the common measures of error such as MAPE, MSE and RMSE. These errors give the core success of the time series models. In this paper, different models are compared by calculating the mentioned errors, each of which is defined as follows.

3.1 Mean Absolute Percentage Error (MAPE)

MAPE can be defined as:

$$\mathrm{MAPE} = \frac{1}{n}\sum_{t=1}^{n}\frac{(Y_t - F_t)^2}{Y_t} * 100 \tag{3}$$

3.2 Mean Square Error (MSE)

MSE can be described as:

$$\mathrm{MSE} = \frac{1}{n}\sum_{t=1}^{n}(Y_t - F_t)^2 \tag{4}$$

3.3 Root Mean Square Error (RMSE)

RMSE can be defined as:

$$\mathrm{RMSE} = \sqrt{\frac{1}{n}\sum\nolimits_{t=1}^{n}(Y_t - F_t)^2} \tag{5}$$

where,

Y_t Actual value
F_t Forecasted value.

4 Dataset Description

Gujarat and Maharashtra are the major cotton-growing states of the country. Majority cotton-producing districts of Gujarat are in the southern part, and districts of Maharashtra are in the middle north part.

There are few weather organizations of India which provides the database as well as weather information. Few of them are Indiawaterportal, India Meteorology Department (IMD) [13], Accuweather, etc. In this paper, the monthly meteorological data from five major cotton-producing districts of Gujarat—Rajkot, Bharuch, Amreli, Sabarkantha and Surendranagar—and also three districts of Maharashtra—Nagpur, Wardha and Akola of 100 years from 1901 to 2001—is taken from Indiawaterportal. The parameters observed are as follows:

(1) Average temperature (°C)
(2) Vapour pressure (hPa)
(3) Reference crop evapotranspiration (mm/day) (Figs. 2, 3 and 4).

Fig. 2 Fluffy growing cotton that is ready to harvest. Reference: www.cottonsource.com

Fig. 3 Map indicating cotton-producing districts of Gujarat. Reference: www.mapsofindia.com

5 Results and Discussion

For experimentation,

Moving average methods for different time intervals are implemented, and it is found that the period is incoherent.

SES works best when the latest data is given maximum weightage.

Using neural network, the data is split into three parts:

For training: Indices [1:1100];
For validation: Indices [1101:1212]; and
For testing: Indices [1213:1224].

The network is optimized for different combinations of activation functions, and it is observed that minimum error is obtained when purelin function is used in hidden layer and tansig in the output layer. We tried to use Levenberg–Marquardt algorithm as it is the fastest algorithm.

We have tried to summarize the variations on different parameters at identical locations using mentioned techniques.

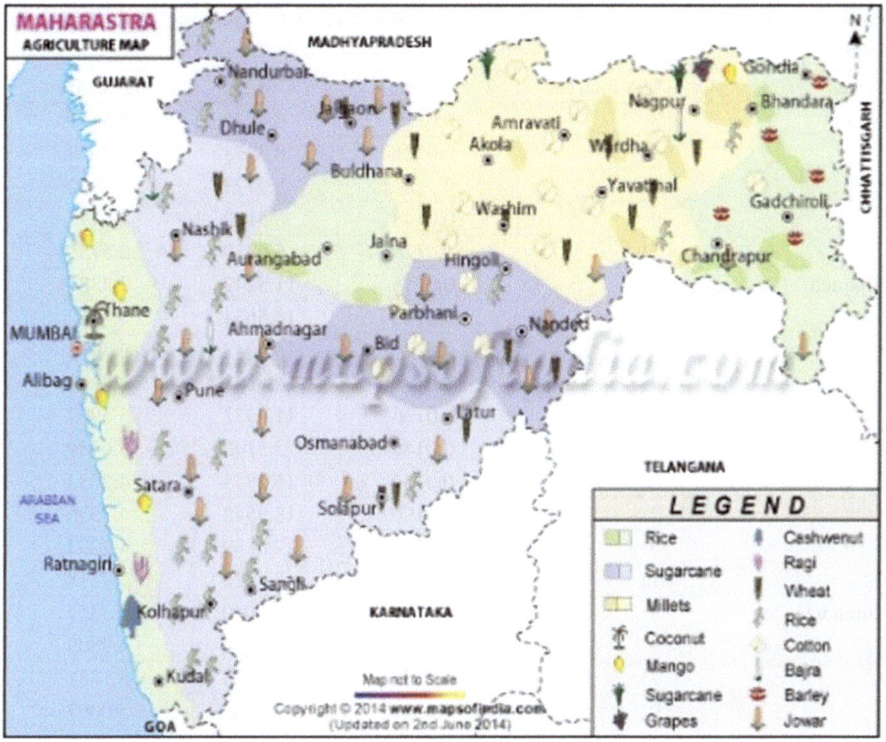

Fig. 4 Map indicating cotton-producing districts of Maharashtra. Reference: www.mapsofindia.com

Tables 1, 2 and 3 show the comparison of moving average, SES and neural networks for five different cotton-producing districts of Gujarat for parameter average temperature.

Tables 4, 5 and 6 show the same for parameter vapour pressure.

Tables 7, 8 and 9 show the same for parameter evapotranspiration.

Tables 10, 11 and 12 show the comparison of moving average, SES and neural networks for three different districts of Maharashtra for parameter average temperature.

Tables 13, 14 and 15 show the same for parameter vapour pressure.

Tables 16, 17 and 18 show the same for parameter evapotranspiration (Figs. 5, 6, 7, 8, 9 and 10; Tables 19 and 20).

Table 1 Comparison of parameter: *Average temperature*; state: *Gujarat*; method: *Moving average*

Place	MA interval	MAPE	MSE	RMSE
Amrell	3	0.0125	*6.4772*	*2.5450*
	5	0.0126	8.4378	2.9048
	7	0.0152	8.9773	2.9962
	9	0.0148	8.5112	2.9174
	11	*0.0110*	6.6330	2.5755
Bharuch	3	0.0145	11.8247	3.4387
	5	*0.0145*	14.5634	3.8162
	7	0.0177	15.0743	3.3972
	9	0.0179	14.6008	3.8211
	11	0.0136	11.4932	3.3902
Rajkot	3	0.0146	*13.3148*	*3.6489*
	5	0.0153	17.6843	4.2053
	7	0.0190	18.9542	4.3536
	9	0.0187	17.8889	4.2295
	11	*0.0142*	13.8520	3.7218
Surendranagar	3	*0.0191*	*18.7959*	*4.3354*
	5	0.0203	24.8093	4.9809
	7	0.0251	26.5026	5.1481
	9	0.0249	25.0967	5.0097
	11	0.0193	19.4295	4.4079
Sabarkantha	3	*0.0198*	*22.6195*	*4.7560*
	5	0.0215	0.0675	5.4834
	7	0.0269	32.2197	5.6762
	9	0.0269	30.4499	5.5181
	11	0.0208	23.5408	4.8519

Table 2 Comparison of parameter: *Average temperature*; state: *Gujarat*; method: *SES*

Place	Alpha	MAPE	MSE	RMSE
Amrell	0.1	1.1404	5.3870	2.3210
	0.2	1.2587	5.5098	2.3473
	0.3	1.2853	5.4021	2.3242
	0.4	1.2642	5.1194	2.2626
	0.5	1.2259	4.7514	2.1798
	0.6	1.1839	4.3605	2.0882
	0.7	1.1393	3.9774	1.9943
	0.8	1.0881	3.6157	1.9015
	0.9	*1.0262*	*3.2820*	*1.8116*

(continued)

Table 2 (continued)

Place	Alpha	MAPE	MSE	RMSE
Bharuch	0.1	1.1768	10.6753	3.2673
	0.2	1.4733	10.8896	3.2999
	0.3	1.4773	10.6378	3.2616
	0.4	1.4219	10.0566	3.1712
	0.5	1.1480	9.3160	3.0522
	0.6	1.2706	8.5305	2.9207
	0.7	1.1897	7.7583	2.7354
	0.8	1.1008	7.0276	2.6510
	0.9	*1.00*	*6.3552*	*2.5210*
Rajkot	0.1	1.4673	11.0730	3.3276
	0.2	1.5610	11.2911	3.3602
	0.3	1.5550	10.9548	3.3098
	0.4	1.4980	10.2362	3.1994
	0.5	1.4272	9.3522	3.0581
	0.6	1.1590	8.4461	2.9062
	0.7	1.2938	7.5648	2.7540
	0.8	1.2260	6.7928	2.6063
	0.9	*1.1501*	*6.0792*	*2.4656*
Surendranagar	0.1	1.9545	16.8082	4.0998
	0.2	2.0591	17.0660	4.1311
	0.3	2.0302	16.4688	4.0582
	0.4	1.9316	15.3000	3.9115
	0.5	1.8134	13.8968	3.7278
	0.6	1.6975	12.4738	3.5325
	0.7	1.5853	11.1459	3.3385
	0.8	1.4701	9.9336	3.1518
	0.9	*1.3465*	*8.8520*	*2.9752*
Sabarkantha	0.1	2.0671	21.408	4.6269
	0.2	2.1652	21.6259	4.6504
	0.3	2.1054	20.7351	4.5536
	0.4	1.9640	19.1225	4.3729
	0.5	1.8013	17.2317	4.1511
	0.6	1.6438	15.3485	3.9177
	0.7	1.4943	13.6008	3.6379
	0.8	1.3467	12.0304	3.4685
	0.9	*1.1951*	*10.6446*	*3.2626*

Table 3 Parameter: *Average temperature*; state: *Gujarat*; method: *Neural network*

Place	Architecture	Transfer function		MAPE	MSE	RMSE
		Layer1	Layer2			
Amrell	1-5-1	Purelin	Tansig	*0.0805*	*0.2203*	*0.0293*
	1-4-1			0.0770	0.2207	0.03190
	1-5-1			0.0765	0.2206	0.03741
	1-10-1			0.0757	0.2207	0.04053
	1-2-1		Logsig	0.4678	1.1324	2.4377
	1-4-1			0.4671	1.1346	2.2294
	1-5-1			0.4729	1.1331	2.5212
	1-10-1			0.4749	1.1328	2.1305
Bharuch	1-2-1	Purelin	Tansig	*0.0951*	*0.3061*	*0.09162*
	1-4-1			0.1052	0.3053	0.0932
	1-5-1			0.1017	0.3063	0.0626
	1-10-1			0.1110	0.3082	0.0727
	1-2-1		Logsig	0.7925	1.7645	3.9686
	1-4-1			0.7908	1.7647	4.0971
	1-5-1			0.7944	1.7641	3.6064
	1-10-1			0.7930	1.7640	3.4386
Rajkot	1-2-1	Purelin	Tansig	*0.1120*	*0.3448*	*0.08126*
	1-4-1			0.1159	0.3464	0.0767
	1-5-1			0.1138	0.3438	0.09224
	1-10-1			0.1132	0.3445	0.084604
	1-2-1		Logsig	0.7561	1.6776	3.5171
	1-4-1			0.7508	1.6736	4.2347
	1-5-1			0.7568	1.6776	4.2323
	1-10-1			0.7535	1.6778	3.9639
Surendranagar	1-2-1	Purelin	Tansig	*0.1546*	*0.4410*	*0.1548*
	1-4-1			0.1598	0.4406	0.15505
	1-5-1			0.1577	0.4407	0.1608
	1-10-1			0.1576	0.4400	0.17048
	1-2-1		Logsig	1.1675	2.1826	5.6167
	1-4-1			1.1677	2.1824	6.0905
	1-5-1			1.1733	2.1835	5.9042
	1-10-l			1.1659	2.1828	7.1715
Sabarkantha	1-2-1	Purelin	Tansig	0.1718	0.4565	0.1560
	1-1-1			0.1608	0.4562	0.17566
	1-5-1			0.1644	0.4560	0.13115
	1-10-l			*0.1477*	*0.4617*	*0.10659*
	1-2-1		Logsig	1.3936	2.6433	8.9727
	1-1-1			1.3993	2.6434	7.8432
	1-5-1			1.3980	2.6433	8.4326
	1-10-l			1.3979	2.6433	9.609

Table 4 Parameter: *Vapour pressure*; state: *Gujarat*; method: *Moving average*

Place	MA interval	MAPE	MSE	RMSE
Amrell	3	*0.0401*	*29.3171*	*5.4145*
	5	0.0540	43.6561	6.9754
	7	0.0623	56.4122	7.5108
	9	0.0572	50.3725	7.0974
	11	0.0461	37.3170	6.1088
Bharuch	3	*0.0229*	*50.6014*	*7.1135*
	5	0.0461	75.7126	8.7013
	7	0.0537	85.1826	9.2294
	9	0.0468	77.0341	8.7769
	11	0.037	59.2497	7.6974
Rajkot	3	*0.0477*	*41.1144*	*6.8640*
	5	0.0631	78.4153	8.8552
	7	0.0716	91.0253	9.5407
	9	0.0656	81.2522	9.0140
	11	0.0530	60.0969	7.7522
Surendranagar	3	*0.0531*	*50.8659*	*7.1320*
	5	0.0741	84.9579	9.2173
	7	0.083	96.6887	9.9342
	9	0.0748	87.9375	9.3775
	11	0.0604	65.1369	8.0707
Sabarkantha	3	0.0626	*62.2430*	*7.8894*
	5	0.0958	100.01	10.0182
	7	0.1018	115.0342	10.7254
	9	0.0852	103.2364	10.1605
	11	*0.0082*	77.0094	8.7755

Table 5 Comparison of parameter: *Vapour pressure*; state: *Gujarat*; method: *SES*

Place	Alpha	MAPE	MSE	RMSE
Amrell	0.1	4.6492	30.8306	5.5525
	0.2	4.7950	30.5001	5.5227
	0.3	4.5539	23.1309	5.3039
	0.4	4.0837	24.5822	4.9580
	0.5	3.5436	20.8025	4.5610
	0.6	3.0312	17.3723	4.1681
	0.7	2.5854	14.5043	3.8084
	0.8	2.2132	12.1993	3.4928
	0.9	*1.9106*	*10.3839*	*3.2224*

(continued)

Table 5 (continued)

Place	Alpha	MAPE	MSE	RMSE
Bharuch	0.1	3.6967	48.755	6.9825
	0.2	3.7063	43.3346	6.9382
	0.3	3.32	46.0762	6.7379
	0.4	2.7145	41.7316	6.46
	0.8	2.0806	37.1555	6.0955
	0.6	1.5447	33.1791	5.7601
	0.7	1.1666	30.1156	5.4878
	0.8	0.9649	27.9930	5.2908
	0.9	*0.9403*	*26.7393*	*5.1710*
Rajkot	0.1	5.3772	49.9033	7.0642
	0.2	5.5604	49.3795	7.0271
	0.3	5.3049	45.5219	6.7470
	0.4	4.7796	39.7359	6.3036
	0.5	4.1606	33.5702	5.7940
	0.6	3.5589	27.9700	5.2387
	0.7	3.0203	23.2764	4.8246
	0.8	2.5540	19.4917	4.4149
	0.9	*2.1565*	*16.4902*	*4.0608*
Surendranagar	0.1	6.1053	54.9654	7.4139
	0.2	6.3345	54.2866	7.3679
	0.3	6.0144	49.9676	7.0688
	0.4	5.3568	43.5562	6.5997
	0.8	4.5843	36.7653	6.0633
	0.6	3.8423	30.6363	5.5350
	0.7	3.1964	25.5507	5.0548
	0.8	2.6653	21.5052	4.6374
	0.9	*2.2490*	*18.3552*	*4.2343*
Sabarkantha	0.1	7.1253	66.5119	8.1555
	0.2	7.5015	65.3247	8.1132
	0.3	7.1445	60.7805	7.7962
	0.4	6.2845	53.2101	7.2945
	0.5	5.2204	45.1565	6.7199
	0.6	4.1729	37.3787	6.1546
	0.7	3.2580	31.8340	5.6422
	0.8	2.5186	27.0269	5.1987
	0.9	*1.9617*	*23.2919*	*4.8262*

Table 6 Comparison of parameter: *Vapour pressure*; state: *Gujarat*; method: *Neural network*

Place	Architecture	Transfer function		MAPE	MSE	RMSE
		Layer1	Layer2			
Amrell	1-2-1	Purelin	Tansig	0.0917	0.3852	0.1484
	1-4-1			0.0887	0.3844	0.1477
	1-5-1			0.0922	0.3848	0.1480
	1-10-1			*0.0806*	*0.3839*	*0.1473*
	1-2-1		Logsig	3.6632	3.3665	11.3268
	1-4-1			3.6644	3.3633	11.3120
	1-5-1			3.6644	3.3631	11.3106
	1-10-1			3.6644	3.3633	11.3116
Kharuch	1-2-1	Purelin	Tansig	0.1909	0.4234	0.1793
	1-4-1			0.1866	0.4201	0.1765
	1-5-1			0.1910	0.4217	0.1778
	1-10-1			*0.1837*	*0.4193*	*0.1758*
	1-2-1		Logsig	3.7281	4.9823	24.8230
	1-4-1			3.7285	4.9824	24.8242
	1-5-1			3.7288	4.9825	24.8249
	1-10-1			3.7275	4.9821	24.8217
Rajkot	1-2-1	Purelin	Tansig	0.0661	0.4748	0.2254
	1-4-1			0.0664	0.4756	0.2262
	1-5-1			0.0631	0.4719	0.2227
	1-10-1			*0.0631*	*0.4745*	*0.2251*
	1-2-1		Logsig	4.2585	4.3336	18.9538
	1-4-1			4.2587	4.3535	18.9525
	1-5-1			4.2573	4.3557	18.9719
	1-10-1			4.2573	4.3557	18.9725
Surendranagar	1-2-1	Purelin	Tansig	*0.0365*	*0.4520*	*0.2043*
	1-4-1			0.0900	0.4525	0.2047
	1-5-1			0.0514	0.4498	0.2023
	1-10-1			0.0682	0.4480	0.2007
	1-2-1		Logsig	3.3293	4.8986	23.9964
	1-4-1			5.3295	4.8999	24.0093
	1-5-1			5.3231	4.9002	24.0117
	1-10-1			3.3293	4.9000	24.0104
Sabarkantha	1-2-1	Purelin	Tansig	*0.1773*	*0.5951*	*0.3541*
	1-4-1			0.2148	0.5907	0.3490
	1-5-1			0.1800	0.5945	0.3534
	1-10-1			0.1781	0.5947	0.3537
	1-2-1		Logsig	7.4741	6.3139	39.8659
	1-4-1			7.4749	6.3138	39.8647
	1-5-1			7.4748	6.3138	39.8645
	1-10-1			7.4743	6.3139	39.8655

Table 7 Comparison of parameter: *Reference crop evapotranspiration*; state: *Gujarat*; method: *Moving average*

Place	MA interval	MAPE	MSE	RMSE
Amrell	3	*0.0166*	0.410	0.6403
	5	0.0130	0.4154	0.6445
	7	0.0192	0.3852	0.6207
	9	0.0244	0.4033	0.6350
	11	0.0213	*0.3324*	*0.5765*
Bharuch	3	0.029	1.0809	1.0397
	5	0.0184	1.2380	1.1127
	7	0.0281	1.2323	1.1101
	9	0.0341	1.2276	1.1080
	11	*0.029*	*0.9773*	*0.9886*
Rajkot	3	0.0239	*0.8110*	*0.9551*
	5	0.0334	1.0131	1.0121
	7	0.0283	1.0105	1.0053
	9	0.0188	1.0244	1.0065
	11	*0.0202*	0.9122	0.9005
Surendranagar	3	*0.0273*	1.6591	1.2881
	5	0.0276	2.0003	1.4143
	7	0.0414	2.0524	1.4326
	9	0.0469	2.0013	1.4147
	11	0.0402	*1.5747*	*1.2549*
Sabarkantha	3	*0.0288*	*2.0970*	1.4481
	5	0.0299	2.5743	1.6045
	7	0.0217	2.6709	1.6343
	9	0.0512	2.5825	1.6070
	11	0.0435	2.0249	*1.4230*

Table 8 Comparison of parameter: *Reference crop evapotranspiration*; state: *Gujarat*; method: *SES*

Place	MA interval	MAPE	MSE	RMSE
Amrell	0.1	1.9379	0.3026	0.5501
	0.2	1.8671	0.3149	0.5611
	0.3	1.7711	0.3152	0.5614
	0.4	1.6931	0.3065	0.5537
	0.5	1.6447	0.2923	0.5406
	0.6	1.6115	0.2745	0.5240
	0.7	1.5725	0.2547	0.5047
	0.8	1.5129	0.2337	0.4534
	0.9	*1.4262*	*0.2124*	*0.4609*

(continued)

Table 8 (continued)

Place	MA interval	MAPE	MSE	RMSE
Bharuch	0.1	2.6672	0.9356	0.9673
	0.2	2.5091	0.9589	0.9792
	0.3	2.2738	0.9351	0.9670
	0.4	2.0533	0.8809	0.9386
	0.5	1.8848	0.8132	0.9018
	0.6	1.7534	0.7421	0.8615
	0.7	1.6469	0.6722	0.8199
	0.8	1.5269	0.6056	0.7782
	0.9	*1.3857*	*0.5433*	*0.7371*
Rajkot	0.1	2.6443	0.7835	0.8351
	0.2	2.5050	0.8017	0.8954
	0.3	2.2905	0.7323	0.8845
	0.4	2.0872	0.7384	0.8593
	0.5	1.9310	0.6831	0.8265
	0.6	1.8139	0.6247	0.7904
	0.7	1.7110	0.5669	0.7529
	0.8	1.6001	0.5116	0.7152
	0.9	*1.4687*	*0.4597*	*0.6780*
Surendranagar	0.1	3.6990	1.4920	1.2215
	0.2	3.5008	1.5191	1.2325
	0.3	3.1668	1.4686	1.2119
	0.4	2.8234	1.3689	1.1700
	0.5	2.5362	1.2495	1.1178
	0.6	2.3069	1.1282	1.0622
	0.7	2.1085	1.0127	1.0064
	0.8	1.9136	0.9059	0.9518
	0.9	*1.7059*	*0.8087*	*0.8993*
Sabarkantha	0.1	3.9901	1.8960	1.3770
	0.2	3.7858	1.9284	1.3387
	0.3	3.4143	1.8581	1.3631
	0.4	3.0234	1.7241	1.3131
	0.5	2.6931	1.5661	1.2514
	0.6	2.4300	1.4076	1.1364
	0.7	2.2054	1.2587	1.1219
	0.8	1.9881	1.1224	1.0595
	0.9	*1.7539*	*0.9996*	*0.9998*

Table 9 Comparison of parameter: *Reference crop evapotranspiration*; state: *Gujarat*; method: *Neural network*

Place	Architecture	Transfer function		MAPE	MSE	RMSE
		Layer1	Layer2			
Amrell	1-2-1	Purelin	Tansig	*0.1415*	*0.0562*	*0.0032*
	1-4-1			0.1412	0.0567	0.0032
	1-5-1			0.1428	0.0563	0.0032
	1-10-1			0.1436	0.0566	0.0032
	1-2-1		Logsig	2.1549	0.4084	0.1668
	1-4-1			2.1552	0.4085	0.1668
	1-5-1			2.1548	0.4084	0.1668
	1-10-1			2.1550	0.4084	0.1668
Bharuch	1-2-1	Purelin	Tansig	0.2545	0.1079	0.1668
	1-4-1			0.2482	0.1078	0.0116
	1-5-1			0.2552	0.1078	0.0116
	1-10-1			*0.2425*	*0.1076*	*0.0116*
	1-2-1		Logsig	3.0160	0.6716	0.4511
	1-4-1			3.0161	0.6718	0.4513
	1-5-1			3.0160	0.6714	0.4508
	1-10-1			3.0161	0.6716	0.4511
Rajkot	1-2-1	Purelin	Tansig	*0.1991*	*0.0975*	*0.0095*
	1-4-1			0.2028	0.0970	0.0095
	1-5-1			0.2012	0.0973	0.0095
	1-10-1			0.2015	0.0172	0.0095
	1-2-1		Logsig	2.8538	0.5951	0.3542
	1-4-1			2.8540	0.5954	0.3545
	1-3-1			2.8539	0.5953	0.3544
	1-10-1			2.8540	0.5953	0.3544
Surendranagar	1-2-1	Purelin	Tansig	*0.4037*	*0.1316*	*0.0173*
	1-4-1			0.4011	0.1316	0.0173
	1-5-1			0.4020	0.1316	0.0173
	1-10-1			0.4246	0.1320	0.0174
	1-2-1		Logsig	4.2508	0.9064	0.8216
	1-1-1			4.2507	0.9063	0.3214
	1-5-1			4.2508	0.9065	0.8217
	1-10-1			4.2508	0.9063	0.8213
Sabarkantha	1-2-1	Purelin	Tansig	0.4356	0.1433	0.0205
	1-4-1			0.4281	0.1433	0.0205
	1-3-1			0.4245	0.1434	0.0206
	1-10-1			*0.4227*	*0.1433*	*0.0205*
	1-2-1		Logsig	4.7100	1.0507	3.3040
	1-4-1			4.71	1.0507	3.3039
	1-5-1			4.71	1.0507	3.3039
	1-10-1			4.71	1.0507	3.3039

Table 10 Comparison of parameter: *Average temperature*; state: *Maharashtra*; method: *Moving average*

Place	MA interval	MAPE	MSE	RMSE
Akola	3	*0.0132*	18.4009	4.2896
	5	0.0116	23.2767	4.8246
	7	0.0169	24.5899	4.9588
	9	0.0203	23.4677	4.8443
	11	0.0222	18.4071	4.2904
	13	0.0260	*13.9550*	*3.7356*
Nagpur	3	0.0112	21.9290	4.6828
	5	*0.0110*	28.7048	5.3577
	7	0.0146	30.7950	5.5493
	9	0.0173	29.0497	5.3898
	11	0.0184	22.6028	4.7542
	13	0.0208	*17.0587*	*4.1302*
Wardha	3	0.0117	20.0899	4.4822
	5	*0.0104*	25.8453	5.0838
	7	0.0155	27.5178	5.2457
	9	0.0186	26.1029	5.1091
	11	0.0201	20.3914	4.5157
	13	0.0231	*15.4170*	*3.9264*

Table 11 Comparison of parameter: *Average temperature*; state: *Maharashtra*; method: *SES*

Place	MA interval	MAPE	MSE	RMSE
Akola	0.1	1.7002	14.5778	3.8416
	0.2	1.5528	14.8728	3.8565
	0.3	1.3481	14.2997	3.7815
	0.4	1.1497	13.3129	3.6487
	0.5	0.9857	12.1862	3.4909
	0.6	0.8548	11.0872	3.3298
	0.7	0.7425	10.0907	3.1766
	0.8	0.6344	9.2250	3.0373
	0.9	*0.5214*	*8.5031*	*2.9160*
Nagpur	0.1	1.6022	18.2086	4.2672
	0.2	1.5090	18.4106	4.2608
	0.3	1.3153	17.6743	4.2041
	0.4	1.1058	16.3479	4.0433
	0.5	0.9228	14.8102	3.8484
	0.6	0.7735	13.2976	3.6466
	0.7	0.6474	11.9146	3.4518
	0.8	0.5319	10.6965	3.2706
	0.9	*0.4187*	*9.6514*	*3.1067*

(continued)

Table 11 (continued)

Place	MA interval	MAPE	MSE	RMSE
Wardha	0.1	1.5933	16.5491	4.0681
	0.2	1.4723	16.7112	4.0879
	0.3	1.2724	16.0511	4.0064
	0.4	1.0696	14.8837	3.8579
	0.5	0.8986	13.5378	3.6794
	0.6	0.7615	12.2160	3.4951
	0.7	0.6489	11.0079	3.3178
	0.8	0.5380	9.9448	3.1535
	0.9	*0.4296*	*9.0360*	*3.0060*

Table 12 Comparison of parameter: *Average temperature*; state: *Maharashtra*; method: *Neural network*

Place	Architecture	Transfer function		MAPE	MSE	RMSE
		Layer1	Layer2			
Akola	1-2-1	Purelin	Tansig	*0.0661*	*0.0874*	*0.2957*
	1-4-1			0.0662	0.0867	0.2945
	1-5-1			0.0667	0.0875	0.2959
	1-10-1			0.0695	0.0881	0.2969
	1-2-1		Logsig	1.7002	6.0358	2.4568
	1-4-1			1.7000	6.0378	2.4572
	1-5-1			1.6988	6.0590	2.4612
	1-10-1			1.6991	6.0515	2.4600
Nagpur	1-2-1	Purelin	Tansig	*0.0885*	*0.1398*	*0.3740*
	1-4-1			0.0921	0.1407	0.3751
	1-5-1			0.0922	0.1426	0.3776
	1-10-1			0.0939	0.1412	0.3758
	1-2-1		Logsig	1.6698	7.4996	2.7385
	1-4-1			1.6698	7.4994	2.7385
	1-5-1			1.6690	7.5027	2.7391
	1-10-1			1.6685	7.5074	2.7400
Wardha	1-2-1	Purelin	Tansig	*0.0831*	*0.1212*	*0.3482*
	1-4-1			0.0721	0.1188	0.3446
	1-5-1			0.0742	0.1190	0.3449
	1-10-1			0.0751	0.1194	0.3456
	1-2-1		Logsig	1.6541	6.8161	2.6108
	1-4-1			1.6548	6.7993	2.6075
	1-5-1			1.6545	6.8052	2.6087
	1-10-1			1.6551	6.7959	2.6069

Table 13 Comparison of parameter: *Vapour pressure*; state: *Maharashtra*; method: *Moving average*

Place	MA interval	MAPE	MSE	RMSE
Akola	3	0.0121	41.4460	6.4379
	5	0.0131	60.1880	7.7561
	7	0.0103	66.9655	8.1832
	9	0.0089	61.3387	7.8319
	11	0.0092	46.7823	6.8398
	13	*0.0107*	*35.0075*	*5.9167*
Nagpur	3	0.0072	69.0969	8.3125
	5	0.0102	82.5055	9.0833
	7	0.0064	85.7072	9.2578
	9	0.0048	81.6048	9.0335
	11	0.0044	66.2720	8.1408
	13	*0.0049*	*51.2735*	*7.1605*
Wardha	3	0.0092	55.0834	7.4218
	5	0.0117	70.0675	8.3706
	7	0.0080	74.4615	8.6291
	9	0.0064	70.0034	8.3668
	11	0.0062	55.5598	7.4538
	13	*0.0071*	*42.4677*	*6.5167*

Table 14 Comparison of parameter: *Vapour pressure*; state: *Maharashtra*; method: *SES*

Place	Alpha	MAPE	MSE	RMSE
Akola	0.1	3.9425	33.5423	6.2082
	0.2	4.3173	33.7171	6.2223
	0.3	4.2557	36.5075	6.0421
	0.4	3.8348	32.3978	5.7357
	0.5	3.2179	23.9712	5.3325
	0.6	2.5573	25.3944	5.0393
	0.7	1.9523	22.4197	4.7349
	0.8	1.4539	20.0743	4.4840
	0.9	*1.0814*	*13.3079*	*4.2788*
Nagpur	0.1	2.3506	53.7187	7.6628
	0.2	2.6392	60.6606	7.7385
	0.3	2.9714	60.1097	7.7530
	0.4	2.7693	57.9919	7.6152
	0.5	2.3429	55.4694	7.4478
	0.6	1.8211	53.3076	7.3012
	0.7	1.3036	51.8508	7.2007
	0.8	0.8505	51.2235	7.1574
	0.9	*0.4837*	*51.5212*	*7.1778*

(continued)

Table 14 (continued)

Place	Alpha	MAPE	MSE	RMSE
Wardha	0.1	1.6022	13.2086	4.2672
	0.2	1.5090	13.4106	4.2608
	0.3	1.3193	17.6743	4.2041
	0.4	1.1058	16.3479	4.0433
	0.5	0.9228	14.3102	3.8484
	0.6	0.7735	13.2976	3.6466
	0.7	0.6474	11.9146	3.4518
	0.8	0.5319	10.6965	3.2706
	0.9	*0.4137*	*9.6514*	*3.1067*

Table 15 Comparison of parameter: *Vapour pressure*; state: *Maharashtra*; method*:* *Neural network*

Place	Architecture	Transfer function		MAPE	MSE	RMSE
		Layer1	Layer2			
Akola	1-2-1	Purelin	Tansig	*0.0507*	*0.1639*	*0.4048*
	1-4-1			0.0466	0.1629	0.4036
	1-5-1			0.0561	0.1653	0.4066
	1-10-1			0.0664	0.1683	0.4102
	1-2-1		Logsig	4.8072	28.0023	5.2917
	1-4-1			4.8071	28.0007	5.2916
	1-5-1			4.8066	27.9927	5.2908
	1-10-1			4.8068	27.9952	5.2910
Nagpur	1-2-1	Purelin	Tansig	*0.2353*	*0.2430*	*0.4929*
	1-4-1			0.2333	0.2441	0.4941
	1-5-1			0.2419	0.2443	0.4941
	1-10-1			0.2338	0.2454	0.4954
	1-2-1		Logsig	3.4611	44.3359	6.6585
	1-4-1			3.4601	44.3333	6.6583
	1-5-1			3.4614	44.3378	6.6587
	1-10-1			3.4615	44.3372	6.6586
Wardha	1-2-1	Purelin	Tansig	*0.1776*	*0.2279*	*0.4774*
	1-4-1			0.1767	0.2269	0.4763
	1-5-1			0.1691	0.2279	0.4774
	1-10-1			0.1763	0.2272	0.4766
	1-2-1		Logsig	4.2893	44.2543	6.6524
	1-4-1			4.2891	44.2524	6.6522
	1-5-1			4.2897	44.2564	6.6525
	1-10-1			4.2895	44.2537	6.6523

Table 16 Comparison of parameter: *Reference crop evapotranspiration*; state: *Maharashtra*; method*: Moving average*

Place	MA interval	MAPE	MSE	RMSE
Akola	3	*0.0357*	1.8560	*1.3624*
	5	0.0307	2.3691	1.5392
	7	0.0510	2.4972	1.5803
	9	0.0647	2.3881	1.5453
	11	0.0738	1.8559	1.3623
	13	0.0860	*1.3861*	1.1773
Nagpur	3	*0.0290*	2.1246	1.4576
	5	0.0249	2.7088	1.6458
	7	0.0448	2.8544	1.6895
	9	0.0570	2.7290	1.6520
	11	0.0639	2.1219	1.4567
	13	0.0860	*1.3861*	*1.1773*
Wardha	3	*0.0135*	2.0082	1.4171
	5	0.0271	2.5765	1.6052
	7	0.0472	2.7229	1.6501
	9	0.0602	2.5985	1.6126
	11	0.0682	2.0175	1.4204
	13	0.0791	*1.5065*	*1.2274*

Table 17 Comparisonof parameter: *Reference crop evapotranspiration*; state: *Maharashtra*; method*: SES*

Place	Alpha	MAPE	MSE	RMSE
Akola	0.1	3.2765	1.6713	1.2928
	0.2	2.8989	1.6833	1.2974
	0.3	2.4425	1.6058	1.2672
	0.4	2.0348	1.4735	1.2139
	0.5	1.7276	1.3229	1.1502
	0.6	1.5082	1.1758	1.0844
	0.7	1.3339	1.0412	1.0204
	0.8	1.1854	0.9214	0.09591
	0.9	*1.0279*	*0.8162*	*0.9033*
Nagpur	0.1	3.1930	1.9275	1.3883
	0.2	2.8242	1.9486	1.3959
	0.3	2.3481	1.8626	1.3648
	0.4	1.9157	1.7113	1.3081
	0.5	1.5900	1.5373	1.2399
	0.6	1.3604	1.3669	1.1691
	0.7	1.1876	1.2104	1.1002
	0.8	1.0358	1.0707	1.0348
	0.9	*0.8840*	*0.9477*	*0.9735*

(continued)

Table 17 (continued)

Place	Alpha	MAPE	MSE	RMSE
Wardha	0.1	3.2792	1.8476	1.3593
	0.2	2.8931	1.8628	1.3648
	0.3	2.4109	1.7765	1.3328
	0.4	1.9774	1.6287	1.2762
	0.5	1.6516	1.4606	1.2085
	0.6	1.4214	1.2965	1.1386
	0.7	1.24.72	1.1465	1.0707
	0.8	1.0932	1.0130	1.0065
	0.9	*0.9381*	*0.08957*	*0.9484*

Table 18 Comparison of parameter: *Reference crop evapotranspiration*; state: *Maharashtra*; method*: Neural network*

Place	Architecture	Transfer function		MAPE	MSE	RMSE
		Layer1	Layer2			
Akola	1-2-1	Purelin	Tansig	*0.4011*	*0.0129*	*0.1134*
	1-4-1			0.3757	0.0125	0.1119
	1-5-1			0.3355	0.0125	0.1119
	1-10-1			0.3365	0.0125	0.1120
	1-2-1		Logsig	4.2695	1.0795	1.0390
	1-4-1			4.2691	1.0801	1.0393
	1-5-1			4.2691	1.0801	1.0393
	1-10-1			4.2694	1.0979	1.0391
Nagpur	1-2-1	Purelin	Tansig	*0.3707*	*0.0146*	*0.1208*
	1-4-1			0.4222	0.0158	0.1256
	1-5-1			0.3766	0.0146	0.1207
	1-10-1			0.3961	0.0147	0.1212
	1-2-1		Logsig	4.5755	1.4155	1.1898
	1-4-1			4.5756	1.4154	1.1897
	1-5-1			4.5755	1.4155	1.1898
	1-10-1			4.5755	1.4156	1.1898
Wardha	1-2-1	Purelin	Tansig	*0.4271*	*0.0141*	*0.1189*
	1-4-1			0.4225	0.0141	0.1136
	1-5-1			0.4163	0.0141	0.1136
	1-10-1			0.4269	0.0141	0.1137
	1-2-1		Logsig	4.5773	1.3325	1.1543
	1-4-1			4.5775	1.3323	1.1545
	1-5-1			4.5775	1.3329	1.1545
	1-10-1			4.5776	1.3326	1.1544

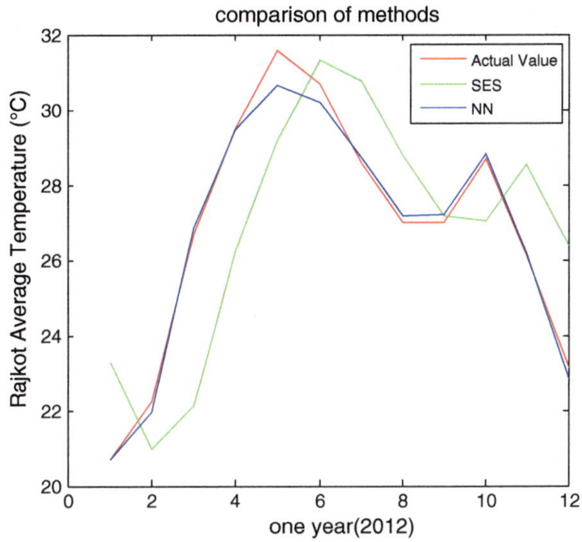

Fig. 5 Comparison of SES and ANN methods for place: Rajkot; parameter: average temperature

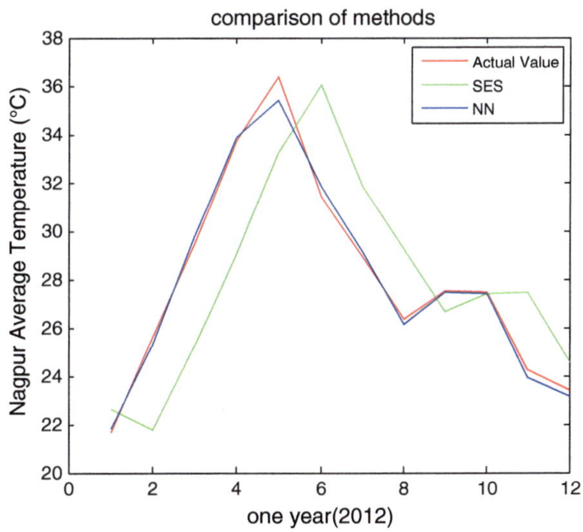

Fig. 6 Comparison of SES and ANN methods for place: Nagpur; parameter: average temperature

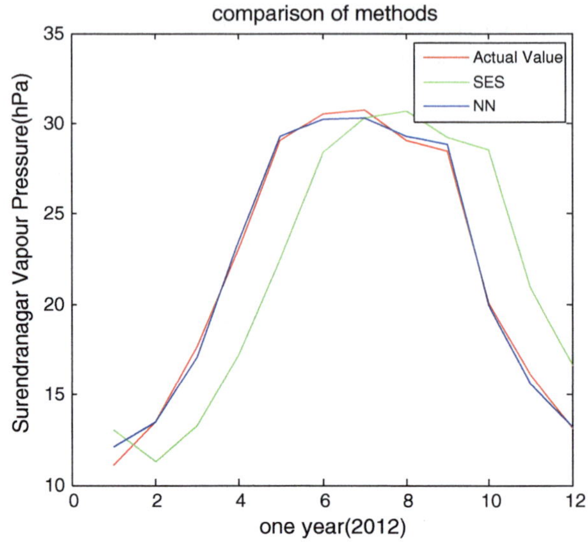

Fig. 7 Comparison of SES and ANN methods for place: Surendranagar; parameter: vapour pressure

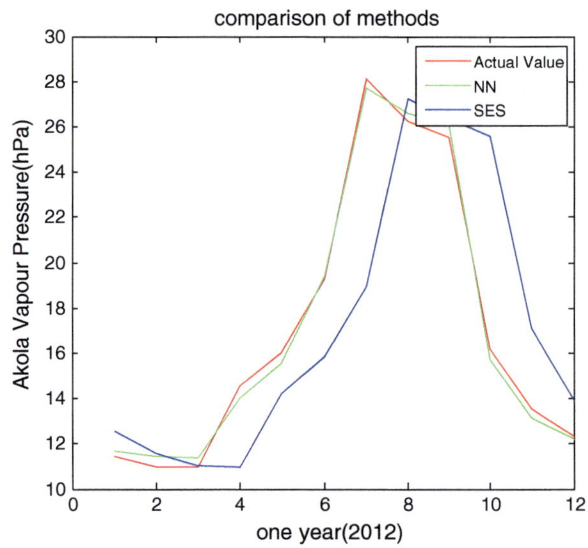

Fig. 8 Comparison of SES and ANN methods for place: Akola parameter: vapour pressure

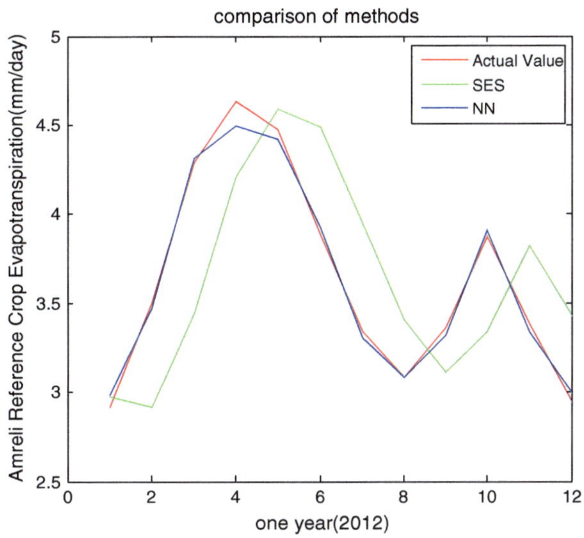

Fig. 9 Comparison of SES and ANN methods for place: Amreli; parameter: reference crop evapotranspiration

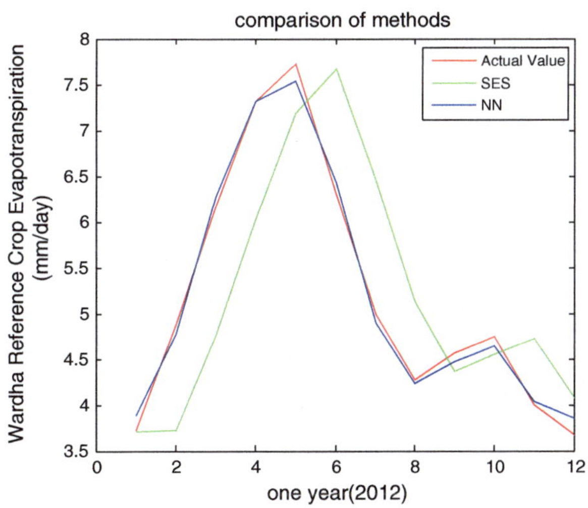

Fig. 10 Comparison of SES and ANN methods for place: Wardha; parameter: Reference crop evapotranspiration

Table 19 Year-wise area, annual production and annual productivity of cotton in Gujarat [1]

Year	Area (million hectares)	Production (million bales)	Productivity per annum
1999–2000	1.516	3.5	392
2000–2001	1.578	2.7	291
2001–2002	1.687	3.3	342

Table 20 Year-wise area, annual production and annual productivity of cotton in Maharashtra [3]

Year	Area (lakh hectares)	Production (lakh bales)	Productivity per annum
1999–2000	32.54	31.26	163
2000–2001	27.93	14.57	89
2001–2002	30.00	27.98	158

6 Conclusion

Prediction of a good crop can be assured if the relevant parameters are rightly predicted well before in time. The monthly predictions for the year 2002 are made based on the past data of the last century for all the five major cotton-producing districts of Gujarat and three districts of Maharashtra. These values are then compared to in situ data as recorded. As the variations are found to be minimum, a good crop is expected, which is reflected in the Article by cotton statistics and news. Moreover, out of all the methods discussed and implemented, it is observed that minimum error is reflected in the readings when neural networks are used as prediction model. It is further observed these parameters alone are not totally sufficient to predict the upcoming crop. Other important factors such as rainfall, soil quality, availability of irrigation facilities play an equally significant role.

With the advent of new mechanisms and technology, many varieties are observed to achieve many evolving characteristics like earliness and adaptability, which are very helpful for growth. Earliness varieties provide for multiple cross-breeding and are known to have excellent pest tolerance with reduced maturity period. Adaptability refers to the characteristic of the crop variety to adapt to changing environmental conditions, specially exhibited by hybrid varieties.

No matter how adaptive the variants become, but forecasting still helps in having an estimate over upcoming weather variations.

References

1. Amar Singh, (2015, Nov 24), "History of cotton in Gujarat", Available: http://www.caionline.in.
2. M. Sabesh, M. Ramesh, A.H. Prakash, G. Bhaskaran, "Is There any Shift in Cropping Pattern in Maharashtra after the Introduction of Bt. Cotton?" Cotton Research Journal, Vol. 6, 2014, pp. 63–70.
3. Performance of Bt. Cotton Cultivation in Maharashtra "Report of State Department of Agriculture" Available: www.envfor.nic.in/divisions/csurv/geac/srmh.pdf.
4. P. Singh, M.S Kairon (2000), "Cotton Varieties and Hybrids", CICR, Bulletin No. 13, Available: http://www.cicr.org.in.
5. Adhikari, Ratnadip, and R.K. Agrawal. "An introductory study on time series modeling and forecasting." arXiv: 1302.6613 (2013).
6. Mahalakshmi, G.S. Sridevi, and S. Rajaram. "A survey on forecasting of time series data." In Computing Technologies and Intelligent Data Engineering (ICCTIDE), International Conference on, pp. 1–8. IEEE, 2016.
7. N.E. Hwa (2007) "Different uses of moving average (MA)," Available: https://www.chartnexus.com/learning/static/pulses_apr2007.pdf.
8. D. Sodha, G. Saha, "Crop management of Agricultural products using Time Series Analysis" IEEE Recent Trends in Electronics Information Communication Technology, pp. 1–4, 2016.
9. Khoshravesh, Mojtaba, Mohammad Ali Gholami Sefidkouhi, and Mohammad Valipour. "Estimation of reference evapotranspiration using multivariate fractional polynomial, Bayesian regression, and robust regression models in three arid environments." Applied Water Science (2015): 1–12.
10. Nema, Manish K., Deepak Khare, and Surendra K. Chandniha. "Application of artificial intelligence to estimate the reference evapotranspiration in sub-humid Doon valley." Applied Water Science (2017): 1–8.
11. Hayati, Mohsen, and Zahra Mohebi. "Application of artificial neural networks for temperature forecasting." World Academy of Science, Engineering and Technology 28, no. 2 (2007): 275–279.
12. https://www.indiawaterportal.org/metdata last accessed on 3rd March, 2017.
13. https://www.imd.gov.in last accessed on 3rd March, 2017.

Enhancement of Data Streaming in Clustering for Uncertain Data

Jeny Ganatra and Chintan Thacker

Abstract Data stream mining presents exciting model from the stream of information. The objective of clustering data stream is to cluster and determine unknown pattern from streaming data. Streaming data generated from imprecise hardware result into uncertainty built into them. In this paper, we propose a novel approach for data stream clustering which reduces the degree of uncertainty and increases the degree of homogeneity. Unlike current procedures whose aim is to cluster data stream, we consider the case where uncertainty of micro-cluster affects the quality of clustering outcome. The proposed algorithm Fuzzy-HCMStream is based on HCMStream algorithm which incorporates the concept of Fuzzy C-Mean algorithm, in account for uncertainty. Our proposed algorithm increases degree of homogeneity, which leads to improvement in clustering quality. The implementation of proposed algorithm is tested with two data sets comprising synthetic and real data set. The Fuzzy-HCMStream algorithm shows improvement in performance compared to the other state-of-the-art algorithms in terms of various indices for evaluating clustering results.

Keywords Data mining · Data stream clustering · Fuzzy clustering
Uncertain data

1 Introduction

Data mining is the methodology for determining stimulating knowledge or pattern from incredible extent of contribution [1]. The current method in knowledge is capable to increase sole calculation of statistics, which is mandatory to discover online as they seem. Data stream is constant generation of sequenced data.

J. Ganatra (✉) · C. Thacker
Computer Engineering Department, HJD-ITER, Kachchh 370430, Gujarat, India
e-mail: jenyganatra9@gmail.com

C. Thacker
e-mail: chintanthacker450@gmail.com

R. Kher et al. (eds.), *Proceedings of the International Conference on Intelligent Systems and Signal Processing*, Advances in Intelligent Systems and Computing 671,
https://doi.org/10.1007/978-981-10-6977-2_13

Instances consist of telephone records, sensor networks, etc. Data stream mining publishes exciting model from the stream of information [2]. These assessments are produced frequently with exciting differ charges.

Due to inexact apparatus or hardware, inexact data are produced which effect into vagueness constructed into them. By considering inexact data, we can increase the result of clustering algorithm in stream mining. With the usage of inexact data stream, we can fortify productivity of data mining procedure [3].

The objective of clustering is to cluster the streaming data into evocative classes and determine unknown pattern from streaming data. There are numerous clustering procedures for static data bank, but they are not appropriate for streaming data [4–6]. Data streams clustering process is capable to identify impulsive pattern, hold clatter, clustering without preceding proficiency about number of groups. An instance of clustering is the pursuing of network data cast-off to study alterations in traffic flow patterns and to identify potential interruptions. In this instance, furthermore, data must be handled as it is formed.

As streaming data originates in vast quantity, we cannot load entire records in memory at an interval. The streaming data cannot be reconsidered due to its growing nature. Fuzzy clustering technique is capable to decrease effort of data via Fuzzy data estimation [7]. In Fuzzy clustering, data section can have its place to numerous clusters (i.e., two or more groups) with certain definite worth of membership. Figure 1 shows an instance of hard clustering and soft (or Fuzzy) clustering.

The Fuzzy clustering problem can be resolved by means of the Fuzzy C-Means procedure. If Fuzzy examination is used, actual fluctuations may be recognized via alterations of degrees of membership. By obstruction, the dynamic Fuzzy clustering contributions to save time and can produce additional consistent outcome.

The remainder of this paper is arranged as follows. In the next section, we overview the clustering algorithms for uncertain data and clustering algorithm based on Fuzzy set theory for data streams. Sect. 2 provides the proposed work for algorithm, and Sect. 3 introduces the GUI and proposed algorithm. Finally, we conclude study.

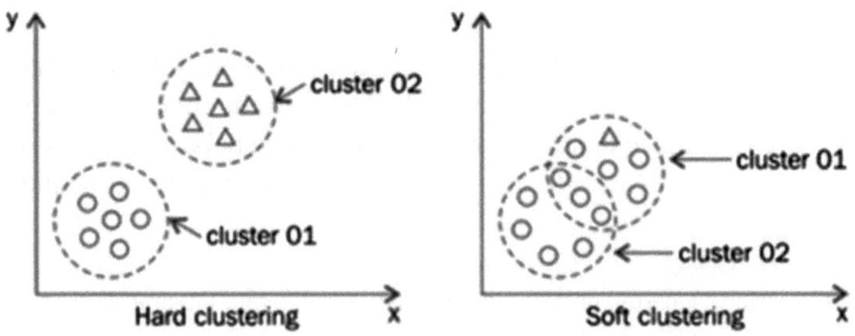

Fig. 1 Hard clustering & Fuzzy clustering

2 Related Work

As per the research scenario, clustering data stream is a significant research topic of stream mining. A lot of researches and projects are developed on data stream mining as well as on data stream clustering [8]. Many successful algorithms are proposed for data stream clustering, for example, CluStream [4] and DenStream [5] which are earliest proposed algorithm. E-Stream and HECES are subsequentially proposed algorithms in order to provide efficient clustering result.

According to research, a number of procedures have been preferred for uncertain flow of data [3]. In the uncertain data clustering, UK-means algorithm [9], M-FDBSCAN algorithm, and UMicro algorithm are proposed.

A Fuzzy method seems to be mainly beneficial for the kind of application, in which the clustering construction is matter to constant fluctuations. In Fuzzy clustering, an object may fit into different groups, and it is conveyed with membership theme [10]. We can see Fuzzy C-Mean as basic clustering method for soft clustering [11–14], as it expresses clustering being assigned to certain clusters with certain definite degree of membership.

HCMStream [6], a data streaming clustering procedure which agrees eliminations of statistics archives at some uninformed interval. It customs hyper-cylindrical micro-clusters produced by assimilation of outmoded spherical micro-clusters, agrees the eliminations of statistics at any random interval. The clustering process is performed in connected manner by incrementally informing adjacency matrix each and every time there is an alteration in the arrangement of the micro-clusters; offline phase is not compulsory. The procedure needs two input variables from end user, namely np and r. Parameter np is the threshold for determining outliers. r is the static radius of spherical and cylindrical micro-clusters. The radius of entirely micro-clusters is conventional at persistent for the entire procedure. This method uses destructor which leads to information loss, and also, uncertainty in micro-clustering is not considered.

The UMicro algorithm [15], considered the deviation of each dimension. This algorithm adds an error representation for each dimension to reflect data uncertainty when dealing with a dimensional uncertain data tuple. This algorithm adopted the two-layer frame of Clustream algorithm. The number of micro-clusters requires to be stated in the algorithm. For each coming data tuple, the similarity between the existing data tuple and new coming micro-cluster needs to be re-calculated. The similarity is the expectation distance between the uncertain data tuple and clustered one. Besides, a pyramidal time frame is proposed in UMicro algorithm to well analyze data stream. However, the outlier of data stream is not considered in this algorithm. There is no good method to deal with outlier whatsoever.

Gloria et al. [16] propose a supplement of DBSCAN algorithm to initiate clusters with Fuzzy density quality. In this method, only density threshold is enrolled and its intention is to identify Fuzzy clusters with estimation of density attribute. This algorithm allows user to specify input parameters for execution, and it suggest a soft constraint to define number of points, approximately.

D-Stream [17] is a foundation for streaming data clustering through a density-based approach. It examines clustering multi-dimensional data in the formation of a stream. This method utilizes online element which plans all input data into mesh and an offline element which calculates the mesh based on weight. It acquires density decaying technique to ensure modification of data stream. It automatically alters clusters without demanding particular. This algorithm is able to cluster high-speed data stream without decaying the nature of clustering. This algorithm is able to identify arbitrary shape clusters with efficiency and accuracy.

3 Proposed Algorithm

Fuzzy-HCMStream clustering algorithm is based on HCMStream clustering algorithm. In our proposed algorithm, Fuzzy clustering technique is used with hyper-cylindrical micro-clustering technique. There are two different kinds of micro-clusters (MC) in HCMStream, i.e., spherical micro-cluster (SM) and cylindrical micro-cluster (CM). A SM is cast off as an essential component to characterize a native cluster. A fresh MC is mainly created as a SM. Then, numerous overlying SMs will procedure a CM. A set of overlying MCs formulates a cluster as shown in Fig. 2.

Thus, incorporating the concept of Fuzzy C-Mean algorithm with HCMStream clustering algorithm results in less degree of uncertainty in micro-cluster, which leads to improvement in clustering quality. Here, SM represents spherical micro-cluster and x represents incoming data element in algorithm described below.

Algorithm 1 Fuzzy-HCMStream

1. Algorithm attempts to combine an incoming element (x) with present CM
2. if x can't be allotted to any CM, it is tested with SM
3. end if
4. if x is not into any present MC, a fresh micro-cluster (SM_{new}) is created to cover x
5. end if
6. When SM_k is revised through addition of a fresh data element
    a. if the population in the MC is intense, we attempt to combine it with present CM
    b. end if

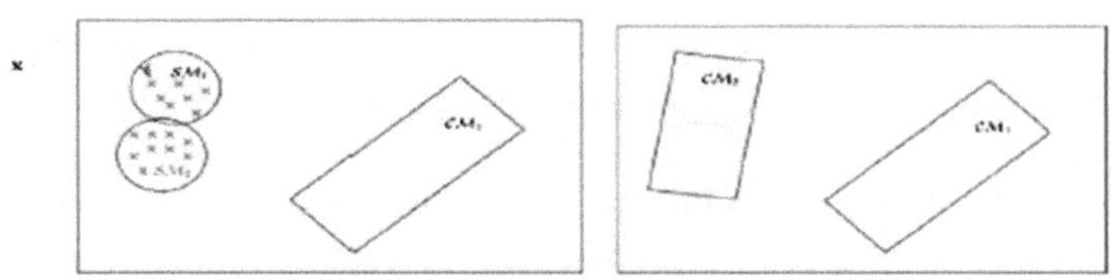

Fig. 2 An example of cluster formation by connecting micro-cluster

7. if SM_k can't be merged with other CM
    a. if it overlays further SM
        (1) The overlying SM are combined jointly & changed by a new CM
        (2) return
    b. end if
8. end if
9. return cluster clst

3.1 Input Parameters

The algorithm entails two input restrictions from user, namely np and r. Restriction np is the threshold designed for defining outliers. On supplementary indicator, MCs whose statistics points are less than np are not consumed in clustering. r is the static radius of SM and CM. The radius of all MCs is established at persistent for the entire procedure.

The parameters np and r play a significant part in defining the clustering performance. When np is huge, added micro-clusters with comparatively less statistics would be accepted as of the clustering. This can assistance extrication of two clusters by rejecting outliers among two clusters.

The radius of micro-clusters r regulates the determination of clustering. Higher clustering determination, resultant as of smaller r, regularly indications to more separate clusters being originate. If r is huge, micro-clusters would develop less dense since they would shelter more zone deprived of statistics opinions. However, if r is too minor, MCs in the similar cluster developed excessively minor to completely attach through all further, which prominent to the solitary cluster violation addicted to trashes.

In adjusting the limits, we usually separate some data points intended for conveying the worth of np and r. First, np is set at little worth such as 5, then preliminary beginning with small r and progressively increasing r, and we can discover the choice of r that revenues the steady number of ultimate clusters. Small r can be selected from this variety to realize high clustering resolve.

3.2 Performance Indices

The data stream clustering outcomes were associated through the class labels providing in the data sets. We have chosen performance indices, namely Adjusted Rand index (AR), Rand index (RI), Hubert's index (HI). These indices reflect the homogeneity of clusters. The principles of these indices commonly span from 0 to 1, except HI. HI can be negative assessment.

Rand index interpretations clustering as a sequence of choices in selecting sets of statistics in the set. Rand index can be deliberate by

$$\mathrm{RI} = (\text{No. of Agreement})/(\text{Total No. of Pairs}) \tag{1}$$

Hubert's index is definite by the alteration of the prospect of contract and the prospect of divergence. It is computed as

$$\mathrm{HI} = (\text{No. of Agreement} - \text{No. of Disagreement})/(\text{Total No. of Pairs}) \tag{2}$$

Adjusted Rand Index (AR) is calculated as

$$\mathrm{AR} = (\text{No. of Agreement} - \text{Expected Index})/(\text{Total No. of Pairs} - \text{Expected Index}) \tag{3}$$

However, the other indices, namely RI and HI capture the comparison founded on including the contract among altogether sets of data substances. RI sincerely tallies the number of assenting sets while Hubert's index comprises the punishment of conflicting pairs in the cunning by withdrawing conflicting sets FP + FN from the supportive sets TP + TN. As an outcome, HI is always fewer than RI; since RI regularly has worth between 0.5 and 1 [6].

4 Experimental Results

Algorithm is implemented in MATLAB. Two data sets: synthetic and real are used. The data set consists of two-dimensional data with changing density, and 7% data are equally and arbitrarily created as noise to the original data set. Forest cover type data set is obtained from UCI machine learning repository. It contains 54 cartographic variables including 44 qualitative binary variables and 10 quantitative variables [18]. We used familiar similarity measures which are frequently used in evaluating clustering outcome, namely Rand Index (RI), Hubert's Index (HI), and Adjusted Rand Index (AR).

Table 1 shows the result of Fuzzy-HCMStream algorithm with different values of r and np. Table 2 shows comparison of outcome of Fuzzy-HCMStream clustering algorithm with HCMStream clustering algorithm. It shows that Fuzzy-HCMStream outperforms the HCMStream clustering algorithm in terms of performance indices which are considered on the basis of homogeneity of micro-cluster. Figure 3 shows comparison of Fuzzy-HCMStream with HCMStream clustering algorithm.

Table 1 Fuzzy-HCMStream clustering result for forest cover data set

PI	$r = 0.15$; np = 5	$r = 0.25$; np = 10	$r = 0.35$; np = 20	$r = 0.50$; np = 30
AR	0.1095	0.1412	0.1728	0.1026
HI	0.3149	0.2965	0.3097	0.2994
RI	0.6575	0.6482	0.6548	0.6497

Table 2 Comparison of Fuzzy-HCMStream with HCMStream for forest cover data set

	Fuzzy-HCMStream algorithm	HCMStream algorithm
PI	$r = 0.25$; np = 30	$r = 0.25$; np = 30
AR	0.1920	0.0885
HI	0.2202	0.1274
RI	0.6101	0.5127

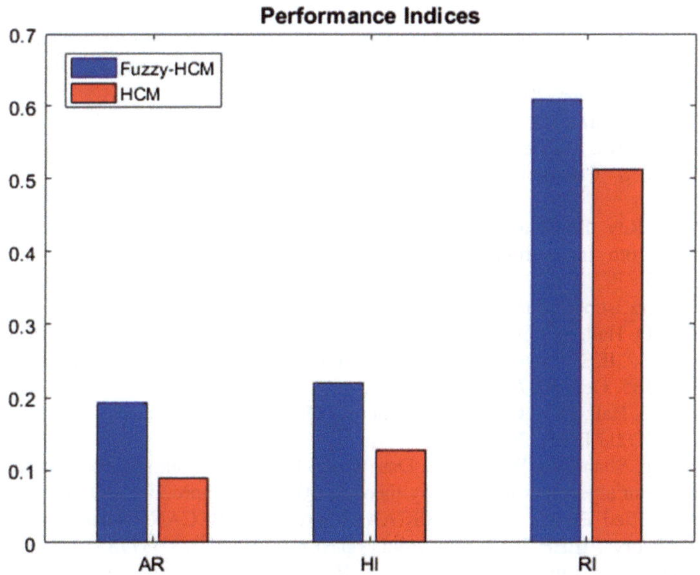

Fig. 3 Comparison of outcome: performance indices for real data set with $r = 0.25$; np = 30

5 Conclusion and Future Work

HCMStream algorithm does not give efficient clustering result when we gradually increase value of r and np. On the other hand, incorporating the concept of Fuzzy clustering in HCMStream clustering gives improved clustering quality when we increase value of r and np. It gives better clustering outcome in terms of homogeneity (i.e., increase degree of homogeneity), which is measured by performance indices, which gives better clustering outcome. In Future work, multi-media streaming data can be used with Fuzzy-HCMStream clustering algorithm.

References

1. Han and Kamber, Book Review: "Data mining – Concepts and Techniques".
2. Mohamed Medhat Gaber, Arkady Zaslavsky and Shonali Krishnaswamy, "Mining Data Streams: A Review", SIGMOD Record, Volume 34, No. 2, June 2005. doi:10.1145/1083784.1083789.
3. Charu C. Aggarwal, "A Survey of Uncertain Data Clustering Algorithms".
4. S. Guha, A. Meyerson, N. Mishra, R. Motwani, L. O'Callaghan, Clustering data streams: theory and practice, IEEE Trans. Knowl. Data Eng. 15 (2003) 515–528. doi:10.1109/TKDE.2003.1198387.
5. F. Cao, M. Ester, W. Qian, A. Zhou, Density-based clustering over an evolving data stream with noise, in: Proceedings of SIAM International Conference on Data Mining, 2004, pp. 328–339.
6. Sirisup Laohakiat, Suphakant Phimoltares, Chidchanok Lursinsap, "Hyper-cylindrical micro-clustering for streaming data with unscheduled data removals", 10.1016/j.knosys.2016.02.004, Elsevier, 2016.
7. Fuzzy Sets and Fuzzy Logic Theory and Applications by George J. Klir and Boyuan.
8. Jonathan A Silva, Elaine R Faria, Rodrigo C Barros, et al., "Data stream clustering, A survey," ACM Computing Surveys, 2013, 46(1). doi:https://doi.org/10.1145/2522968.2522981.
9. Ngai Wang Kay, Ben Kao, Chun Kit Chui et al., "Efficient clustering of uncertain data," in Proc. of 6th International Conference on Data Mining, Hong Kong, China, 2006, pp. 436–445.
10. http://citeseerx.ist.psu.edu/viewdoc/download?doi=10.1.1.381.2407&rep=rep1&type=pdf.
11. P. Hore, L. O. Hall, and D. B. Goldgof, "A Fuzzy C Means Variant For Clustering Evolving Data Streams," IEEE International Conference on Systems, Man and Cybernetics, Montreal, pages 360–365, Oct. 2007.
12. P. Hore, L. O. Hall, D. B. Goldgof, W. Cheng, "Online Fuzzy C Means", 978-1-4244-2352-1/08/$25.00 ©2008 IEEE, 2008.
13. Baoju Zhang, Shan Qin, Wei Wang, Dan Wang, Lei Xue, "Data stream clustering based on Fuzzy C-Mean algorithm and entropy theory", 10.1016/j.sigpro.2015.10.014, 2015.
14. Abir Smiti, Zied Eloudi, "Soft DBSCAN: Improving DBSCAN Clustering method using fuzzy set theory", IEEE 2013. https://doi.org/10.1109/HSI.2013.6577851.
15. Yue Yang, Zhuo Liu, Zhidan Xing, "A Review of Uncertain Data Stream Clustering Algorithms", 2015. https://doi.org/10.1109/ICICSE.2015.30.
16. Gloria Bordogna, Dino Ienco, Irstea, Montpellier, "Fuzzy Core DBScan Clustering Algorithm", Springer, 2014.
17. Yixin Chen, Li Tu, "Density-Based Clustering for Real-Time Stream Data", 13th ACM SIGKDD International Conference on Knowledge Discovery and Data Mining, 2007. https://doi.org/10.1145/1281192.1281210.
18. UCI Machine Learning Repository. http://archive.ics.uci.edu/ml/, 1999.

Performance Analysis of Word Spotting Techniques Using HOG and Shape Descriptor on Gujarati Script

Himanshu M. Kathiriya and Mukesh M. Goswami

Abstract The current paper represents the performance analysis of word spotting techniques in Gujarati script using two of the well-known feature extraction techniques, namely HOG and shape descriptors (SD). The experiments were performed using dynamic time warping (DTW) on a small-sized word image database collected from machine printed book images and the initial results are encouraging.

Keywords Word spotting · Gujarati script · Document image retrieval

1 Introduction

Information retrieval (IR) is a task to retrieve relevant information from the documents. It is easier to retrieve information from a digital text document whereas in document images it is an area of research. Mainly two approaches are used for IR in document images, namely recognition based and recognition free. The recognition-based approach converts the document image into a machine editable text document using OCR and uses standard text processing techniques for IR. However, OCR restricts transformation of poor quality document images. Also, the word-level accuracy of OCR systems for Indian scripts has remained low because of the issues of characters and glyphs segmentation [1]. Therefore, many researchers are motivated to explore recognition free approach.

Over the years, various techniques have been evolved to query the document images using the physical layout, logical and structural information, and other extracted contents [2]. Word (keyword) spotting from the document image is one of the frequently used techniques for IR task in which the query word image is directly matched with document word images without explicitly recognizing the characters.

H. M. Kathiriya (✉) · M. M. Goswami
Dharmsinh Desai University, Nadiad, India
e-mail: hm.kathiriya@gmail.com

M. M. Goswami
e-mail: mukesh.goswami@gmail.com

R. Kher et al. (eds.), *Proceedings of the International Conference on Intelligent Systems and Signal Processing*, Advances in Intelligent Systems and Computing 671,
https://doi.org/10.1007/978-981-10-6977-2_14

Gujarati is a prominent Indian script having rich cultural heritage and the literature. There are around 39,000 machine printed Gujarati books already available online by DLI [3]. However, the OCR technology for Gujarati script is still in progress. While it will take some time for the Gujarati OCR to reach maturity, word spotting-based document image retrieval can be a viable intermediate solution to search and retrieve documents from large repositories like DLI. The current paper presents word-spotting experiments on a small-sized Gujarati word image database using well-known histogram of oriented gradients (HOG) and shape descriptor (SD) features. The remaining of the paper is organized as follows, Sect. 2 gives brief literature review followed by the discussion on feature extraction and matching in Sect. 3. The experimental setup and results are discussed in Sect. 4 followed by conclusion in Sect. 5.

2 Related Work

Numerous amount of work has been reported for major Indian scripts like Devanagari, Bengali, and Telugu. The approaches suggested for word spotting in Devanagari and Sanskrit script include features like moment function [4], hybrid features [5–7], and SD [8]. The highest accuracy reported for Devanagari script was using hybrid features. The technique also performs well for Telugu script in comparison with SIFT [9]. SIFT and SD were also used for the other Indian scripts like Bangla and Malayalam and reported an acceptable results. To the best of our knowledge, there is very less amount of work reported so far for the other Indian scripts like Gurumukhi, Gujarati, Kannada, Dogri.

3 Feature Extraction and Matching

The word images are divided into fixed sized blocks to extract features. As the height of the word image in the database is fixed (50 pixels), the number of the blocks in the word image will very depend on the width of the block (ranging from 30 to 50 pixels). Figure 1 shows the division of word image into blocks. It can be seen that a block may not represent a single character. The block-wise representation of word images facilitates incremental matching in which the entire word image may not be same but partially matching.

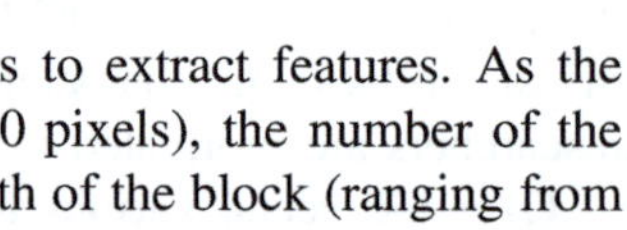

Fig. 1 Dividing word image into fixed size of blocks

In the next step, the HOG and SD features are extracted from each block. The HOG technique divides the block image into fixed number of small connected regions known as cells and counts the occurrences of gradient orientation in localized portions of interest (ROI). A histogram of nine different gradients is calculated for each cell in an image, and finally, all histograms are concatenated to generate a final feature vector [10, 11]. On the other hand, SD is categorized as contour-based SD which uses only the boundary information of shape extracted by conventional edge or contour detection approach. A logical grid of constant size is applied on boundary image and the boundary points which are intersecting with the logical grid are selected as descriptor points. The feature vector is generated by finding the relative arrangement of the descriptor points in the grid using the summation of log-polar histogram computed at each descriptor point [8].

After computing feature descriptor for each block on a word image, we compute a similarity score between two word images using feature matching technique. The current experiment focuses on incremental matching instead of exact matching. Thus DTW-based algorithm is used to match block-wise features of two word images. Distance between *i*th and *j*th block feature vector, *d(i, j)* is computed using Euclidian distance feature vector of each block of a word image is compared with every block of other word image, and distance matrix for two word images is calculated as described in [5] using Eq. (1). Figure 2 shows the distance matrix calculated for two word images. The distance between two word images is normalized in between 0 and 1.

$$D(i,j) = \min\begin{cases} D(i-1,j-1) \\ D(i,j-1) \\ D(i-1,j) \end{cases} + d(i,j) \qquad (1)$$

where *d(i, j)* is Euclidian distance of *i*th and *j*th block feature vector.

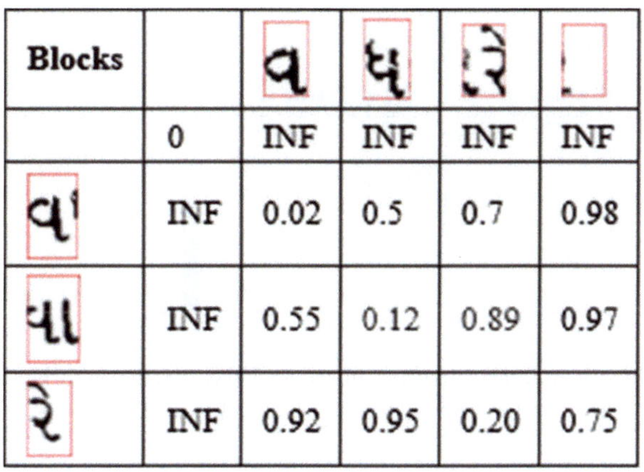

Blocks		વ	ધ	રે	
	0	INF	INF	INF	INF
વ	INF	0.02	0.5	0.7	0.98
ધા	INF	0.55	0.12	0.89	0.97
રે	INF	0.92	0.95	0.20	0.75

Fig. 2 Distance matrix to compute sequential matching score using DTW

4 Experiments and Results

The experiments are performed on a small-sized printed Gujarati word image database collected from machine printed Gujarati document images available on DLI. The database consists of 280 word images of size 3–7 characters manually grouped based on shape similarity. In order to show the discriminating strength of the features, the shape similarity matrix is computed that shows the distance between every pair of word images in the database. Figure 3a, b shows the optimum shape similarity matrix for HOG and SD features, respectively. It was observed that for both HOG and SD features increasing the size of the block within group similarity is improved but across group similarity has remained unchanged. However, for SD features increasing the number of bins in log-polar histogram by varying radial and angular bins, the result was improved for both within and across group similarities (Fig. 3b).

To perform the word retrieval experiments, a random query image is selected from the database and matched against all other images. The word images having similarity score more than some threshold are retrieved and ranked according to the similarity score. The precision (PR) and recall (RC) values were computed based on the relevant and retrieved images to measure the performance. Figure 4a, b shows the PR and RC versus query radius (threshold) graph for HOG and SD, respectively. The highest PR of 68% was recorded at 50% RC for SD features, whereas for HOG it is 57% only. Figure 5 shows the visual results for the word image retrieval experiments using SD. It is evident that the proposed approach also retrieves the morphological variants of the query image as well as handles structural noise like cuts and blobs.

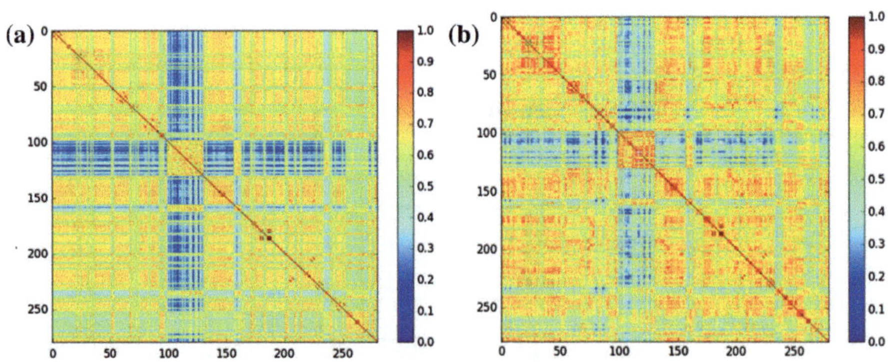

Fig. 3 **a** Similarity matrix image for HOG descriptor for block size (50, 50). **b** Similarity matrix image for shape descriptor with log-polar histogram of 35R and 36T

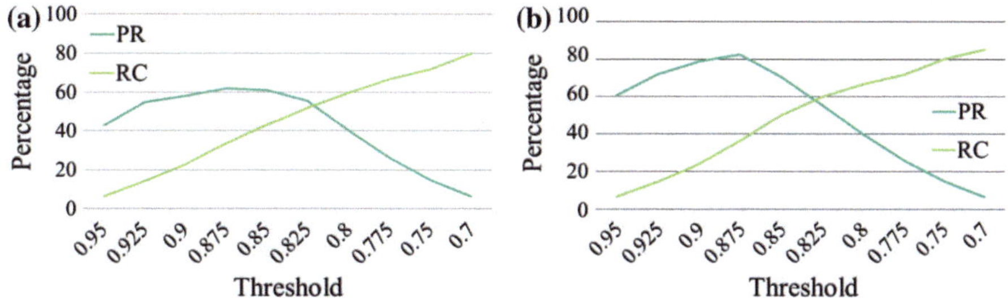

Fig. 4 **a** Precision & recall versus various thresholds for HOG descriptor. **b** Precision & recall versus various thresholds for HOG descriptor

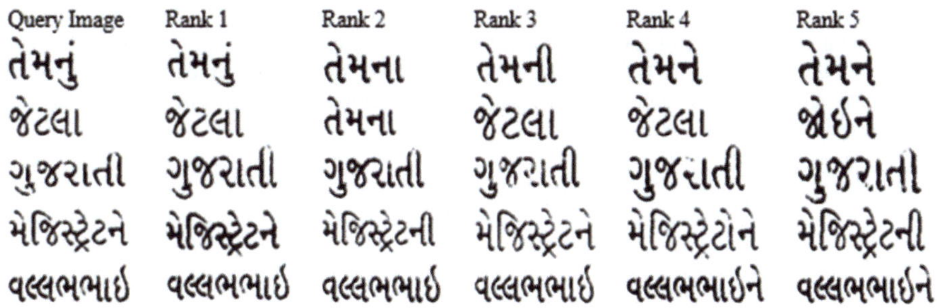

Fig. 5 Visual results for the word image retrieval experiments using SD

5 Conclusion

Based on the analysis of the literature for word spotting techniques in Indian script, two feature extraction techniques, namely HOG and SD, were selected and word-matching experiments were performed on printed Gujarati script. It is evident from the results that the SD features gives better results than HOG. However, the method requires further improvement to match the state of the art. The work can be extended in future by considering a larger database with a combination of multiple features to improve the performance.

References

1. S. Kompalli, et al. "Challenges in OCR of Devanagari documents.", Proceedings of the Eighth International Conference on Document Analysis and Recognition, (2005):1–5.
2. B. Varghese and S. Govilkar, "A Survey on Various Word Spotting Techniques for Content Based Document Image Retrieval" International Journal of Computer Science and Information Technologies, Vol. 6.3 (2015):2682–2686.
3. "DLI Home." [Online]. Available: http://www.dli.ernet.in/, [Accessed: 15-Oct-2016].

4. A. Bhardwaj, D. Jose, and V. Govindaraju, "Script Independent Word Spotting in Multilingual Documents", International Joint Conference on Natural Language Processing, (2008):48–54.
5. M. Meshesha and C. V. Jawahar, "Matching word images for content-based retrieval from printed document images", International Journal of Document Analysis and Recognition, Vol. 11.1 (2008):29–38.
6. S. N. Srihari, H. Srinivasan, C. Huang, and S. Shetty, "Spotting words in Latin, Devanagari and Arabic scripts", Vivek: Indian Journal of Artificial Intelligence, Vol. 16.3 (2006):2–9.
7. A. Kumar, C. V Jawahar, and R. Manmatha, "Efficient Search in Document Image Collections" Asian Conference on Computer Vision, (2007):586–595.
8. E. Hassan, S. Chaudhury, and M. Gopal, "Word shape descriptor-based document image indexing : a new DBH-based approach", International Journal on Document Analysis and Recognition, Vol. 16.3 (2013):227–246.
9. R. Shekhar and C. V Jawahar, "Word Image Retrieval using Bag of Visual Words", 10th International Workshop on Document Analysis Systems, (2012):297–301.
10. A. Kovalchuk, L. Wolf, and N. Dershowitz, "A Simple and Fast Word Spotting Method", 14th International Conference on Frontiers in Handwriting Recognition, (2014):3–8.
11. A. J. Newell and L. D. Griffin, "Multiscale histogram of oriented gradient descriptors for robust character recognition", International Conference on Document Analysis and Recognition, (2011):1085–1089.

Investigation of Surface Degradation in Polymer Dielectrics with Micro-fillers

Nitin V. Nair and R. R. Patel

Abstract Dielectrics used in high voltage applications are prone to partial discharge. Recent studies have indicated that addition of fillers in epoxy polymers lead to considerable improvement in dielectric properties. This paper illustrates an experimental study on how incorporation of micro-fillers lead to improvement of surface properties on application of voltage stress via surface degradation experiment. For the experimental study, Al_2O_3 micro-fillers were dispersed in Bisphenol-A type epoxy resin. Surface discharge was performed on both neat epoxy and micro-filled specimen with 4% w/w loading for comparison. Both effect of time and voltage level was noted. High resolution microscopic photographs were used to compare eroded area in both specimen. Simulation of electrical field intensity was performed on ANSYS Maxwell to illustrate the field distribution around the periphery of electrode.

Keywords Surface discharge · Al_2O_3 · Micro-composites · Epoxy Degradation · Ansoft maxwell · Microscopic analysis

1 Introduction

Epoxy resins are the most popular thermoset polymers for high voltage applications including outdoor and indoor line insulators, transformer bushings and PCB coatings. Being around the scene for almost 5 decades, it has withstood the test of time due to its robustness, low cost and superior dielectric properties compared to other polymers [1, 2].

Surface degradation is an anomaly that occurs when the dielectric strength of the material is strong enough to withstand the voltage stress without causing breakdown and the effect is observed on the surface of the dielectric in form of tracks and

N. V. Nair · R. R. Patel (✉)
Department of Electrical Engineering, G H Patel College of Engineering and Technology, Anand, India
e-mail: riteshpatel@gcet.ac.in

R. Kher et al. (eds.), *Proceedings of the International Conference on Intelligent Systems and Signal Processing*, Advances in Intelligent Systems and Computing 671,
https://doi.org/10.1007/978-981-10-6977-2_15

turbid areas. Surface discharge causes degradation and volume erosion, for polymers with weak surface resistivity.

Researches show that addition of fillers enhance dielectric properties and surface characteristics. Epoxy micro-fillers, as they are known after curation, exhibit better properties than its neat counterpart in reduction of surface degradation in a long run which may prevent eventual treeing and erosion when exposed to chemicals [3]. Various authors [5–8] have proved that using fillers (micro or nano) lead to less material erosion compared to neat epoxy when exposed to surface discharges. This was investigated by Maity et al. [4] using Laser Surface Profilometry (LSP). Rafaey et al. [9] observed greater degradation area when voltage magnitude was increased.

This paper intends to investigate the surface properties of neat and micro-composite epoxy matrices. Two specimen were prepared for the experimentation which consisted of neat epoxy and composite with 4% w/w filler loading. Addition of fillers reinforces the dielectric and stronger inter-atomic bonds that prevent volume erosion.

2 Experimental

2.1 Materials

Bisphenol A type Epoxy resin named Araldite CY230 (Density 1.15 gm/cm^3 at 25 °C) and Amide Hardener Aradur HY951 (Density 0.98 at 25 °C) was selected for this study. This resin is widely popular for its low viscosity and insulation properties manufactured by Huntsman Pharmaceuticals Pvt. Ltd. Al_2O_3 were selected as micro-fillers.

2.2 Specimen Preparation

For neat epoxy, base resin about a 100 ml was mixed with hardener HY951 in a 10:1 proportion. Micro-composite specimen was prepared by adding 4% w/w micro-fillers to the base resin in a glass beaker and stirring for 20 min with a glass rod. The mixture was kept in an ultrasonic bath for 180 min for proper dispersion. For gelation, specimens were poured into an acrylic mold of the size 200 mm 90 mm × 3 mm and kept at room temperature for 24 h. Specimens were also post-cured for 24 h and kept another 24 h for cooling in a temperature controlled oven before experimentation.

2.3 *Test Setup*

The surface degradation experiments were carried out in accordance to IEC-60343. The set-up consists of The IEC (b) electrode configuration consists of a rod-plane electrode assembly as shown in Fig. 1. The electrodes are made of stainless steel and the diameter of the rod electrode is 6 mm, with 1 mm end curvature.

Experimentation was done for the following degradation time and voltage levels.

1. 10 kV for 3 h
2. 15 kV for 3 h
3. 15 kV for 6 h

3 Results and Discussions

The tests were performed on both neat and micro-composite matrix. The areas affected by voltage stress were observed under high-resolution microscope manufactured by Olympus Corporation Ltd. to observe the effect of voltage magnitude and time shown in Fig. 2 degraded for 3 h under 10 kV and 15 kV and 6 h for 15 kV respectively. The effect of degradation was significantly less towards the outer periphery of the degraded region i.e. areas farther from the edge of the electrode as tabulated in Table 1.

From the results in Table 1, voltage magnitude and time of voltage application can be correlated with degradation. The effect of partial discharge is rigourous at region closer to the electrode edge. In case of neat epoxy, increase in time does not increase the area of degradation. Rather, there is increase volume erosion within the periphery of degraded region as obtained by Mostafa et al. [9].

Fig. 1 IEC (b) type electrode

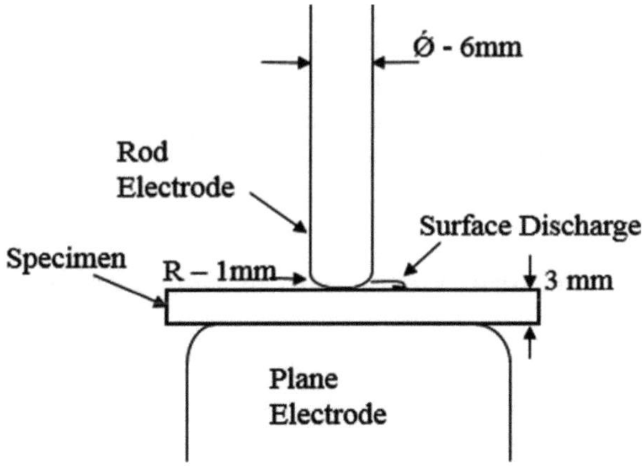

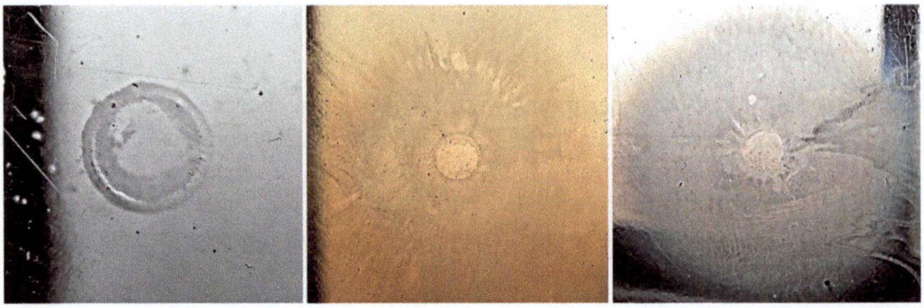

Fig. 2 Photographs of degraded area 10 kV for 3 h, 15 kV for 3 h and 15 kV for 6 h for neat matrix

Table 1 Degraded area for neat epoxy

Time and magnitude of voltage application	Radius of degraded region from the edge of the electrode (mm)
10 kV for 3 h	2
15 kV for 3 h	15
15 kV for 6 h	15

Table 2 Degraded area for micro-composite

Time and magnitude of voltage application	Radius of degraded region from the edge of the electrode (mm)
10 kV for 3 h	8
15 kV for 3 h	14
15 kV for 6 h	20

Surprisingly, the visible tracks obtained for the voltage 15 kV for 6 h in micro-composite was found to be higher as tabulated in Table 2. But, the size of the craters on further investigation was found to be less in micro-composites at 4% w/w as compared to the neat epoxy for the same voltage level and time. The craters seen in the microscopic images indicate volume erosion (Figs. 3, 4 and 5).

4 Possible Mechanism of Degradation

The effect and possible mechanism of surface degradation can be illustrated in Fig. 6. For neat epoxy, the field distribution is uniform. As for micro-composite matrix, when field distribution is uniform, both area and volume of degradation is less, given that particle distribution is also uniform. Due to the absence of any

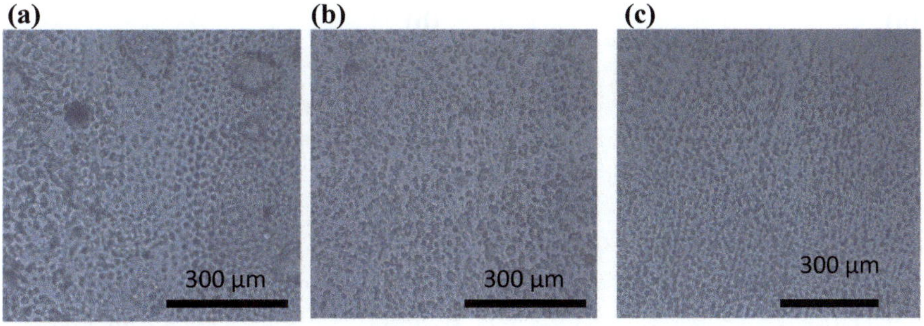

Fig. 3 Degraded area in neat epoxy under 40× resolution 15 kV for 6 h: **a** 5 mm, **b** 8 mm and **c** 12 mm from the electrode edge

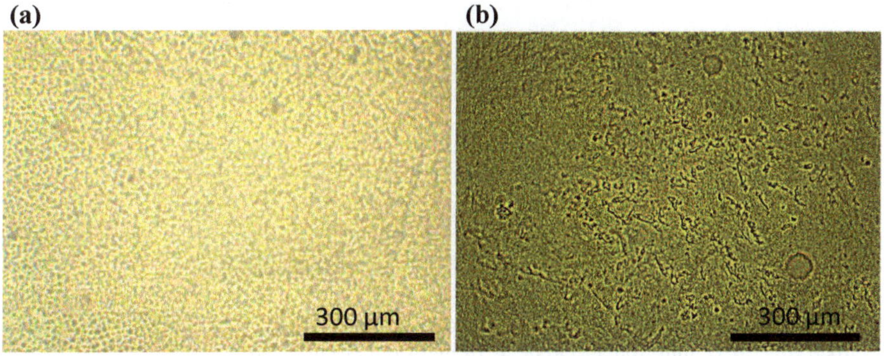

Fig. 4 Degraded 10 kV for 3 h: **a** neat epoxy, **b** micro-composite

reinforcements unlike in micro-composites, neat epoxy degrades significantly more in comparison forming deeper craters indicating higher volume erosion.

The most likely hypothesis in micro-composite is shown in Fig. 7, as the volume eroded and depth of erosion were not quantified. First, if the volume of material eroded is significantly less as compared to the neat matrix, the reinforcement and stronger inter-atomic bonds subjects the leakage current only onto the surface preventing its penetration into the volume of the material. Second, the visible result show higher degradation for micro-composite. If the volume erosion more or less remains the same, then the obtained result is caused either by agglomeration or higher particle content than required. The possible explanation can be confirmed after performing the experiments on particle concentration less than 4% w/w.

The simulation of experimentation is done in Ansoft Maxwell. The result obtained shows the electric field distribution indicating greater field density along the area subjected to voltage stress. It reduces gradually farther away from the edge of the electrode as illustrated in Fig. 8. Electric field intensity as expected, changes with voltage magnitude.

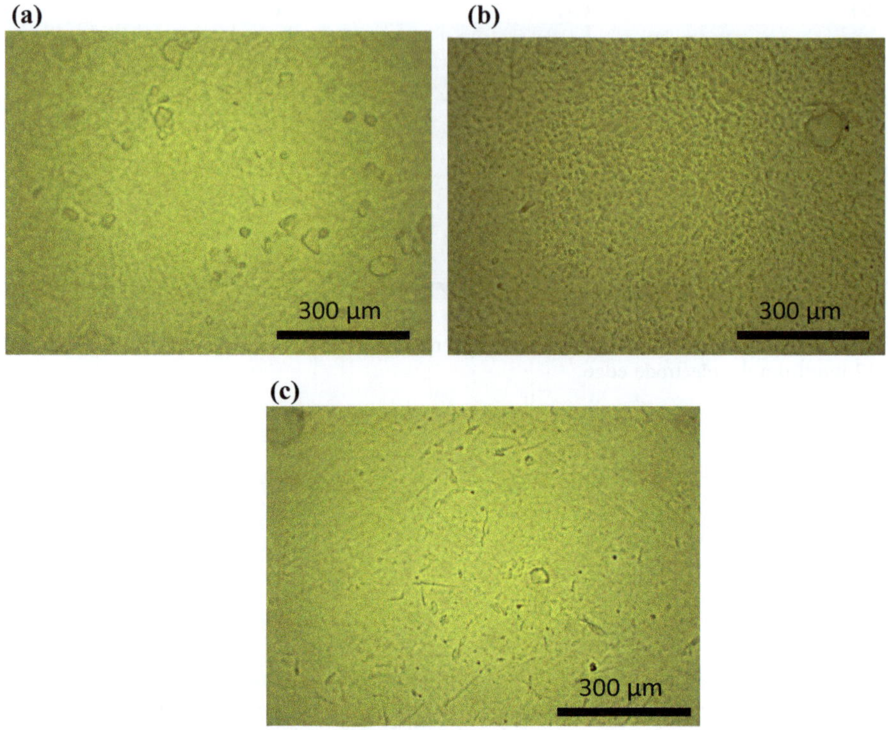

Fig. 5 Degraded area of micro-composite under 40× resolution 15 kV for 6 h: **a** 5 mm for electrode edge, **b** 12 mm from electrode, **c** 18 mm from electrode edge

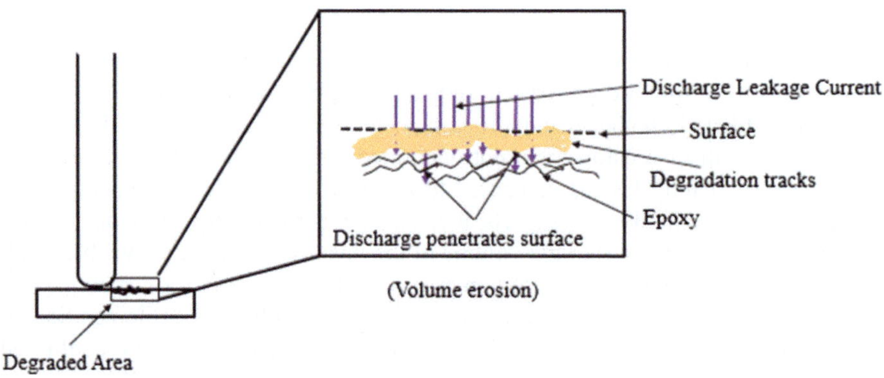

Fig. 6 Degradation mechanism for neat epoxy

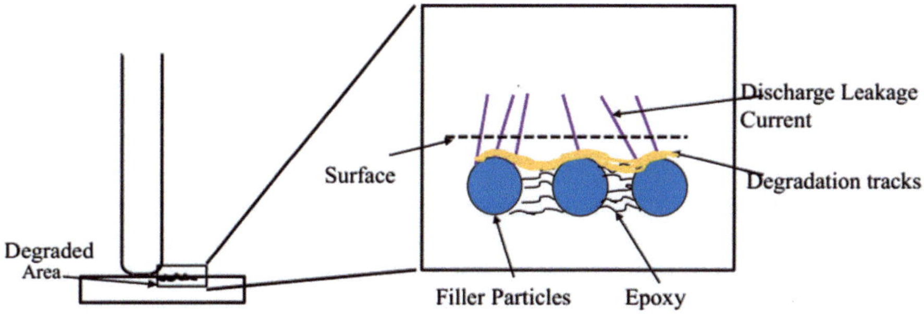

Fig. 7 Degradation mechanism and field distribution for micro-composites evenly distributed fillers

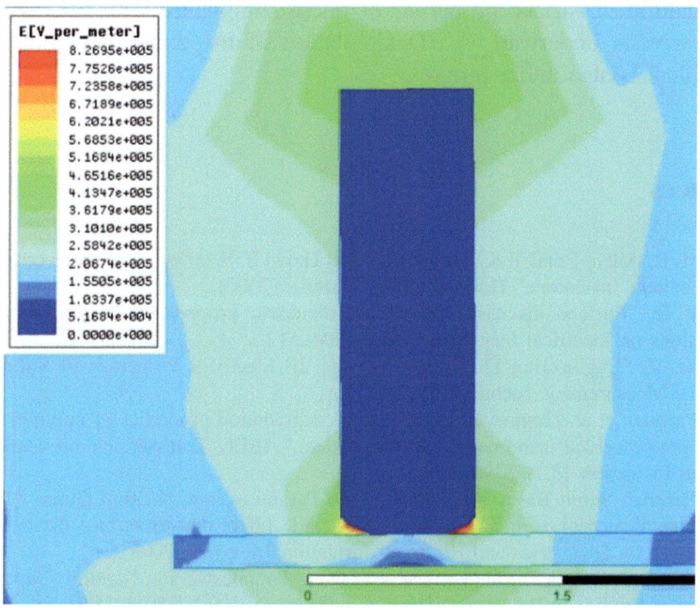

Fig. 8 Simulation of field distribution during surface degradation in Ansoft Maxwell for 15 kV

It is still to be investigated whether properties improve with change in concentration. Amount of filler loading also result in change of relative permittivity and thermal conductivity which may contribute to either worsening or improvement of the surface and dielectric properties.

5 Conclusions

In an attempt to improve surface properties of epoxy dielectrics, micro-fillers were reinforced into the polymer by ultrasonic dispersion. 4% w/w filler loading was selected and the following results were noted.

1. Area and turbidity of degraded area increase with voltage magnitude and time respectively in both neat and micro-composite matrix respectively.
2. Visible tracks formation in micro-composite with 4% w/w filler loading was higher. But further investigation showed less volume erosion as compared to neat epoxy.
3. For 6-h of degradation, observation indicated greater degradation along with increase in periphery of degraded region.

Investigation of surface properties of micro-composites with filler loading less than 4% w/w is underway. A 3D simulation of the experimentation with and without fillers is also under progress.

References

1. Bernstorf, R. Allen, Randall K. Niedermier, and David S. Winkler. "Polymer compounds used in high voltage insulators." Hubbell power Systems (2000).
2. Schifani, R. "Surface discharge effects on dielectric properties of epoxy resins." IEEE Transactions on Electrical Insulation (1983): 504–512.
3. Bartnikas, R. "Engineering Dielectrics Volume 2B-Electrical Properties of Solid Insulating Materials: Measurement Techniques." (1987).
4. Maity, Parimal, et al. "Improvement in surface degradation properties of polymer composites due to pre-processed nanometric alumina fillers." IEEE Transactions on Dielectrics and Electrical Insulation 15.1 (2008).
5. Maity, Parimal, Sumit Basu, Venkitanarayanan Parameswaran, Nandini Gupta. "Degradation of polymer dielectrics with nanometric metal-oxide fillers due to surface discharges." IEEE Transactions on Dielectrics and Electrical Insulation 15, no. 1 (2008).
6. Maity, Parimal, Sumit Basu, Venkitanarayanan, Parameswaran, Nandini Gupta. "Surface Degradation Studies in Polymer Dielectrics with Nano- sized Fillers."IEEE 8th International Conference on Properties & applications of Dielectric Materials, pp. 171–174. IEEE, 2006.
7. Wu, X., Wang, Y., Xie, L., Yu, J., Liu, F., & Jiang, P. (2013). "Thermal and electrical properties of epoxy composites at high alumina loadings and various temperatures". Iranian Polymer Journal, 22(1), 61–73.
8. Iizuka, Tomonori, Yuqing Zhou, Tomoaki Maekawa, Toshikatsu Tanaka, and Kohei Tatsumi. "Surface analysis of epoxy nanocomposite insulator materials eroded by partial discharge. IEEE Conference on Electrical Insulation and Dielectric Phenomena (CEIDP), pp. 703–706. IEEE, 2014.
9. Refaey Mostafa, and Josef Kindersberger. "Factors of influence on surface erosion of epoxy resin exposed to electrical discharges." IEEE Transactions on Dielectrics and Electrical Insulation (2014): 1198–1206.

Characterization of Power Cable Using Various Diagnostic Techniques

R. R. Patel and Vidushi Badmera

Abstract The increased need of continuous high-quality electrical energy combined with various issues related to underground cable makes the condition assessment of the cable very necessary. The main contribution of this paper is to establish the correlation between Tanδ Tip-up and partial discharge performance of newly manufactured medium voltage cable samples of 11 and 33 kV, of crossed-linked polyethylene (XLPE), at power frequency. A complete sequence of type test is performed which includes partial discharge test, Tanδ test, heat cycle test, bending test, impulse test and HV AC withstand test. From the results obtained, the Tanδ Tip-up value and partial discharge value are used and performance ranking is done. On the basis of performance ranking, a plot is obtained which correlates the above-mentioned features and shows which sample has better performance as compared to other.

Keywords Tanδ Tip-up · Partial discharge · XLPE · Type test

1 Introduction

Continuous, reliable and high-quality supply of electrical energy is the heart of our economy. The engineers of the world work hard to make this power supply reliable and safe. Distribution- and transmission-class cable systems play key role in achieving this goal and when we talk about underground cable! The insulation becomes the centre of attraction.

The underground cable from day one of installation is subjected to thermal, electrical and mechanical stress, foreign particle ingress and variations of

R. R. Patel (✉) · V. Badmera
Department of Electrical Engineering, G H Patel College of Engineering and Technology, Vallabh Vidyanagar, Anand 388120, Gujarat, India
e-mail: riteshpatel@gcet.ac.in

V. Badmera
e-mail: vidushibadmera@yahoo.com

R. Kher et al. (eds.), *Proceedings of the International Conference on Intelligent Systems and Signal Processing*, Advances in Intelligent Systems and Computing 671,
https://doi.org/10.1007/978-981-10-6977-2_16

temperature and humidity. All these increase the working temperature of the cable. The heating accelerates the chemical reactions in the insulation which results in degradation of the dielectric character of the insulation. This process has an avalanche effect. Hence, the changing electrical character of the insulation increases the loss factor. So if the dissipation factor Tanδ is monitored periodically, then it becomes possible to predict and/or avoid the catastrophic failure of the electrical equipment. Dissipation factor is the energy taken by the flow of current through the resistive component of the insulation.

The most widely used diagnostics technologies include dissipation factor (Tanδ) and partial discharge measurements; therefore, they are the main focus. The discussion is based on data obtained from tests performed in Polycab Wires Pvt. Ltd.

2 Experimental

Type test is performed on all the manufactured cables before supplying on commercial basis. A sequence of type test was performed on four, newly manufactured, cable samples as described in Table 1. These tests were performed in high-voltage laboratory of Polycab Wires Pvt. Ltd. Type test was performed in a fixed sequence which is listed in Table 2 [1].

2.1 Partial Discharge Test

Partial discharge test was performed on four cable samples listed in Table 1 as per IS: 10810 part 49 [2].

In this test, 1.73 times the rated voltage was applied to the four samples using HV AC series resonant test set. The value of partial discharge was noted for each sample in partial discharge recorder. The results obtained clearly indicate that the value of partial discharge for all the four samples does not exceed 5 pC. Hence, it is within the limit set by IS 7098 part 2 (Fig. 1 and Table 3).

Table 1 Cable sample description

Sample ID	Insulation	Voltage class (kV)	Number of cores, cross-sectional area	Length (m)
1	XLPE	11/11 kV (Ue)	3C X 240 mm^2 A2xFY FRLSH	10
2	XLPE	11/11 kV (Ue)	3C X 300 mm^2 A2xFY FRLSH	10
3	XLPE	19/33 kV(e)	1C X 300 mm^2 A2xWaY FR	10
4	XLPE	19/33 k(e)	1C X 630 mm^2 A2xWaY FR	10

Table 2 Sequence of type test

List of tests	Specifications	
Partial discharge Test	Less than 5 pC at $1.73V_0$	
Bend test	Mandrel size 20 OD ± 5%	
Partial discharge test	Less than 5 pC at $1.73V_0$	
Tanδ function of voltage	At $0.5V_0$ V_0 $2V_0$	0.004
	Increment in Tanδ $0.5V_0$ to $2V_0$	0.002
Tanδ function of temperature at 2 kV ambient and 100 °C	0.004	
Heat cycle test	At least 2-h heating At least 3-h cooling 20 cycles	
Partial discharge test	Less than 5 pC at $1.73V_0$	
Tanδ function of voltage	At $0.5V_0$ V_0 $2V_0$	0.004
	Increment in Tanδ $0.5V_0$ to $2V_0$	0.002
Impulse test	10 shots +ive 10 shot –ive of 95 kV for samples 1 & 2 170 kV for samples 3 & 4	
Four-hour HV AC withstand test	$4V_0$	

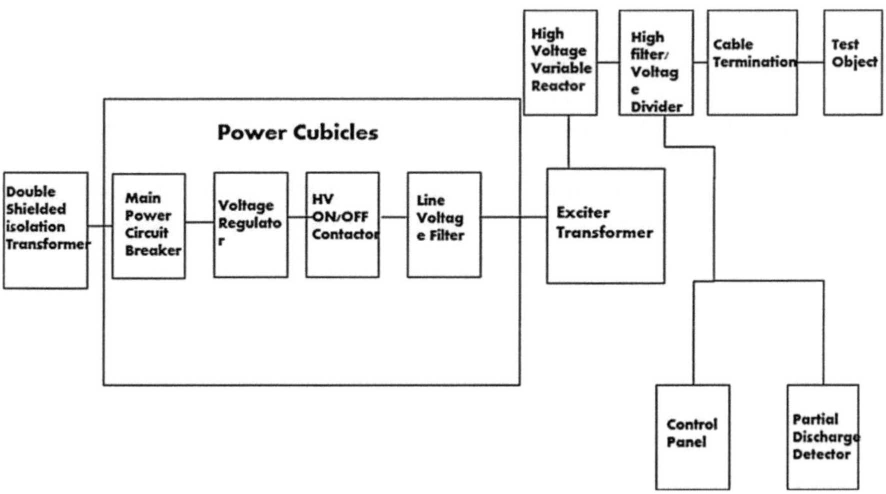

Fig. 1 Block diagram for HV AC series resonance test set

Table 3 Applied voltage for partial discharge test

Cable sample	Applied voltage (kV)
1 and 2	10.98
3 and 4	32.96

2.2 Dielectric Power Factor Test

This test was performed on four cable samples listed in Table 1. The test is performed based on IS 10810 part 48 [3]. HV AC series resonant test set along with Tanδ measuring instrument is used to obtain the value of Tanδ. The value of Tanδ was noted for $0.5V_0$, V_0 and $2V_0$. Thereafter the cable samples were heated at 100 °C, and Tanδ value was noted as 2 kV.

The results obtained show that the value of Tanδ is not exceeding 0.004. Hence, the limit set by IS 7098 part 2 is met.

2.3 Heat Cycle Test

Heat cycle test was performed on the four cable samples as listed in Table 1. The test is performed on the basis of IS 10810 part 49 [4]. The maximum continuous operating temperature at normal condition is 90 °C. The cable samples were heated at 100 °C temperature for 2 h. Then, it is cooled for 3 h. Twenty such cycles are conducted to check the adequacy of manufacturing technique.

2.4 Bending Test

Bending test was performed on the four cable samples as given in Table 1 and as per IS 10810 part 50 [5]. In this test, 10-m-long cable samples are wound on the cylinder of diameter, 10 times the outer diameter of cable. Then, it is unwound and wound again. Bending is carried out at reasonably uniform speed between 15 and 30 s per half cycle. Three such cycles were conducted to check the adequacy of manufacturing technique.

2.5 Impulse Test

In this test, 10 shots of positive voltage and 10 shots of negative voltage were applied on the four cable samples. The voltage level is listed in Table 4. These voltage levels are as per IS 7098 part 2. The test is performed as per IS 10810 part 47 [6].

Table 4 Impulse voltage level

Sr. No	Voltage class (kV)	Applied voltage (kV)
1	6.35/11 (e)	75
2	11/11 (Ue)	95
3	19/33 (e)	170

2.6 *Four-Hour HV AC Withstand Test*

This was conducted on the same four cable samples as per IS 10810 part 45 [7]. In this test, $4V_0$ was applied to the core of the conductor continuously for four hours.

3 Result and Discussion

Type test included a number of tests which includes

1. Partial discharge test
2. Bend test
3. Tanδ test as a function of voltage
4. Tanδ test as a function of temperature
5. Heat cycle test
6. Impulse test
7. Four-hour HV test

These tests were performed in a fixed sequence. The results obtained are given in Table 5.

3.1 *Performance Ranking*

Performance ranking was performed on the basis of series of electrical tests performed under type test. The diagnostic ranks are simply assigned by looking at the value of feature under consideration. For example, if the mean value of Tanδ is considered at V_0, then, the cable sample that has highest value (here sample 4) is selected as the poorest performer, thus assigned the rank of one. The performance continues until the best performer is identified. The same procedure is repeated for another diagnostic feature (Table 6).

A graph is plotted of partial discharge performance rank (*C*) versus Tanδ Tip-up rank (*E*) as shown in Fig. 2. Accuracy of the diagnostic feature is directly related to how far the dots are from the dashed lines. Similarly, a graph of Tanδ as function of voltage (*H*) versus Tip-up (*I*) can also be plotted. But in this paper, partial discharge after bending (*C*) and Tanδ Tip-up after bending test (*E*) are selected for performance ranking.

Table 5 Results of type test

Sample ID	1	2	3	4
PD before bending (pC)	1.8	1.9	2.3	2.4
PD after bending (pC)	2.0	2.3	2.4	2.6
Tanδ as a function of voltage	(1) At $0.5V_0$ 0.00012 (2) At V_0 0.00021 (3) At $2V_0$ 0.00072 (4) Increment $0.5V_0$–$2V_0$ 0.00060	(1) At $0.5V_0$ 0.00015 (2) At V_0 0.00034 (3) At $2V_0$ 0.00047 (4) Increment $0.5V_0$–$2V_0$ 0.00032	(1) At $0.5V_0$ 0.00021 (2) At V_0 0.00056 (3) At $2V_0$ 0.00105 (4) Increment $0.5V_0$–$2V_0$ 0.00084	(1) At $0.5V_0$ 0.00063 (2) At V_0 0.00110 (3) At $2V_0$ 0.00172 (4) Increment $0.5V_0$–$2V_0$ 0.00109
Tanδ at 2 kV as function of temperature (100 °C)	At ambient 0.00090 At 100 °C 0.00017	At ambient 0.00090 At 100 °C 0.00027	At ambient 0.00015 At 100 °C 0.00005	At ambient 0.00043 At 100 °C 0.00015
PD after HCT (pC)	2.3	2.5	2.8	3.6
Tanδ as a function of voltage	(1) At $0.5V_0$ 0.00015 (2) At V_0 0.00030 (3) At $2V_0$ 0.00065 (4) Increment $0.5V_0$–$2V_0$ 0.00050	(1) At $0.5V_0$ 0.00017 (2) At V_0 0.00039 (3) At $2V_0$ 0.00099 (4) Increment $0.5V_0$–$2V_0$ 0.00082	(1) At $0.5V_0$ 0.00019 (2) At V_0 0.00049 (3) At $2V_0$ 0.00092 (4) Increment $0.5V_0$–$2V_0$ 0.00073	(1) At $0.5V_0$ 0.00030 (2) At V_0 0.00074 (3) At $2V_0$ 0.00132 (4) Increment $0.5V_0$–$2V_0$ 0.00102
Impulse test	Withstood	Withstood	Withstood	Withstood
4-h HV AC withstand test	Withstood	Withstood	Withstood	Withstood

Table 6 Performance ranking table

A	B	C	D	E	F	G	H	I
1	4	4	4	3	2	4	4	4
2	3	3	3	4	1	3	3	2
3	2	2	2	2	4	2	2	3
4	1	1	1	1	3	1	1	1

A—Sample ID
B—Partial discharge before bending (pC)
C—Partial discharge after bending (pC)
D—Tanδ as a function of voltage
E—Tip-up $0.5V_0$–$2V_0$
F—Tanδ at 2 kV as a function of temperature
G—Partial discharge after heat cycle test
H—Tanδ as a function of voltage
I—Tip-up $0.5V_0$–$2V_0$

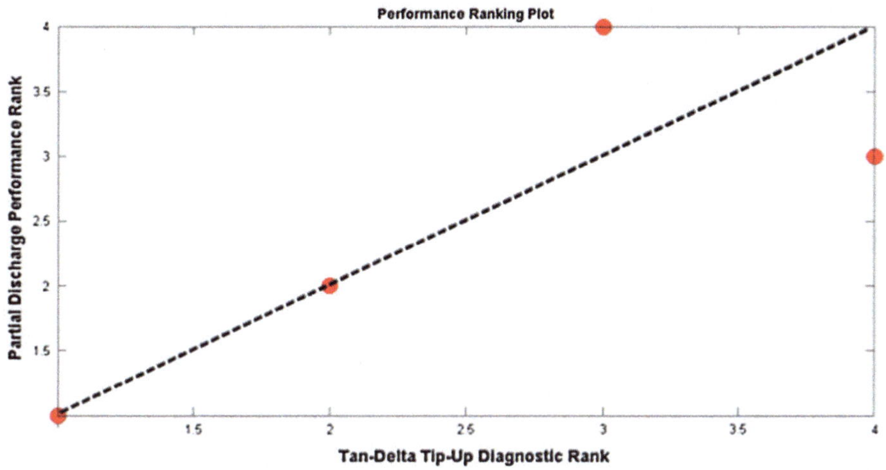

Fig. 2 Performance ranking plot

4 Observations and Conclusions

4.1 Observations

The performance ranking plot can be interpreted as follows:

The samples in the left corners are the worst performers; that is, (e.g. rank 1) the one having higher value of Tanδ Tip-up and higher value of partial discharge after bending, which indicates that their performance will be poor as compared to other cables in the group.

On the other hand, the upper-right corner of the plot contain better performers; that is, (e.g. rank 4) the one that have lower value of Tanδ Tip-up and lower value of partial discharge after bending, which indicates that their performance will be better as compared to other cables in the group.

Dashed line can be thought of as the perfect correlation between any two diagnostic features.

4.2 Conclusions

Type test was conducted in a fixed sequence as listed in Table 2 on four cable samples as listed in Table 1. Results obtained were used to do performance ranking. Performance ranking leaded to the results that the performance of the cables can be easily judged on the basis on partial discharge performance rank versus Tanδ Tip-up diagnostic rank. Hence, graph in Fig. 2 clearly indicates that the performance of the cable improves as we move from left end towards right end.

Performance ranking table clearly shows the rank of each sample for every diagnostic feature. Hence, the performance of each sample can be judged and compared.

Acknowledgements The authors are grateful to the engineers of Polycab Wires Pvt. Ltd. for providing help to perform the tests.

References

1. Indian Standard, "Cross-linked polyethylene insulated thermoplastic sheathed cables-specification" Part 2 for working voltages from 3.3 kV up to and including 33 kV
2. Indian Standard "Methods of test for cables" Part 46 Partial Discharge Test
3. Indian Standard "Methods of test for cables" Part 48 Dielectric Power Factor Test
4. Indian Standard "Methods of test for cables" Part 49 Heating Cycle Test
5. Indian Standard "Methods of test for cables" Part 50 Bending Test
6. Indian Standard "Methods of test for cables" Part 47 Impulse Test
7. Indian Standard "Methods of test for cables" Part 45 High Voltage Test
8. Steven Boggs and John Densley "Fundamentals of Partial Discharge in Context of Field Cable Testing" IEEE Electrical Insulation Magazine Vol. 16, No. 5; September/October 2001
9. John Densley "Ageing Mechanisms and Diagnostics for Power cables- An Overview" IEEE Electrical Insulation Magazine Vol. 17, No. 1; January/February 2001
10. J. C. Hernández-Mejía, R. Harley and N. Hampton, and R. Hartlein "Correlation between Tanδ Diagnostic Measurements and Breakdown Performance at VLF for MV XLPE Cables" IEEE Transactions on Dielectrics and Electrical Insulation Vol. 16, No. 1; February 2009 16

Radial Feeder Protection by Definite Time Overcurrent Relay

Param Mehta and Vijay H. Makwana

Abstract This paper describes the modeling of definite time overcurrent relay using MATLAB/Simulink tool. The model is developed as a Simulink block which accepts input from the user to feed relay settings. The model is validated on Simulink for different values of pickup values and time delays. A hardware model implementation is carried out using ATMega328 microcontroller and ACS712 current sensor. The performance of the hardware model is tested on a radial feeder network with the backup protection relays operating at a time delay of 0.5 s. The results are satisfactory and accurate and therefore can be used in digital relays and applied in relay systems. In addition to this, a hardware model with different inverse definite minimum time characteristics is included along with different interfacing elements such as keypads, LCD, LEDs. The hardware model designed is ready to use for any overcurrent applications and shows accurate and satisfactory behavior.

Keywords Power system protection · Power system relaying · Power system faults · Microcontrollers

1 Introduction

Overcurrent relay is simple and widely used in protection strategy in distribution circuits, maintaining high degree of serviceability and security [1]. The relay works on the basic principle that when current exceeds a threshold value or pickup value, it trips with or without time delay, tripping the associated circuit breakers [2]. The overcurrent relays with fixed time delays are called as definite time overcurrent relay [3].

P. Mehta (✉) · V. H. Makwana
G.H. Patel College of Engineering and Technology, Vallabh Vidhyanagar, Anand 388120, Gujarat, India
e-mail: theparammehta@gmail.com

V. H. Makwana
e-mail: vijaymakwana@gcet.ac.in

R. Kher et al. (eds.), *Proceedings of the International Conference on Intelligent Systems and Signal Processing*, Advances in Intelligent Systems and Computing 671,
https://doi.org/10.1007/978-981-10-6977-2_17

Various time delays are covered by the following equation:

$$t = K/((\mathrm{PSM})^{\alpha} - 1) \tag{1}$$

where t is the time of operation of relay, PSM = I/I_p (I is the fault current and I_p is the pickup current), and K, α are the constants which determine relay characteristics.

For achieving selectivity and backup protection, delay in operation of relay is an important requirement. If $k \to 0$ and $n \to 0$, then definite time characteristics are obtained. With definite time delaying arrangement, a time-graded system is yielded. The definite overcurrent relay starts timing out once it exceeds the preset value, termed as 'pickup.' The pickup settings range from 50 to 200% of the rated current, and the time settings range from 0.1 to 1, 1 to 10, 3 to 30, 6 to 60, etc. [4].

The overcurrent protection is one of the basic strategies for protecting power systems. This paper describes the modeling of definite time overcurrent relay using MATLAB/Simulink tool. Furthermore, hardware implementation of definite time overcurrent relay using ATMega328 microcontroller board and the result are compared for radial feeder network. A comparative result is carried out for simulation and hardware implementation.

2 Definite Time Overcurrent Relay Model

2.1 Modeling of Overcurrent Relay Using MATLAB/Simulink

MATLAB/Simulink is one of the most powerful and efficient tools for modeling power system. The relay model includes the use of various libraries included in Simulink. The definite time overcurrent relay measures the current and compares with the predefined threshold values, and after fixed time delay issues trip signal to the circuit breaker. Figure 1 shows the definite time overcurrent relay block modeled using MATLAB/Simulink. The relay model should be fed with input current '*i*', and the output terminal issues trip signal.

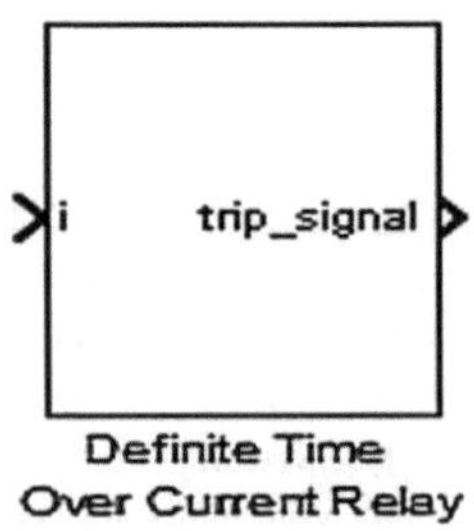

Fig. 1 Definite time overcurrent relay Simulink block

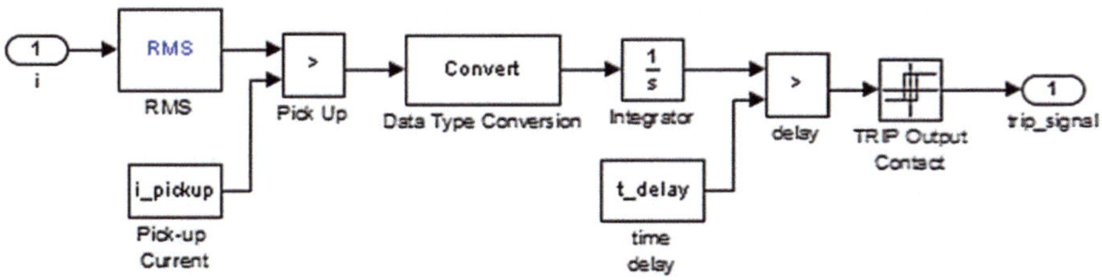

Fig. 2 Definite time overcurrent relay Simulink block subsystem

The RMS block outputs the RMS value of the input current signal which is utilized to compare with the threshold value. If the fault current is increased beyond the threshold value, then the relational comparator generates active HIGH signal, which is integrated to reflect time delay. The integrated signal is compared with the time dial settings. When the integrated time increases beyond the time threshold, the relational comparator generates active HIGH signal. The relay block ensures that the relay does not reset by itself. Figure 2 shows the subsystem of the definite over current relay block [3].

A. **Block Mask Parameter Dialog Box**

The block mask parameter dialog box is designed in such a way that the user can input relay settings very easily. It takes input from the user like pickup value and time delay value. Then, this setting is passed to the corresponding pickup current block and time delay block. Figure 3 illustrates the block mask parameter dialog box of the proposed definite time overcurrent relay block.

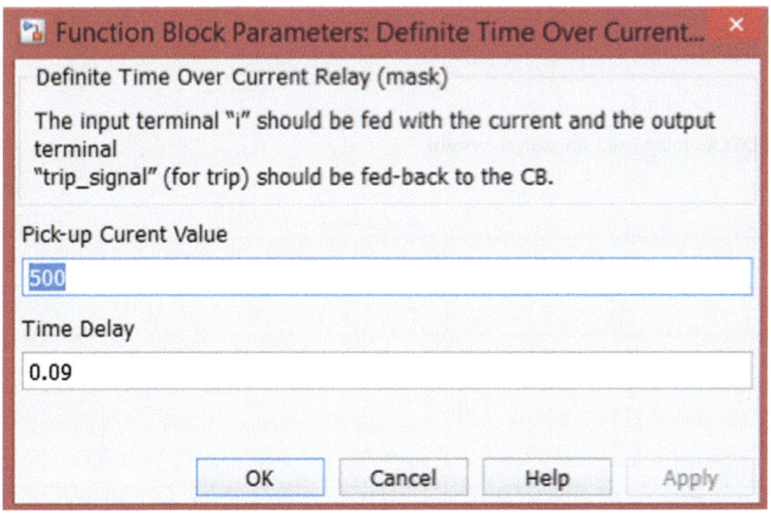

Fig. 3 Definite time overcurrent relay Simulink block subsystem mask

2.2 Single-Phase System Protection Using Definite Time Overcurrent Relay Model

Figure 4 shows the single-phase system protected by the definite time overcurrent relay block. The system is supplied by a single-phase AC voltage source. Similarly, the transmission line is shown by equivalent series RL branch and load at the end. The current is fed to the input terminal of the overcurrent relay model, while the output terminal is fed back to the circuit breaker via a NOT block as the circuit breaker takes active low (i.e., the control signal connected to the block input port must be 0 (zero) to open the breaker).

To simulate the fault condition, a switch is used with other end connected to the ground. This fault can be simulated by configuring switching times. Figure 5 shows

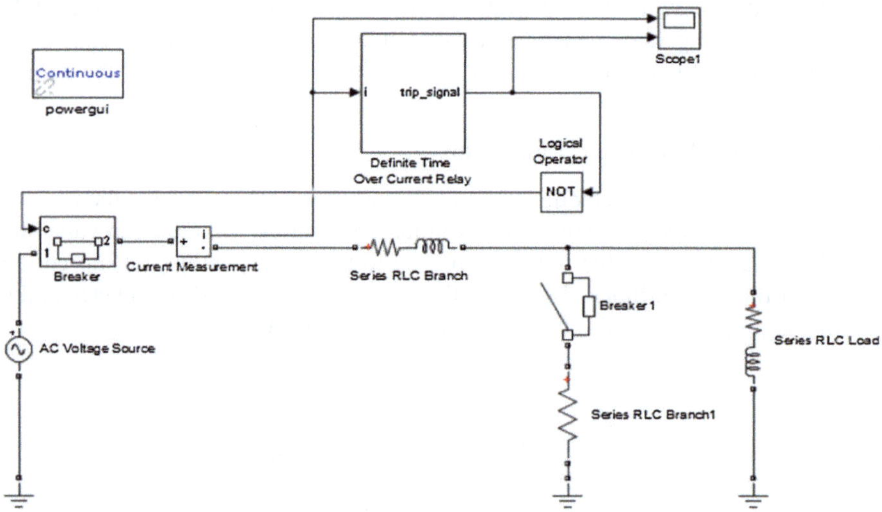

Fig. 4 DTOC relay used for power system

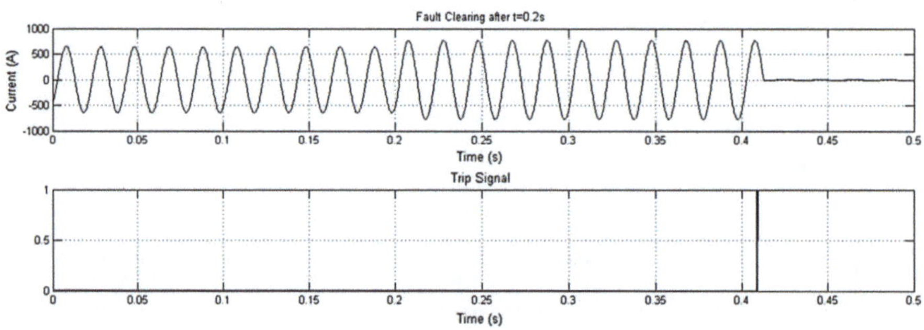

Fig. 5 DTOC relay sending a trip signal at t = 0.4 s, when fault ocuurs

the performance of the overcurrent relay model in fault conditions. At time $t = 0.2$ s fault occurred, and exactly after 0.2 s, i.e., at time $t = 0.4$ s relay issued tripping signal to the circuit breaker.

3 Relay Model Using ATMega328P Microcontroller

Arduino microcontroller is an open source platform which is powered by ATMega328P, a low-power CMOS 8-bit microcontroller based on the AVR-enhanced RISC architecture [5]. It is equipped with the input and output ports which forms the basic block for the modeling of a digital relay.

The block diagram of the proposed relay model can be simplified into three major bocks as shown in Fig. 6.

- The input transducer block essentially comprises of current sensor that samples the input current by generating magnetic field in which the Hall IC converts into a proportional voltage. This sampled data is sent to the microcontroller's analog pin.
- The analog to digital conversion circuit features 10-bit successive approximation ADC. The ADC contains a sample and hold circuit which ensures that the input voltage to the ADC is held at a constant level during conversion.
- The decision-making block analyzes the current data and uses definite time overcurrent relay principles to issue trip signal.

3.1 Current Measurement with ACS712 Sensor

The sensor consists of a precise, low-offset, linear Hall circuit. The current flowing through sensor generates a magnetic field in which the Hall IC converts into a proportional voltage. The ACS712 outputs an analog signal, V_{OUT}, which varies linearly with the primary sampled current, IP, within the range specified. Capacitor is recommended for noise management. Figure 7 shows the sensor package using ACS712 IC [6]. The output voltage is sent to analog pin input A0 of the microcontroller.

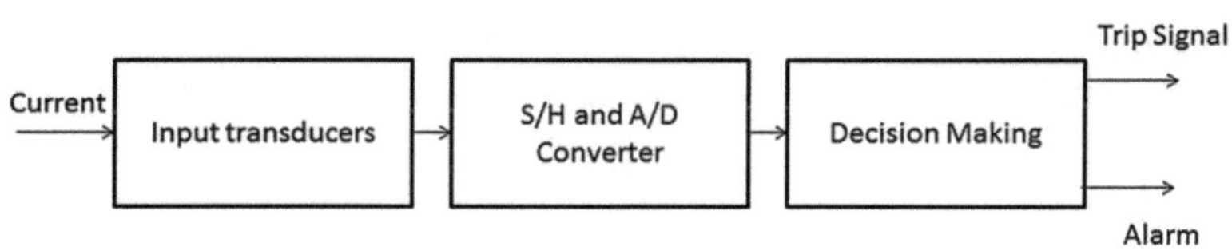

Fig. 6 Basic block diagram of a digital relay

Fig. 7 ACS712 current sensor package

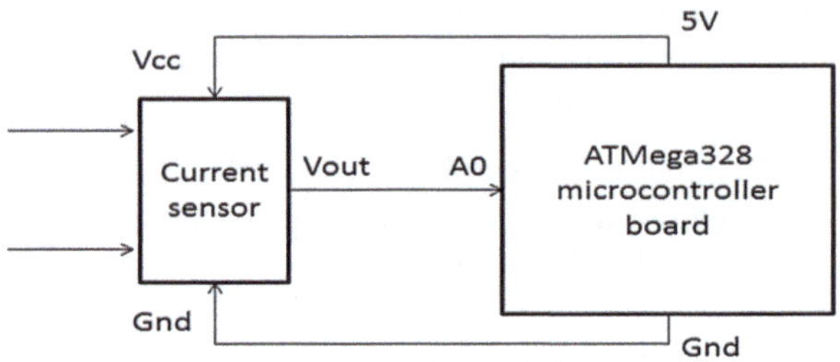

Fig. 8 Interfacing of current sensor with microcontroller

Figure 8 shows the interface circuit of current sensor with microcontroller. The input samples fed to the analog pin A0, which is converted to digital bits and then it is converted to current data using the following equation:

$$I = ((A \times (5/1024)) - 2.5)/0.234 \quad (2)$$

where

I is the sampled current value and
A is the ADC value of sample.

Here in (2), 2.5 is the offset voltage. At supply voltage of 5 V, the sampled data is measured with an offset of 2.5 V. Thus, we have to subtract that amount from the original sampled value. The offset voltage depends on the supply voltage to the current sensor, i.e., in this case it is 5 V thus an offset voltage of 2.5 V. The plot for output voltage for a unit change in input current is shown in Fig. 9, whose slope is 0.234 which is used in (2). The sampled data plot is shown in Fig. 10, which is a sinusoidal curve. Table 1 shows the sampled current data value, and the average number of samples per cycle is 34, which is sufficient for modeling of an over-current relay.

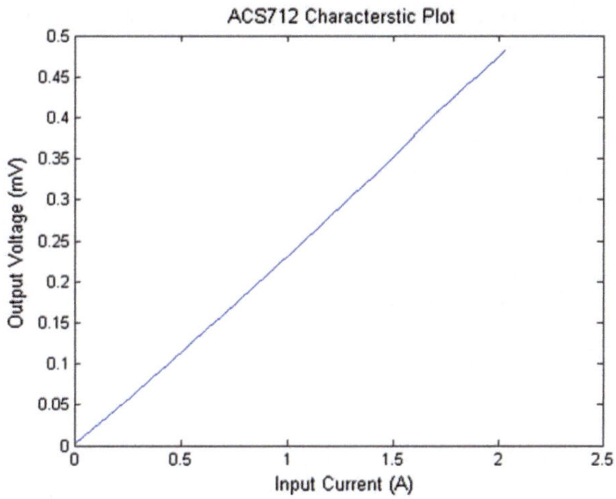

Fig. 9 ACS712 current sensor characteristics

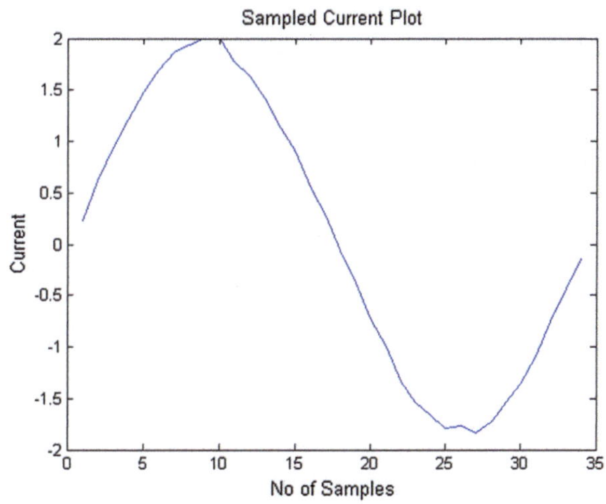

Fig. 10 ACS712 current sensor sampled current waveform

Table 1 Instantaneous current sample data

Current sample data				
0.23	2	0.27	−1.79	−0.44
0.61	2	−0.06	−1.77	−0.15
0.94	1.77	−0.38	−1.84	
1.21	1.63	−0.73	−1.73	
1.48	1.42	−1	−1.54	
1.69	1.13	−1.34	−1.36	
1.86	0.9	−1.54	−1.09	
1.94	0.56	1.67	−0.75	

3.2 Modeling of Definite Time Overcurrent Relay Using Microcontroller

This is a crucial part of the system where the sampled data should be processed and then process the data according to generic description of an overcurrent relay and issue trip signal [7–10]. The proposed design uses the microcontroller with ATMega328P as a core, capable of doing such task. The complete process is described as follow.

The sampled data is obtained by microcontroller and is converted to digital signal using A/D converter of the microcontroller. The flowchart of the proposed relay algorithm is shown in Fig. 11.

- Next after getting samples, it calculates RMS value of the current.
- The current values compare it against the pickup value of the relay setting. If the calculated RMS current is higher than the pickup value, then microcontroller starts counting time, from the moment of initiation of fault.

If the fault time is higher than the time delay, then the trip command is issued, which makes digital pin 12 HIGH.

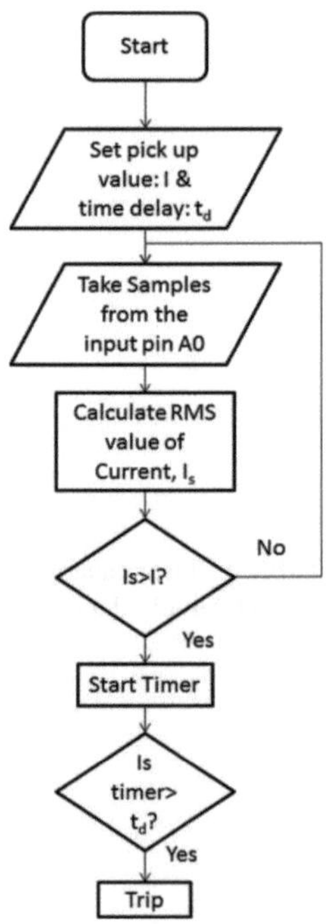

Fig. 11 Flowchart of the proposed algorithm

4 Results—Radial Feeder Protection Using Definite Overcurrent Relay

Figure 12 shows the basic setup of a radial protection scheme using definite overcurrent relay. The definite time overcurrent principle for a radial feeder network is illustrated. It shows the time delays of the relays in the protection system of the network. The microcontroller is fed with the relay settings such that it provides backup to other relay with a fixed time delay of 0.5 s. Table 2 shows the relay time of operation of relays used for the protection of radial feeder network when fault was initiated at time t = 1.0 s, at different locations, i.e., F1, F2, F3, etc. The tabulated results clearly indicate that the relays operated satisfactorily with desired accuracy. In the entire discussion,

FL Fault location
t_C Calculated time of operation in seconds
t_A Actual time of operation in seconds.

$\%E$ is the percentage error given by, $\%E = \left(\frac{t_C - t_A}{t_C}\right) \times 100$

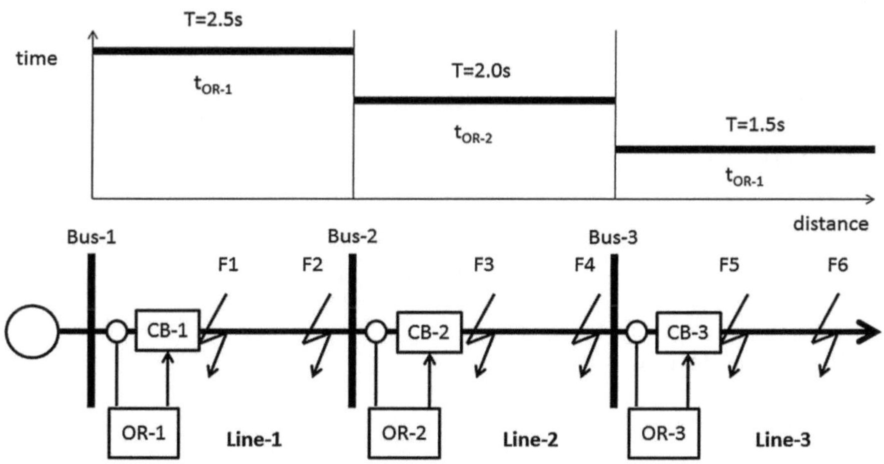

Fig. 12 Radial feeder protection scheme

Table 2 Hardware results

	Section-I (Relay-1)			Section-II (Relay-2)			Section-III (Relay-3)		
FL	t_c	t_A	$\%E$	t_c	t_A	$\%E$	t_c	t_A	$\%E$
F1	1.50	1.51	0.67	–	–	–	–	–	–
F2	1.50	1.51	0.67	–	–	–	–	–	–
F3	2.00	2.02	1.00	1.50	1.57	4.67	–	–	–
F4	2.00	2.02	1.00	1.50	1.57	4.67	–	–	–
F5	2.50	2.54	1.60	2.00	2.09	4.50	1.50	1.54	1.78
F6	2.50	2.54	1.60	2.00	2.09	4.50	1.50	1.54	1.78

5 Inverse Definite Minimum Time (IDMT) Overcurrent Relay

IDMT overcurrent relay works on the principle that when the current increases beyond a threshold value, than after a time delay, calculated by a mathematical equation described in [11], sends the tripping signal. The mathematical equation for different inverse characteristics is as follows:

IEC Standard Inverse

$$t = T \times \left(\frac{0.14}{\left(\frac{I}{I_s}\right)^{0.02} - 1} \right) \tag{3}$$

IEC Very Inverse

$$t = T \times \left(\frac{13.5}{\left(\frac{I}{I_s}\right)^{1} - 1} \right) \tag{4}$$

IEC Extreme Inverse

$$t = T \times \left(\frac{80}{\left(\frac{I}{I_s}\right)^{2} - 1} \right) \tag{5}$$

ANSI/IEEE Moderately Inverse

$$t = T \times \left(\frac{0.0515}{\left(\frac{I}{I_s}\right)^{0.02} - 1} + 0.114 \right) \tag{6}$$

ANSI/IEEE Very Inverse

$$t = T \times \left(\frac{19.61}{\left(\frac{I}{I_s}\right)^2 - 1} + 0.491 \right) \quad (7)$$

ANSI/IEEE Extreme Inverse

$$t = T \times \left(\frac{28.2}{\left(\frac{I}{I_s}\right)^2 - 1} + 0.122 \right) \quad (8)$$

Schneider Electric Long Inverse

$$t = T \times \left(\frac{120}{\left(\frac{I}{I_s}\right)^1 - 1} \right) \quad (9)$$

Out of these seven different inverse characteristics, one can select the desired characteristic as per application.

6 Hardware Model

The hardware model shown in Fig. 13 comprises of a special enclosure housing all the interfacing elements like LCD, LEDs, keypads, current sensor, relay driving circuits connected to the ATMega2560 microcontroller board.

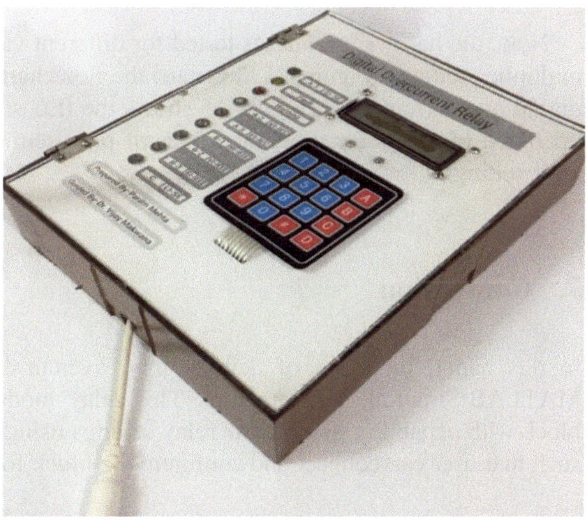

Fig. 13 Hardware model of IDMT overcurrent relay

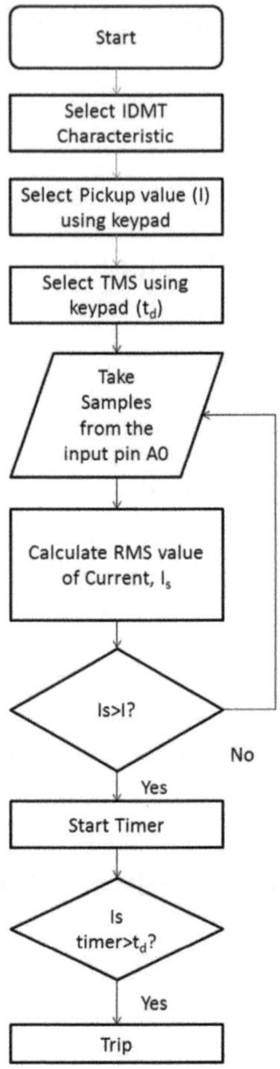

Fig. 14 Flowchart of the proposed IDMT overcurrent relay model

Now, the hardware model is tested for different values of plug settings and time multiplier settings. Figure 14 illustrates the flowchart of the algorithm of proposed digital overcurrent relay. Figure 15 shows the IEC standard inverse plot of time of operation of relays for different values of plug setting multipliers and time multiplier settings.

7 Conclusion

In this paper, modeling of definite time overcurrent relay is carried out using MATLAB/Simulink software tool. This relay model is designed as a Simulink block with capability to input the relay settings using a mask parameter dialog box, such that user can connect and configure the block for the modeling of a protection

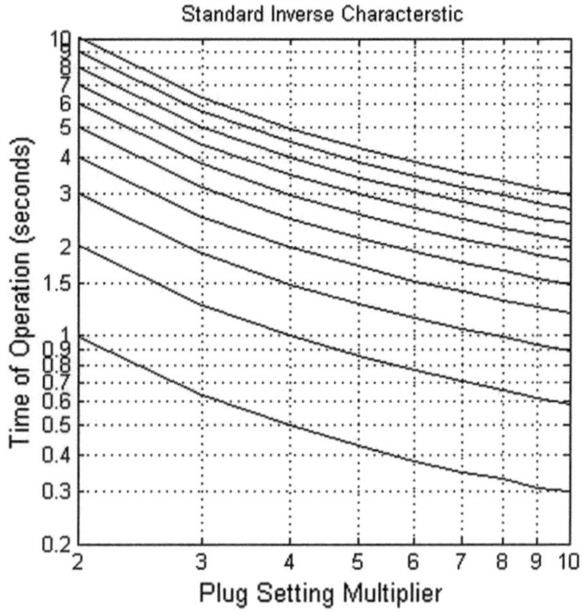

Fig. 15 Standard inverse characteristic curves for different values of TMS

scheme for power system network. The relay model was validated using a single-phase model for different values of plug settings and time delay. Furthermore, an implementation of hardware model was carried out using ATMega328 microcontroller and ACS712 current sensor. The hardware model was tested on a radial feeder network, and time of operation of relays for different faults at different locations is noted, operations of relay being satisfactory and accurate. Similarly, a hardware model with different interfacing elements is also designed which includes seven different inverse characteristics. The flowchart of the proposed relay as well as the plot of standard inverse characteristic curve is also illustrated.

Acknowledgements Special acknowledgments for G H Patel College of Engineering and Technology, Vallabh Vidhyanagar, India, for letting us to use switchgear and protection laboratory which helped us to do this research work.

References

1. Y. L. Goh, A. K. Ramasamy, A. A. Z. Abidin and F. H. Nagi, "Modelling of overcurrent relay using digital signal processor," *2010 IEEE Symposium on Industrial Electronics and Applications (ISIEA)*, Penang, 2010, pp. 367–370.
2. J. C. Tan, P. G. McLaren, R. P. Jayasinghe and P. L. Wilson, "Software model for inverse time overcurrent relays incorporating IEC and IEEE standard curves," *IEEE CCECE2002. Canadian Conference on Electrical and Computer Engineering. Conference Proceedings (Cat. No.02CH37373)*, 2002, pp. 37–41 vol. 1.

3. M. Kezunovic, J. Ren, and S. Lotfifard, *Design, modeling and evaluation of protective relays for power systems*. Springer, 2016.
4. B. A. Oza, N. Nair, R. P. Mehta, and V. H. Makwana, *Power system protection and switchgear*. TataMc-Graw Hill, 2010.
5. Atmel, *8-bit AVR Microcontrollers - ATMega328/P*, 2011.
6. A. M. LLC, *ACS712 Fully Integrated, Hall Effect-Based Linear Current Sensor IC with 2.1 kVRMS Isolation and a Low-Resistance Current Conductor*, 2013.
7. M. Weeks, *Digital signal processing using MATLAB & wavelets*. Jones & Bartlett Learning, 2010.
8. M. A. Ali and F. M. Bendary, "Design of prototype non directional overcurrent relay micro-controller-based," in *22nd International Conference and Exhibition on Electricity Distribution (CIRED 2013)*, June 2013, pp. 1–4.
9. S. Chan and R. Maurer, "Modeling overcurrent relay characteristics," *IEEE Computer Applications in Power*, vol. 5, no. 1, pp. 41–45, Jan 1992.
10. A. Nasiruzzaman, "A student friendly toolbox for power system analysis using matlab," *Matlab-Modelling, Programming and Simulations*, pp. 67–86, 2010.
11. S. Electric, MiCOM P125, P126 and P127 *Directional or Non Directional Relay Manual*, 2011.

Conditional Analysis of Dust Accumulated on Solar Photovoltaic Panel

Kaushal B. Jani and Jatinkumar J. Patel

Abstract The paper presents analysis of impact of dust on photovoltaic solar panel located at CSIR, Bhavnagar city, Gujarat, India. Proposed work is used to analyze the losses due to accumulation of dust on the surface of photovoltaic panel. Due to dust, radiation changes have been observed using the array tester system (SAT 100V 30A). The accumulation of the dust on the surface of polycrystalline PV panel shows reduction in power generation compared to clean module. It is observed that 8–10% decrement has been noticed in the given constraints and efficiency has been reduced by 5–10%. Result analysis has been discussed in detail with MATLAB/SIMULINK-based environment as well as practical setup at research laboratory for the different time duration.

Keywords Solar photovoltaic panel · Dust accumulation · MATLAB/SIMULINK Condition monitoring

1 Introduction

Electricity demand in year 2050 is predicted to lie between 8 and 26 TW with an average growth of about 3.75–17.81%. Fossil fuel, e.g., natural gas, coal, and oil, is still the dominating source of worldwide energy [11]. Nevertheless, due to increasing prices of fossil fuels and also due to the alarming increase of environmental issues related to fossil fuel, renewable energy is getting more attention. It has been projected that by 2040 more than 50% of the global energy is going to be in the form of renewable energy [5]. Solar energy, being a major source of renewable energy, is crucial in meeting future energy demand.

In this paper, we have included the normal use of resources. Study of the environmental effect of dust and the exhaustion of non-renewable resources. Due to

K. B. Jani (✉) · J. J. Patel
Electrical Department, G H Patel College of Engineering & Technology, V.V. Nagar, Anand, India
e-mail: janikaushal@yahoo.com

R. Kher et al. (eds.), *Proceedings of the International Conference on Intelligent Systems and Signal Processing*, Advances in Intelligent Systems and Computing 671,
https://doi.org/10.1007/978-981-10-6977-2_18

these reasons, there is an increase in worldwide interest for sustainable energy production and energy saving. Among all, the technologies involved in the generation of viable and common energy, a key role is played by photovoltaic (PV) cells, wind generators, biomass plants and fuel cells. In the particular area, photovoltaic systems can be considered one of the universal results with hopeful margins of development while ensuring the generation of energy with low environmental effect [5].

A regular cleaning of PV panels is suggested in order to maximize the efficiency. However, this method can be difficult to apply in complex systems or panels located in particular sites. In case, such activities lead to a major increase in its costs. In order to enhance the maintenance of the system, some parameters have to be overseen [6]. Renewable natural resources are the source to generate renewable energy, such as solar radiation, wind, rainfall, tides, geothermal heat. To convert solar radiation into electricity, a photovoltaic cell is required.

The photovoltaic effect is a physical phenomenon of converting the energy carried by optical electromagnetic radiation into electrical energy. The main application has been in space vehicles for many years, instead terrestrial application of PV developed very slowly, nevertheless PV fascinated not only the researchers but also the general public. Solar PV system is also used where the electricity cannot reach with line that is area easily generates electricity using solar photovoltaic system (Fig. 1).

The clearance of dust and sand on cell surfaces may be uniform or non-uniform depending on the mass of the PV arrays and the ground of the location. Smaller panels may have the drops, and uniform dust accumulation in panel performance may be same for many panels in the array [6]. However, the shape of dust addition may not be equal for all the panels in the module, whether they are located in large area or at large distance. Therefore, the drops in panel performance due to dust removal may not be same over the total array and hence it is more difficult to predict the atmospheric effect like rainfall, temperature, for an installation of PV system.

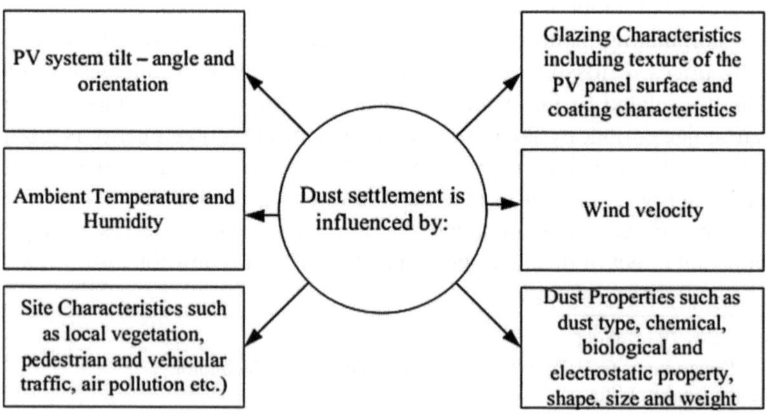

Fig. 1 Factors influencing dust accumulation [6]

On the other hand, it is seen that wind and rainfall often help us to clear the dust on PV surfaces. As it is verified by the experimental setups, that at high wind speed the tendency of dust deposition increases on PV surface and is the reason of drops in panel power generation. Rainfall is one of the sources to remove the dust particle from PV system but if this drop evaporates from PV surface without wearing off, it creates grip of dust on PV system. It is necessary to investigate the effect of the dust particles on PV system to find out the losses and also to set the maintenance schedule for better performance.

The effect of dust on PV system can be tested by comparing the performance of *I–V* characteristics of the panels with and without dust clearance on their surfaces. The *I–V* characteristics can be reflected to investigate the photovoltaic element for every PV element, which produces a unique *I–V* curve whose shapes depend on the conditions of surface, temperature, cell functioning temperature, mismatch of loss, and other balance system-related losses.

This approach has been adopted in the present study to find out the effect of dust accumulation on PV operating characteristics, while keeping the panels unprotected identical ambient parameters but with differing quantities of dust accumulation. The paper presents the measurement setup and the experimental procedure. It also presents a statistical analysis of the results and a detailed discussion.

2 System Introduction

2.1 *Simulation Setup*

Here, analysis has been carried out for PV system of 75 W connected with isolated *R-L* load based on simulation and real-time setup at CSIR-CSMCRI, Bhavnagar. A MATLAB-based simulated model of PV system with input variation of different values of temperature and radiation has been carried out as shown in Fig. 1. We have considered maximum power (P_{max}) 75 W, maximum voltage (V_{max}) 17.06 V, maximum current (I_{max}) 4.40, the open-circuit voltage (V_{oc}) 21.78 V, and short-circuit current (I_{sc}) 4.73 A (Fig. 2).

The performance of simulation characteristics (short circuit and open circuit) has been achieved and indicated in Fig. 4.

2.2 *Experimental Setup*

The experimental analysis was based on PV panels which were kept at same ambient temperature (29–35 °C) and radiation (278–985 W/m^2). The result is recorded for two cases, i.e., with dust particle and without dust particle, and the

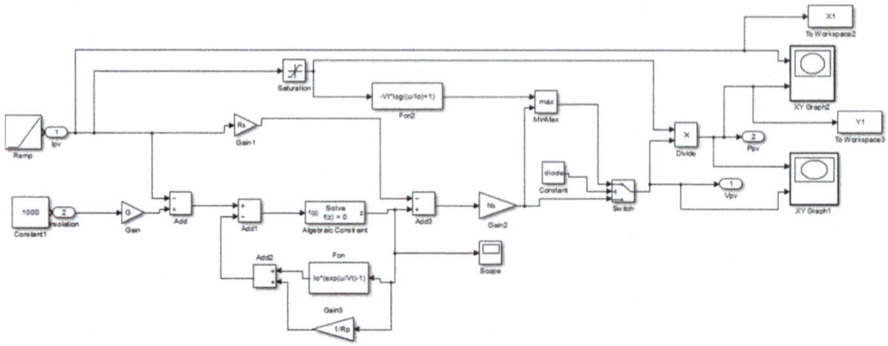

Fig. 2 MATLAB simulation of photovoltaic panel

Fig. 3 Experimental setup of polycrystalline PV panel

setup was installed on the rooftop at the Central Salt & Marine Chemicals Research Institute, Bhavnagar (latitude 21.76°N, longitude 72.15°E). The test setup consists of two identical open-stand-mounted polycrystalline silicon photovoltaic panels installed parallel and sloped at 21 °C to the ground facing southward. The arrangement for the experimental work is shown in Fig. 3. The detailed specifications for PV panels are listed in Table 1. The *I–V* characteristics for the panels were plotted and listed using an array tester. The cell surface temperatures are measured using Mextech DT–8811 Digital Infrared and Contact Thermometer.

As mentioned specifications of 75 W Polycrystalline silicon based martial made by Kotak Urja Ltd. in Table 1. Here, as shown in Fig. 3, two panels with same ratings are situated side by side with same environmental but different operational conditions. Here, one panel is fully cleaned, while the other is the one accumulated with dust particle of 0.5 mm thickness.

Table 1 PV panel specification

Dimensions (in mm)	
PV crystal panel	1190 × 545
Module dimensions	125 × 125
Cell dimensions	36
Maximum power (*W*)	75 W
Voltage at maximum power (*V*)	17.06 V
Open-circuit voltage (V_{oc})	21.78 V
Current at maximum power (*A*)	4.40 A
Short-circuit current (I_{sc})	4.73 A

3 Results Analysis

3.1 Simulation Analysis

Simulation-based analysis is for different radiating conditions that are elaborated in Fig. 4. It is observed that short-circuit current (I_{sc}) and open-circuit voltage (V_{oc}) are normal *I–V* characteristics of polycrystalline PV panel. It has been notice that there is accountable amount of decrement in dusty panel as compared to clean panel for *I*sc is 22% for the radiation level 435 W/m^2, 15% for 756 W/m^2 while for 9% for 856 W/m^2.

Furthermore, ratio of values of open-circuit voltage V_{oc} (dusty)/V_{oc} (clean) has been shown in Table 2 for input intensity range of 290–856 W/m^2. The open-circuit voltage (V_{oc}) of the dusty panel is slightly lower than that of the clean panel at all light intensities. Similarly, parameters like short-circuit current (I_{sc}) and power output are obtained as per Tables 3 and 4, respectively.

3.2 Real-Time Analysis

As described in experimental setup, output of 75 W polycrystalline PV panel creating prototype of different environmental conditions like clean and dusty surface (which includes various constrains) has been established at Terrace of Solar Lab at CSIR-CSMCRI, Bhavnagar.

The comparison of *I–V* characteristics of clean panel and dusty panel is shown in Fig. 5. The *I–V* characteristic curves obtained from the panels which are parallel to each other with or without dust particle under different atmospheric conditions at different time periods from different intensity levels, like 435, 848, and 925 W/m^2 have been plotted in Fig. 5.

Due to similarities, graphical analysis is done for initial, final, and average value of given set of readings.

From the results achieved by practical setup, the following observations can be drawn.

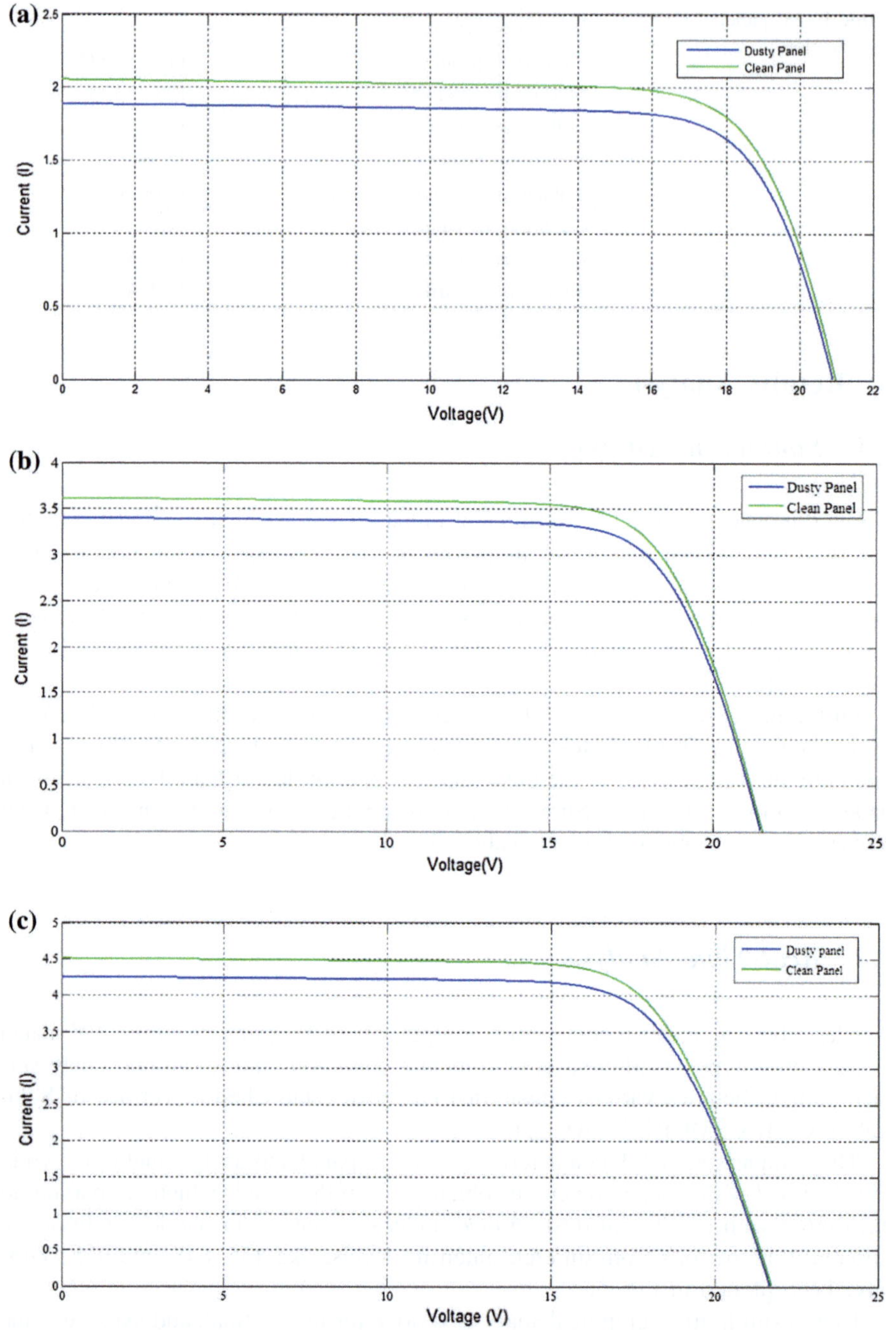

Fig. 4 Performance characteristics of polycrystalline panel. **a** For 435 W/m^2. **b** For 756 W/m^2. **c** For 856 W/m^2

Table 2 Open-circuit voltage losses

Intensity (W/m^2)	Open-circuit voltage (V)		V_{oc} (dusty)/V_{oc} (clean) average (%)
	Clean	Dusty	
345	20.77	20.65	99.42
435	20.98	20.90	99.61
525	21.18	21.07	99.48
756	21.50	21.43	99.67
856	21.69	21.58	99.49
985	21.72	21.65	99.67

Table 3 Short-circuit current losses

Intensity (W/m^2)	Short-circuit current (A)		I_{sc} (dusty)/I_{sc} (clean) average (%)
	Clean	Dusty	
345	1.64	1.51	92.07
435	2.05	1.89	92.19
525	2.52	2.30	91.26
756	3.61	3.40	94.18
856	4.10	3.72	90.73
985	4.50	4.25	94.44

Table 4 Power output losses

Intensity (W/m^2)	Average power output (W)		Power output losses %
	Clean	Dusty	
345	34.0628	31.1815	8.45
435	43.009	39.501	8.15
525	53.3736	48.461	9.20
756	75.615	72.862	3.64
856	88.929	80.2776	9.72
985	97.74	92.0125	5.85

1. Dust has a minor adverse effect on short-circuit current produced by the panels. The clean panel consistently produced a higher current output compared to the dusty panel.
2. The Voc of the dusty panel is observed to be lower in comparison with that of the clean panel at a particular solar intensity as shown in Table 5. This ratio remains close to 100% for all solar insulations. It is therefore proposed that dust does not affect system voltage significantly.

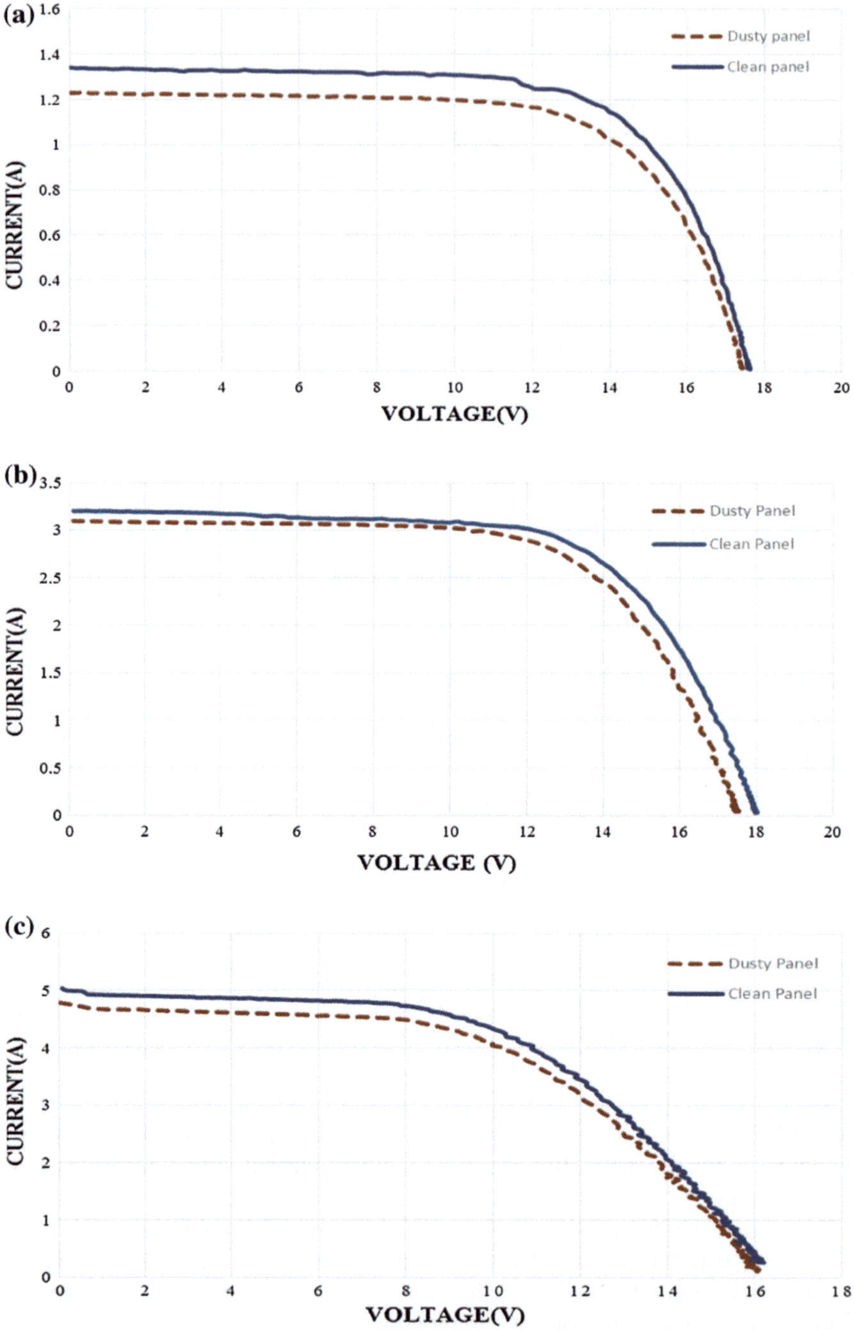

Fig. 5 Performance characteristics of polycrystalline panel. **a** For 435 W/m^2. **b** For 756 W/m^2. **c** For 856 W/m^2

Table 5 Open-circuit voltage losses

Intensity (W/m^2)	Open-circuit voltage (V)		V_{oc} (dusty)/V_{oc} (clean) average (%)
	Clean	Dusty	
345	17.09	16.8	98.30
435	17.98	17.6	97.88
525	7.87	7.6	96.56
756	17.62	17.19	97.55
856	16.65	16.32	98.01
985	16.53	16.14	97.64

3. From graph, we observed dust accumulation thickness and the power losses related to dust accumulation on photovoltaic panels. Here in real-time experimental setup where the dust thickness was found to be 1.8 g/m^2, the power loss due to dust was 8–12% of the power generation (Tables 6 and 7).
4. There is main change in short-circuit current (I_{sc}) on panels in the presence of dust. Hence, showing in Table 8 the real-time experiment analysis of the two panel losses for accumulation of dust. When the clean panel efficiency of power generation is high compare to the dusty panel.

The power generation is more in clean panel by 8–12% as compared to the dusty panel. And the cell temperature of the both panel is different. Maximum cell Temperature 54 °C in clean panel. The density of collected dust was found to be 0.35 g and 0.28 m^2 surface area which resulted in 5–6% of power losses as compared to dust-free PV system. This reduction can be observed better in a large PV plant as compared to a small PV system.

The proposed work suggests remedies to reduce this effect on solar PV system and improvement of power generation. Photovoltaic panels are constantly exposed to various types of weather and thus become a target for dirt, dust, industrial residues, atmospheric pollution, bird droppings, etc. The presence of these elements on the panel's surface prevents the sun's rays from filtering onto the panel's photovoltaic cells completely, reducing the solar performance and therefore efficiency. Accumulated dust reduce efficiency and may damage the material the panels are made from. The following care can be taken to avoid dust accumulation thus improving the performance of PV panel.

1. **Daily cleaning of PV panel surface with water**

The agents responsible for the deterioration of the surfaces also include the chemicals used during cleaning. These form an excellent attraction for dust and accelerate the deterioration process, with negative repercussions on the appearance and above all, the function of the structure.

Cleaning is, therefore, an indispensable part of the maintenance of these structures, and the frequency of the cleaning depends on the geographic area and climate. Industrial areas may require more thorough maintenance than rural areas.

Table 6 Short-circuit current losses

Intensity (W/m^2)	Short-circuit current (A)		I_{sc} (dusty)/I_{sc} (clean) average (%)
	Clean	Dusty	
345	2.94	2.58	87.75
435	3.17	3.08	97.16
525	3.41	3.08	90.32
756	3.42	3.32	97.07
856	3.79	3.54	93.40
985	4.93	4.7	95.33

Table 7 Power output losses

Intensity (W/m^2)	Cell temperature (°C)		Average power output (W)		Power output losses%
	Clean	Dusty	Clean	Dusty	
345	34	33	32.25	26.86	16.71
435	36	35	37.64	35.56	5.52
525	45	45	39.27	35.56	9.44
756	51	50	38.84	36.62	5.71
856	46	45	38.87	34.79	10.49
985	54	55	44.08	41.07	6.82

Table 8 Comparison of simulation and real-time data

Intensity (W/m^2)	Simulation results			Experimental results		
	V_{oc} (dusty)/V_{oc} (clean) average (%)	I_{sc} (dusty)/ I_{sc} (clean) average (%)	Power output losses (%)	V_{oc} (dusty)/V_{oc} (clean) average (%)	I_{sc} (dusty)/ I_{sc} (clean) average (%)	Power output losses (%)
345	99.42	92.07	8.45	98.30	87.75	16.71
435	99.61	92.19	8.15	97.88	97.16	5.52
525	99.48	91.26	9.20	96.56	90.32	9.44
756	99.67	94.18	3.64	97.55	97.07	5.71
856	99.49	90.73	9.72	98.01	93.40	10.49
985	99.67	94.44	5.85	97.64	95.33	6.82

If we clean PV panel with water, panel corrosion and aging effect will increase. Due to this, the panel life decreases and also power generation decreases gradually.

2. **Placing PV panel at the location having green grass**

Advantage to select such as location is that dust particle will decrease. But, due to the green grass, the humidity will play an adverse role on the performance of PV panel.

4 Conclusions

The proposed work is considered for dust accumulation on photovoltaic system performance. Here, the panels' performance was investigated under the similar conditions of ambient temperature and insolation with the help of *I–V* characteristics when one of the panels was dusty and other one was cleaned of dust.

The proposed work presents result analysis of dust-accumulated panel in MATLAB/SIMULINK environment as well as experimental setup of PV panel which are kept side by side having different atmospheric temperatures 28–33 °C and different time periods like 1 month, 2 months, and 6 months. The presence of dust particles on PV panel will be approximately 5–10% deviation, and due to this, there will be losses in power generation. Through the diagnosis of power losses in the presence of dust particle in PV panel, we concluded that in the absence of dust particles the power generation from PV panel will be more and losses will be less compared to dust particles present in PV panel.

Acknowledgements The author(s) of proposed work would like to thank Dr. (Mrs.) Subarna Maiti, (Senior Scientist), and her team (PDEC Solar Department) CSIR (Central Salt & Marine Chemicals Research Institute, Bhavnagar, Gujarat.) for providing technical and infrastructure support.

References

1. A. Chatterjee, Abir, Ali Keyhani, and Dhruv Kapoor. "Identification of photovoltaic source models." IEEE Transactions on Energy conversion 26.3 (2011): 883–889.
2. A. Rao, R. Pillai, M. Mani, and P. Ramamurthy, "Influence of dust deposition on photovoltaic panel performance," Energy Procedia, vol. 54, pp. 690–700, 2014.
3. Akram, Mohd Nafis, and Saeed Lotfifard. "Modeling and Health Monitoring of DC Side of Photovoltaic Array." IEEE Transactions on Sustainable Energy6.4 (2015): 1245–1253.
4. Costa, Suellen CS, Antonia Sonia AC Diniz, and Lawrence L. Kazmerski. "Dust and soiling issues and impacts relating to solar energy systems: Literature review update for 2012–2015." Renewable and Sustainable Energy Reviews 63 (2016): 33–61.
5. M. Catelani, L. Ciani, L. Cristaldi, M. Faifer, M. Lazzaroni, and M. Rossi, "Characterization of photovoltaic panels: the effects of dust," in Energy Conference and Exhibition (ENERGYCON), 2012 IEEE International, 2012, pp. 45–50.

6. Mani, Monto, and Rohit Pillai. "Impact of dust on solar photovoltaic (PV) performance: Research status, challenges and recommendations."Renewable and Sustainable Energy Reviews 14.9 (2010): 3124–3131.
7. Nahar, N. M., and Jagdish P. Gupta. "Effect of dust on transmittance of glazing materials for solar collectors under arid zone conditions of India." Solar & wind technology 7.2–3 (1990): 237–243.
8. Parikh, Abhishek, et al. "Solar Panel Condition Monitoring System based on Wireless Sensor Network."
9. Said, S. A. M. "Effects of dust accumulation on performances of thermal and photovoltaic flat-plate collectors." Applied Energy 37.1 (1990): 73–84.
10. Stauffer, Y., et al. "Power monitoring based photovoltaic installation fault detection." 2015 International Conference on Renewable Energy Research and Applications (ICRERA). IEEE, 2015.
11. Xu, Xiaoli, Huan Wang, and Yunbo Zuo. "Method for diagnosing photovoltaic array fault in solar photovoltaic system." Power and Energy Engineering Conference (APPEEC), 2011 Asia-Pacific. IEEE, 2011.
12. V.U. Hoffmann and A. Goetzberge, Photovoltaic Solar Energy Generation.
13. V. F. Pires, J. Fernando, and A. Silva, "Teaching nonlinear modeling, simulation, and control of electronic power converters using MATLAB/SIMULINK," IEEE Trans. Educ., vol. 45, no. 3, pp. 253–261, Aug. 2002.
14. Kimber A., 2007. The Effect of Soiling on Photovoltaic Systems Located in Arid Climates, Proceedings of the 22nd European Photovoltaic Solar Energy Conference, Milan, Italy.
15. F. Jiang, A. Wong, April 2005. Study on the performance of different types of PV modules in Singapore, IEEE 7th International Power Engineering Conference, Niigata, Japan.
16. U.S. Department of Energy, The History of Solar.
17. Weidong Xiao, July 2003. A Modified Adaptive Hill Climbing Maximum Power Point Tracking (MPPT) Control Method For Photovoltaic Power System, The University of British Columbia.
18. Introduction to Photovoltaic Systems, Renewable Energy, the Infinite Power of Texas, SECO Fact Sheet No. 11.
19. E. Allen, N. LaWhite, Y. Yoon, J. Chapman, and M. Ilic, "Interactive object-oriented simulation of interconnected power systems using SIMULINK," IEEE Trans. Educ., vol. 44, no. 1, pp. 87–94, Feb. 2001.

Modeling of Three-Phase Distribution Transformer for Analysis of Core and Ohmic Losses Using FEM

Priyal V. Patel and Sameer B. Patel

Abstract This article presents modeling of three-phase 315 kVA-distribution transformer. The analysis of magnetic flux density, core loss, and ohmic loss of the transformer is done using 3D finite element method (FEM). Finite element method is one of the most efficient methods used to analyze magnetic fields. The result shows the comparison between practical value and the value obtained from 3D FEM.

Keywords Transformer · Core loss · Ohmic loss · Finite element method (FEM)

1 Introduction

Transformer plays a key role in the interconnection of power systems. Without the transformer, it would not be possible to use electric power. In many different ways, it is used today. Also, transformers occupy prominent positions in the electric power system, being the key links between power-generating stations and points of electric power utilization [1]. The efficiency of a transformer is highest among all the machines; it is around 95–98% [2]. But the major problem associated with a transformer is a hot spot formation due to increase in temperature at particular portion. The heating problem is due to increase in losses. Mainly, losses are divided into two parts: load loss and no-load loss. The no-load loss is the constant loss; they do not change with the change in load, whereas load loss changes with the change in load. Core loss is a constant loss. Generally in core, losses occurs in joints. [3].

Core loss is divided into two parts: eddy current loss and hysteresis loss. To reduce core loss laminated core is used and by increasing numbers of laminations it can be reduced. Numerical methods are used to estimate and study the leakage and magnetic fields. One of the accurate and popular method is finite element method

P. V. Patel (✉) · S. B. Patel
Electrical Department, GCET College, Anand, Gujarat, India
e-mail: priyalp2305@gmail.com

R. Kher et al. (eds.), *Proceedings of the International Conference on Intelligent Systems and Signal Processing*, Advances in Intelligent Systems and Computing 671,
https://doi.org/10.1007/978-981-10-6977-2_19

(FEM). There are many methods to reduce stray loss, such as topological change and shielding.

2 Finite Element Method

FEM is powerful and versatile numerical technique for handling problems involving complex geometries. It is also referred as finite element analysis (FEA). It subdivides a large problem into smaller, simpler parts that are called finite elements. The simpler equations that model these finite elements are then assembled into a larger system of equations that model the entire problem.

There are many advantages of FEM over other numerical methods; one is that formulation is independent of problem's geometrical complexity. Anisotropic, non-uniform, and nonlinear media can be incorporated into the solution scheme in FEM. Many commercial software are available that are based on FEM. A flowchart of finite element analysis is shown below [4] (Fig. 1).

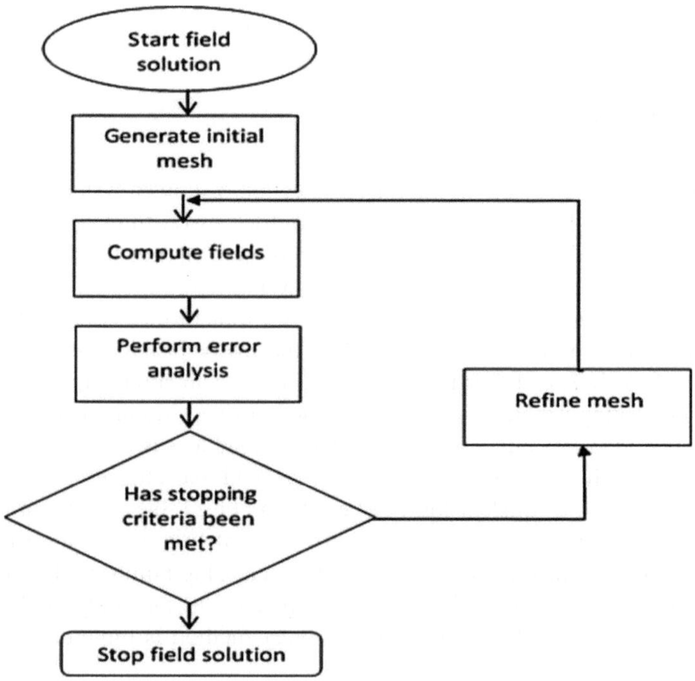

Fig. 1 FEM flowchart

3 Distribution Transformer Modeling

The modeling of 315 kVA-distribution transformer has been done for the analysis of magnetic fields using 3D FEM. The core is made of CRGO steel; it is laminated to reduce eddy current loss. Initially, stepped core is designed as shown in Fig. 2. Further, L.V and H.V winding made of copper is designed according to the specifications as shown in Fig. 3.

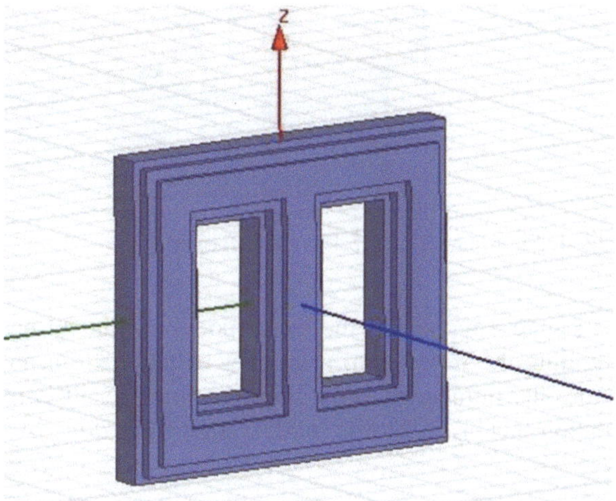

Fig. 2 Core design of 315 kVA-transformer

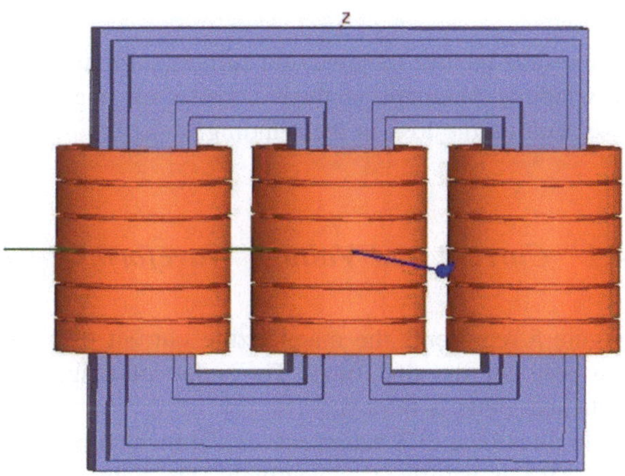

Fig. 3 Model of 315 kVA-distribution transformer

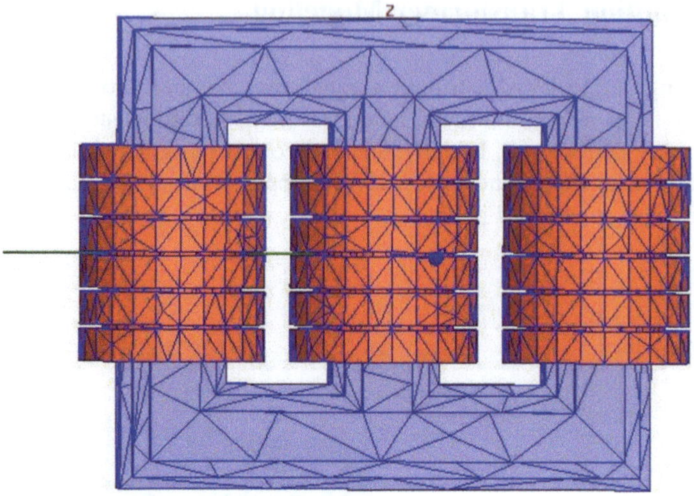

Fig. 4 Mesh model of 315 kVA-transformer

Figure 4 represents the mesh model of transformer. Mesh generation consists of dividing the model into a number of tetrahedron units to get accurate results [5]. More the number of cells, more accurate result is produced, but the time of simulation increases. In this model, analysis is done by considering 500 meshes.

4 Results

The results include magnetizing current, current, and voltage of L.V winding and H.V winding. Magnetic analysis consists of magnetic flux density in the core, core loss, and ohmic loss in the winding.

4.1 Current and Voltage Waveforms

Figure 5 represents the magnetizing current of transformer required to magnetize the core for the first time. The value of magnetizing current is very less. The obtained value is 0.69 A.

Figures 6 and 7 represent the branch current of L.V and H.V windings of 315 kVA-transformer, respectively. The value of branch current is nearly the rated value. The obtained value of L.V branch current and H.V branch current is 386 A and 8.38 A, respectively.

Figures 8 and 9 represent the waveform of L.V and H.V winding peak voltages of 315 kVA-transformer, respectively. Initially, the voltage is increasing thereafter

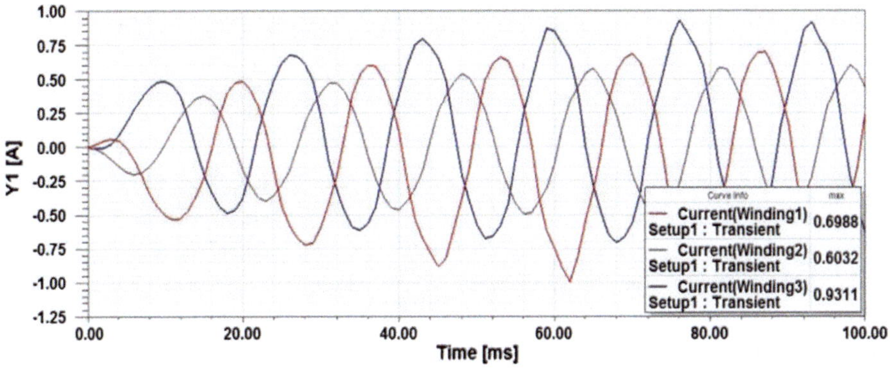

Fig. 5 Magnetizing current of 315 kVA-distribution transformer

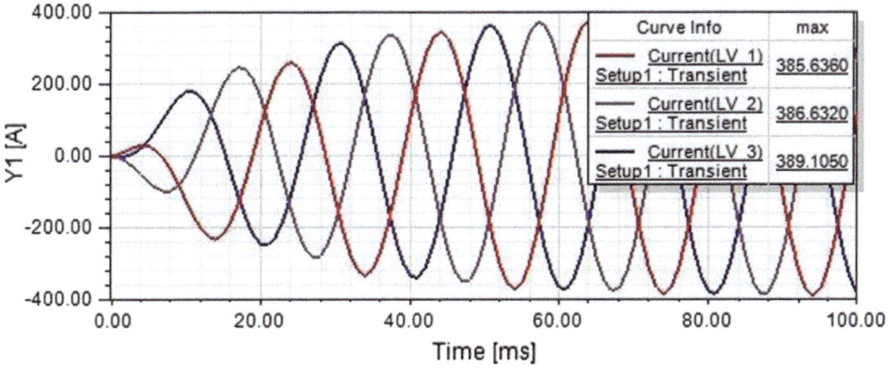

Fig. 6 Branch current of L.V winding

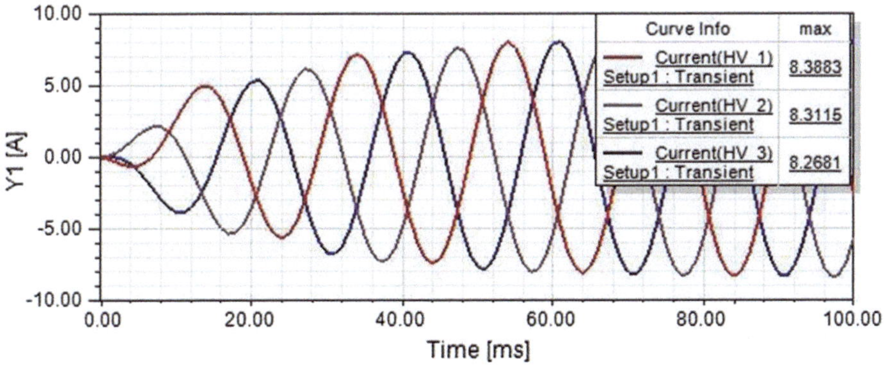

Fig. 7 Branch current of H.V winding

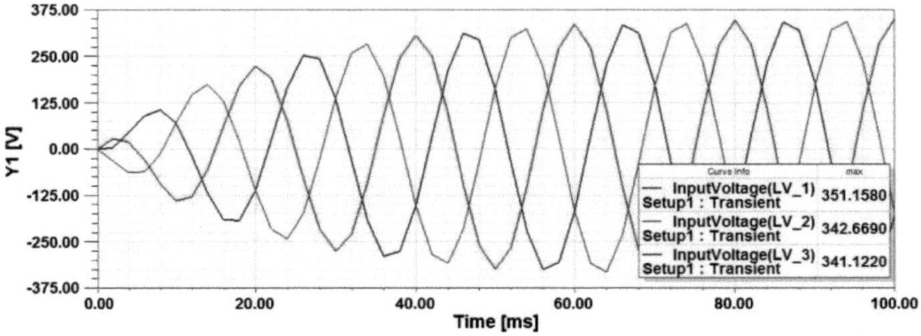

Fig. 8 L.V winding peak voltage

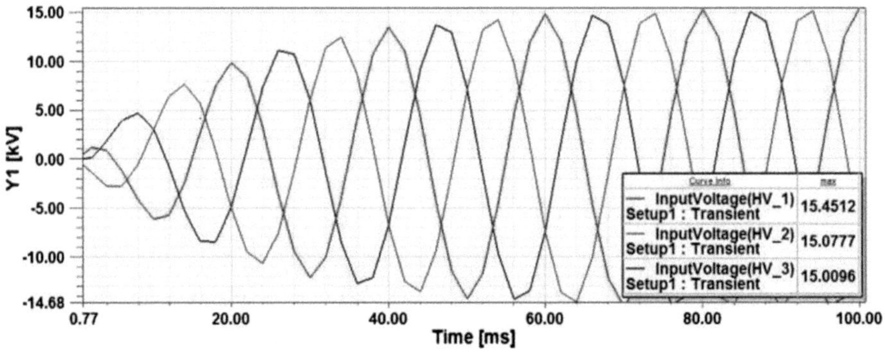

Fig. 9 H.V winding peak voltage

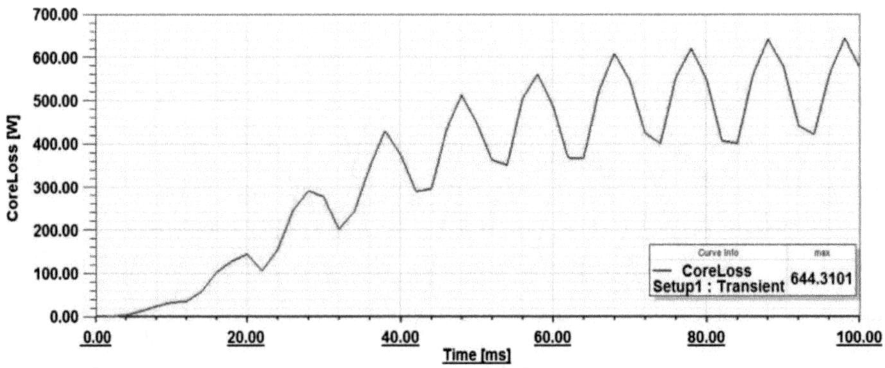

Fig. 10 Core loss versus time graph

it becomes stable. The obtained value of L.V and H.V winding peak voltage is 351 V and 15 kV, respectively.

The core loss in a transformer is the constant loss; it will not change with the change in load. Figure 10 shows the core loss versus time graph. It can be seen from the graph that initially core loss is increasing with time, but after few cycles it becomes stable. The obtained value of core loss is 644.31 W.

4.2 *Magnetic Field Analysis*

Figure 11 represents the magnetic flux density distribution of 315 kVA-distribution transformer. The maximum flux density value obtained is 1.75 T. It can be observed from the figure that flux density is maximum at the corners; it is due to the turning of flux lines which lead to high value of flux density. This will lead to increase in the loss.

Figure 12 shows the ohmic loss in the winding of the transformer. The obtained value is 1.917e + 005 watts per meter cube. The total volume of all the three winding is 0.018 m^3. So the ohmic loss in watts is 3450.

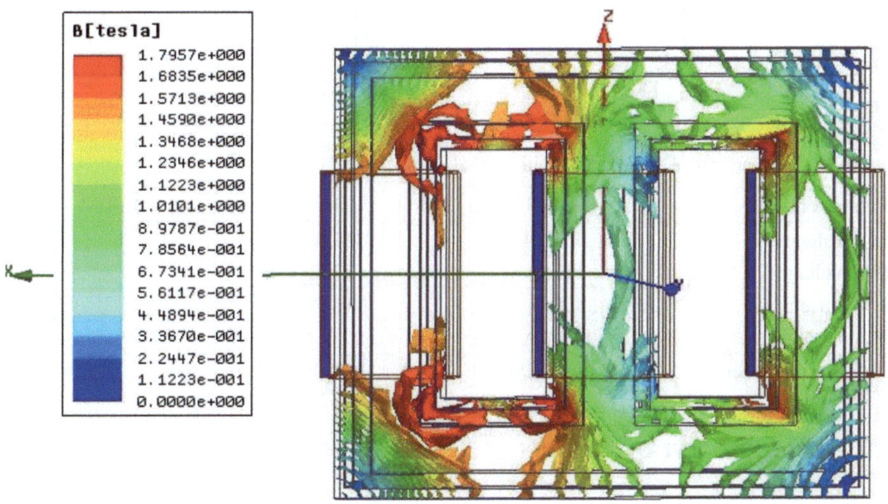

Fig. 11 Magnetic flux density distribution in 315 kVA-transformer

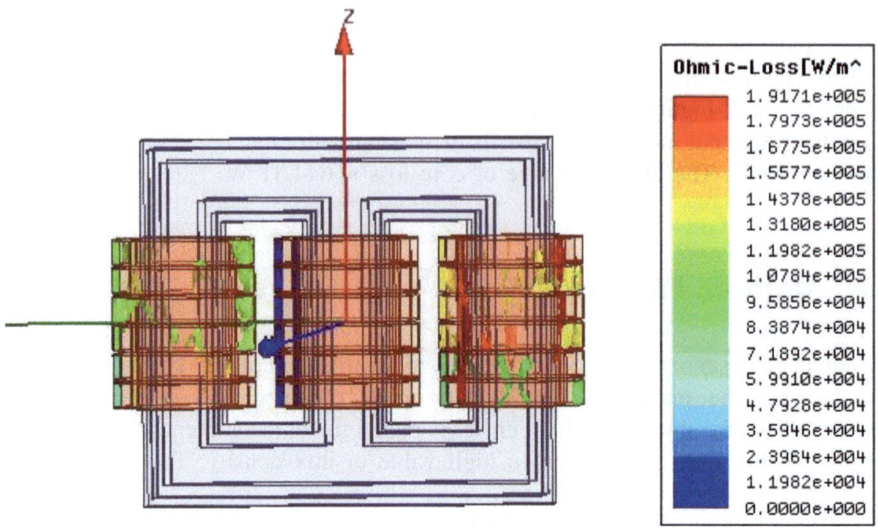

Fig. 12 Ohmic loss in the windings of 315 kVA-transformer

Table 1 Comparison of core and ohmic loss value with experimental value

Loss quantity	Experimental value (W)	Value obtained using 3D FEM (W)
Core loss	689	644.31
Ohmic loss	3722	3450

Table 2 Comparison of parameters with actual value

Parameters	Actual value	Value obtained using 3D FEM
L.V voltage	353.54 V	351 V
H.V voltage	15556 V	15302 V
L.V current	420 A	386 A
H.V current	9.54 A	8.36 A
Magnetic flux density	1.71 b/m^2	1.79 b/m^2

4.3 Comparison of Results

Table 1 shows the comparison of the result of core loss and ohmic loss with experimental values.

The above results show that the percentage variation of core loss from experimental value is 6.5% and variation of ohmic loss is 7%.

Table 2 represents the comparison of transformer parameters with their actual values. L.V and H.V voltage, current, and magnetic flux density values are compared; it can be seen that all the values are very close to that of actual value.

5 Conclusion

From the magnetic flux density plot, we can conclude that the flux density is maximum at the corners. The core loss versus time graph represents that the core loss becomes constant after a few cycles. From the comparison table, it can be concluded that the FEM is very accurate as all the values are very close to the experimental values.

References

1. Pavlos S. Georgilakis, "Spotlight on Modern Transformer Design", Springer-Verlag London Limited 2009.
2. R.K. Agrawal, "Principles of electrical machine design", S.K. Kataria & Sons.
3. Olivares-Galván, Juan C., Pavlos S. Georgilakis, and Rodrigo Ocon-Valdez. "A review of transformer losses." Electric Power Components and Systems 37.9 (2009): 1046–1062.
4. Kirar, Mukesh, et al. "Study of stray losses reduction through Finite Element Method." 2013 Annual IEEE India Conference (INDICON) 2013.
5. Olivares, Juan Carlos, et al. "Reducing losses in distribution transformers." IEEE transactions on power delivery 18.3 (2003): 821–826.

Day-Ahead Wind Energy Forecasting Using Feed Forward Neural Network

Ankitkumar M. Makwana and P. R. Gandhi

Abstract Wind energy injection is intermittent in nature and is one of the most affecting factors in maintaining deviation limits. This paper proposes an artificial neural network-based day-ahead short-term wind energy forecasting using feed forward neural network. Different training algorithms were studied, and mean absolute percentage error of each has been calculated. The results show that Levenberg–Marquardt back propagation provides a better accuracy than other training algorithms.

Keywords Wind energy forecasting · ANN · Feed forward neural network MATLAB

1 Introduction

Wind energy is one of the most promising sources for electricity generation. The utilization of wind energy has become increasingly popular due to its free availability and environmental friendly. Renewable energy sources are growing remarkably since the last two decades among which wind energy is the fastest growing source of energy. Also renewable energy sources are considered as best alternate source for conventional energy.

The factor to be considered in wind energy is its reliability as wind is uncertain. Wind speed is the most difficult weather parameter to forecast. The intermittent behavior of wind speed causes difficulty in forecasting wind generation [1]. To deal with such challenging nature of wind generation, also increasing wind energy

A. M. Makwana (✉) · P. R. Gandhi
Electrical Department, Sardar Vallabhbhai Patel Institute of Technology, SVIT, Vasad, Gujarat, India
e-mail: ankitkumarmakwana97@gmail.com

R. Kher et al. (eds.), *Proceedings of the International Conference on Intelligent Systems and Signal Processing*, Advances in Intelligent Systems and Computing 671,
https://doi.org/10.1007/978-981-10-6977-2_20

penetration and to overcome its impact, a realistic forecast is vital and prerequisite for reliable operation. Moreover, real-time wind generation and operational availability of wind mills are also crucial part for more accurate wind energy forecasting. For better operation of power system, a good forecasting has to be made by transmission system operator [2]. A physical approach for wind forecasting technique uses wind pattern along with wind turbine power curve for forecasting output power of wind. Mathematical model can be used by forecaster based on physical approach [3]. Physical models can be used which considers terrain, obstacles, temperature, and pressure to estimate wind power generation.

Currently, various methodologies have been developed to predict the wind power and wind speed. Methodology like time series modeling such as ARMA and ARIMA provide a good forecasting. To forecast for short- and medium-term interval of time artificial neural network (ANN) has proved very useful. ANN provides accurate result with minimum errors.

ANN is inspired from biological nervous system. ANN is composed of highly interconnected processing elements working in union to solve specific problems. ANN is used in specific applications such as pattern recognition and data classification through learning process. Thus, ANN is an information processing system. In this information processing system, the elements called neurons which process the information. ANN consists of many nodes and connecting synapses. Nodes operate in parallel and communicate with each other through connecting synapses [4]. The signals are transmitted by connected links. This links possess an associated weight, which are multiplied with incoming signal. The output signal is obtained by applying activation to the net input. In recent years, ANN is used for forecasting and for solving complex problems. The advantage of ANN includes modeling of both linear and nonlinear systems by means of learning with training data [5].

Transmission system operator currently uses day-ahead wind energy forecast to predict the power to be delivered for each time blocks of the next day. This forecast ensembles day-ahead commitments of generation resources [6]. By using day-ahead forecasting, it helps the system operator to meet with the demand of power.

Due to variability and uncertainty in wind power generation, it becomes difficult for power system operator to maintain the balance between generation and demand. A reliable prediction of day-ahead forecast helps the power system operator to make good decision in critical situation and also to maintain grid stability of the system.

Development of wind generation has increased in state of Gujarat. It has 52 pooling station for wind generation. Wind farms are located at the coastal regions, and also offshore wind projects are being developed. As wind energy injection is intermittent in nature, an accurate forecasting should be made.

Section 2 describes wind generation characteristics, wind pattern of one region. Section 3 presents the methodology implemented in wind energy forecasting. Section 4 shows the forecast values and is compared with actual values and Sect. 5 provides conclusion.

2 Wind Generation Characteristics

Gujarat has a good wind potential. The wind installed capacity of Gujarat is 4086 MW as on July 2016 and currently ranks 3rd in installed capacity across India. Various wind farms which is located at coastline of Gujarat produce large amount of wind power. So it gives challenge to system operator to meet between generation and demand. The power obtained from renewable energy sources should be used instantly at the time it is produced. Initially, it is needed to understand the behavior of wind generation before developing wind energy forecasting methodology. Annual-, month-, day-wise scenarios of actual wind generation as well as seasonal scenario need to be observed to identify the behavior of wind generation. Wind generation is also affected by heavy rain. Sudden rise and fall of wind may lead to wide variation in wind generation.

Seasonal variation of Gujarat also affects wind generation. Gujarat has primarily three seasons: winter, from November to March; summer, from March to June; and monsoon, from July to September. As Gujarat has a long coast line, sea breezes and it influences the wind characteristics in this region.

Wind power output depends on weather parameters such as wind speed, temperature, pressure, humidity, and air density [7]. The wind power equation is given by

$$P = 0.5\,\rho A v^3 \tag{1}$$

where ρ is air density in kg/m^3, A is the swept area of turbine blade, v is the wind speed in m/s, and P is the power in W. Weather parameters play important role in wind energy forecasting model. Wind speed is one of the major parameters to be considered as a slight variation in wind speed causes change in wind power. Wind power curve depends on three key points: cut-in speed, speed below which turbine not produce power; rated speed, speed at which rated power of turbine is produced; and cutoff speed, defined as the speed beyond which the turbine is not allowed to deliver power. The production of wind power depends humidity, and the relative humidity of air depends on amount of water vapor in air, which determines air density. Temperature and pressure directly influence wind power as they are related to air density. Also geographic location also affects wind generation, if wind power plant are installed where the availability of wind speed, temperature, pressure, humidity is not available at desired range wind power plant will not give a better output.

Wind generation trend remains same during particular season, but it does not mean that the same pattern will be followed the entire season. Wind generation may have change of pattern on hourly basis.

Figure 1 shows the wind pattern of one region of Jamnagar, Gujarat. From the figure, a wide variability in wind energy of three days can be seen.

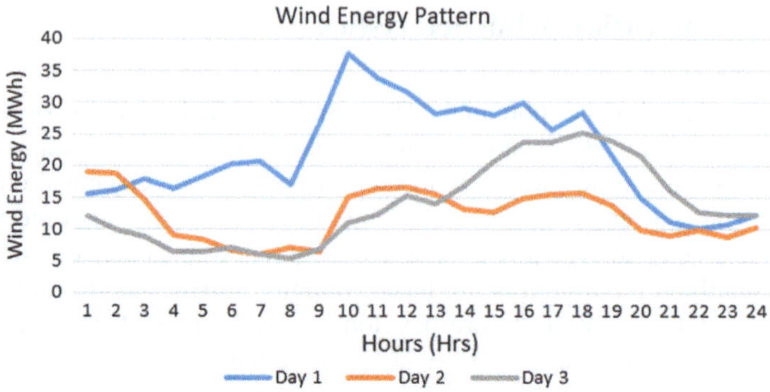

Fig. 1 Wind energy pattern of three days

Fig. 2 Three-layer feed forward neural network

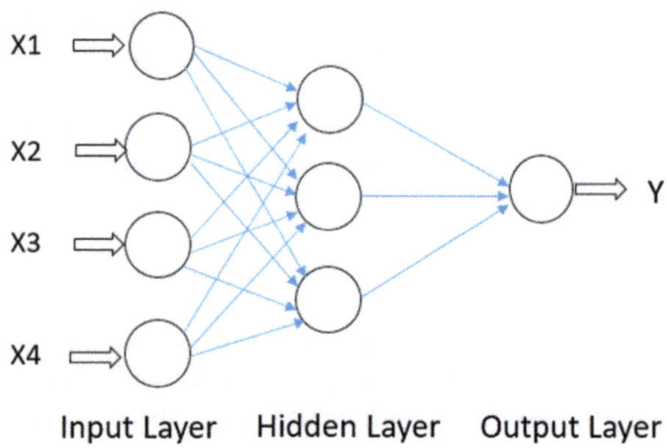

Wind energy pattern also changes from one region to another as it is related to wind speed. Due to this type of variability, it becomes difficult to manage grid operations.

3 Methodology Implemented

ANN-based methods are good choice to study forecasting problems. ANN has ability to learn from the past data which is very useful for future prediction. Various ANN techniques like feed forward neural network (FFN), radial basis function neural network (RBFN), and Elman recurrent neural network can be used for forecasting.

In this paper, wind energy forecasting is done using feed forward neural network. Feed forward neural network is formed by interconnection of several layers. A three-layered feed forward neural network is shown in Fig. 2.

FFN often has more than one hidden layers followed by output layer. Multiple layers allow network to learn nonlinear relationship between input and output.

Analysis was done on one of pooling stations of Gujarat. The wind farm of WWIL is considered for forecasting. The wind farm has installed capacity of 319 MW. It is connected to Tebhda substation. Wind power and weather data were provided by SLDC GETCO. Location of wind farm is at Jamnagar, so hourly weather dataset was provided.

Various weather data which affect the wind power are to be considered. Weather dataset is considered as the input data, and generated wind power is given as Target data. After specifying, the inputs and target data were normalized to a range of $[-1, 1]$. Normalization is to bring data in a uniform scale. It is helpful for better training performance. The equation of normalization used is as below:

$$pn = \text{min}pn + (\text{max}pn - \text{min}pn) \times (p - \text{min}p)/(\text{max}p - p\text{min})$$

After normalization, feed forward neural network has to be created by specifying number of inputs and target data and hidden layer. Various parameters such as training algorithm, learning rate, number of epochs are to be specified. The neural network is then trained by using different types of training algorithm.

Short-term wind energy forecasting procedure is as below:

Weather parameters are comprised of wind speed, temperature, pressure, humidity. The parameters selected as weather data are related with wind power. From Eq. (1), wind power is related with cube of wind speed and is also related with air density. Air density is related with pressure and temperature, so all these parameters are used in weather data. Wind power data are considered for the investigation. Hourly dataset of months of July and August was used for detail study. A multilayered feed forward neural network consisting of input layer, hidden layer, and output layer is used. The number of neurons in hidden layer is selected on a trial-and-error basis. Two types of training algorithms are used and compared. The neural network is trained on hourly dataset of months of July and August, and prediction was made for day ahead. The network is trained using Levenberg–Marquardt algorithm. The LM algorithm tries to reduce mean square error, i.e., minimizing the performance function in form of sum of squares.

$$\text{MSE} = \frac{1}{n}\sum_{q=1}^{n}(Tq - Aq) \tag{2}$$

Tq = target value
Aq = actual value

The second algorithm used for training the network is gradient descent with momentum back propagation. Gradient descent with momentum depends on two parameters: 1. learning rate which is similar to simple gradient descent and 2. the momentum constant in which the value is to be set between 0 and 1.

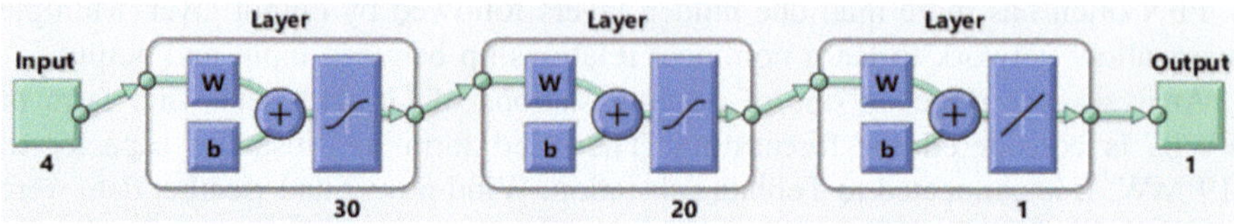

Fig. 3 Developed network

Another algorithm is gradient descent back propagation. The weights and biases are updated in the direction of negative gradient of performance function. The learning rate is multiplied times the negative of the gradient to determine the changes to the weights and biases [8].

The network is ready for training; during the training, weights and bias are adjusted. After training, the neural network was tested on new dataset, and error has been computed.

The network is developed and trained using MATLAB R2015a. Figure 3 shows the feed forward neural network consisting of four input parameters followed by two hidden layers, and one output layer has been created. First hidden layer consists of 30 neurons, and second hidden layer consists of 20 neurons. The output layer consists of single neuron. The transfer function of first layer and second layer in 'tansig' i.e. hyperbolic tansigmoid. It is selected as it gives high convergence. The output layer has transfer function 'purelin.'

Error analysis can be computed by mean absolute percentage error (MAPE) and is given by

$$\text{MAPE} = \frac{1}{n}\sum_{i=1}^{n} Zai - Zpi/Zai \tag{3}$$

where

Zai is actual value
Zpi is forecasted value
n is number of periods

4 Result and Discussion

Feed forward neural network has been developed, network was first trained on hourly data of two months, and prediction was made for day ahead. Three types of training algorithm were considered: LM back propagation, gradient descent with momentum back propagation, and gradient descent back propagation. MAPE for all the three training algorithms is obtained.

1. Training algorithm: LM back propagation

Levenberg–Marquardt back propagation was implemented on the developed neural network. The learning rate was set at 0.01. Parameters were divided into random division consisting of training samples used for analysis is 70%, validation and testing of network at 15% each (Fig. 4).

The best validation performance is at 0.033959 at epochs 5 as LM back propagation is fastest and gives a stable convergence.

Figure 5 depicts the regression analysis on the test data. Regression analysis clarifies in terms of training, testing, and validation of the developed model. It can also be seen from Fig. 5 test and validation have a value greater than 0.96.

Also the actual and predicted plots of wind power generation using feed forward neural network trained with LM back propagation on hourly basis are shown in Fig. 6. MAPE with LM back propagation is 13.08%.

2. Training algorithm: gradient descent with momentum back propagation

In this algorithm, learning rate was same and momentum constant of the parameter was considered as 0.8. Data division was random as above case. The best validation performance was obtained at 0.0035504 at epoch 286 (Fig. 7).

Regression analysis was also done as shown in Fig. 8. MAPE obtained is 28.75%. The validation of the trained algorithm obtained is $R = 0.92479$, and testing result is $R = 0.96434$.

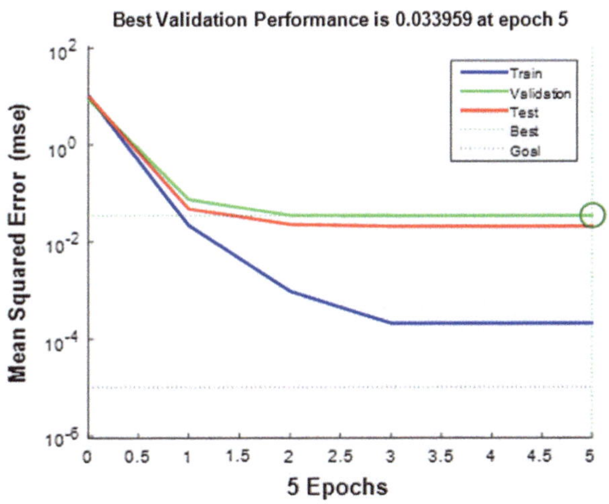

Fig. 4 Performance plot

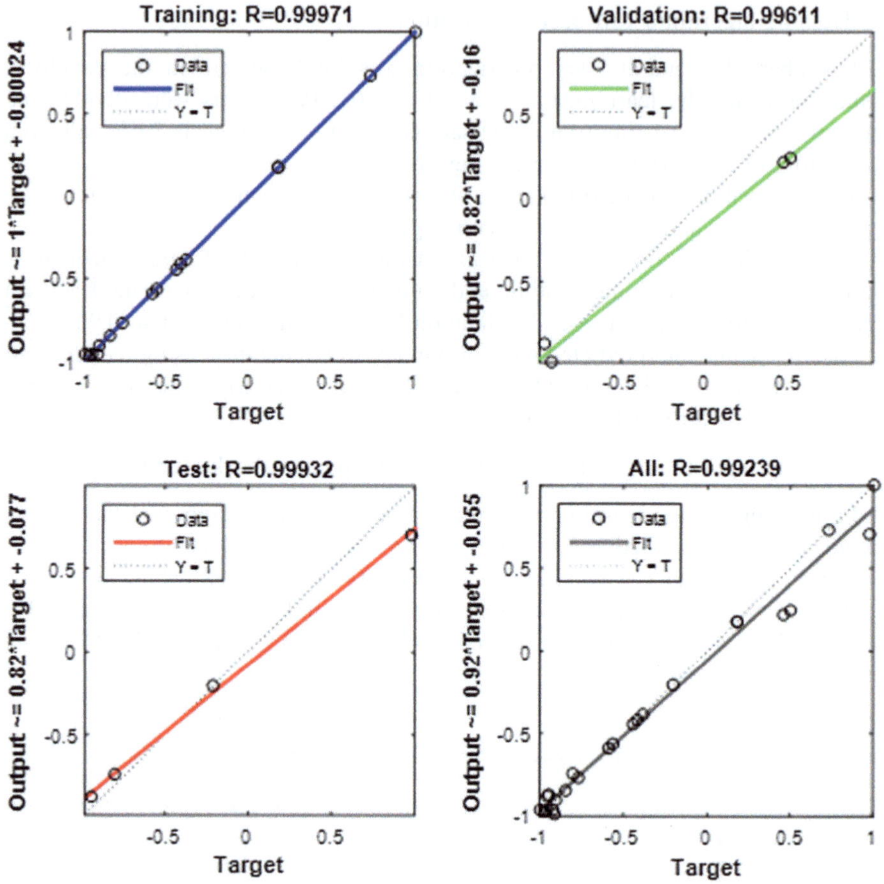

Fig. 5 ANN regression analysis

Figure 9 shows actual and predicted outputs of wind power generation using GDM back propagation. MAPE obtained is 28.75%.

3. Training algorithm: gradient descent back propagation

Gradient descent back propagation combines learning rate with momentum training. The momentum constant was taken as 0.8. The training algorithm took 115 iterations. Data division was random, i.e., 70% data were used for training, rest 15% for testing and validation. The learning rate was same as previous case. In Fig. 10, performance plot best validation is obtained at 0.07787 at epoch 98.

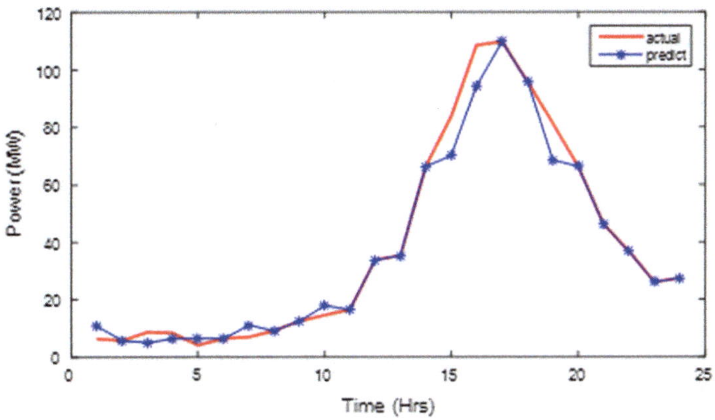

Fig. 6 Comparison of actual and predicted wind power using LM back propagation

Fig. 7 Performance plot

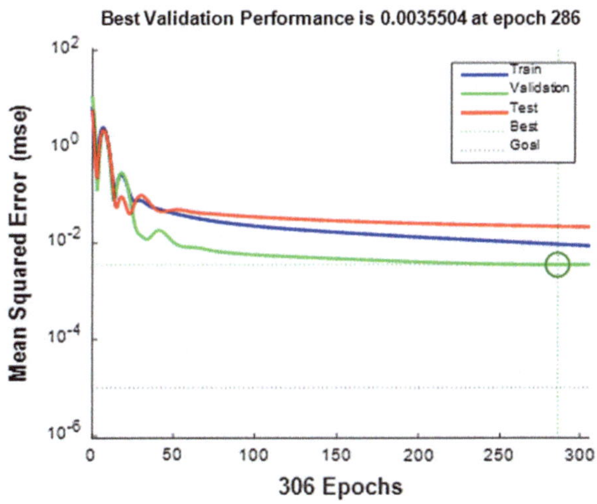

Regression plot is shown in Fig. 11. The testing and validation value for regression analysis is more than 0.95.

Figure 12 depicts actual and predicted results of wind power generation with GD back propagation. MAPE of above algorithm is 18.89%.

Table 1 indicates the comparison of various training algorithm.

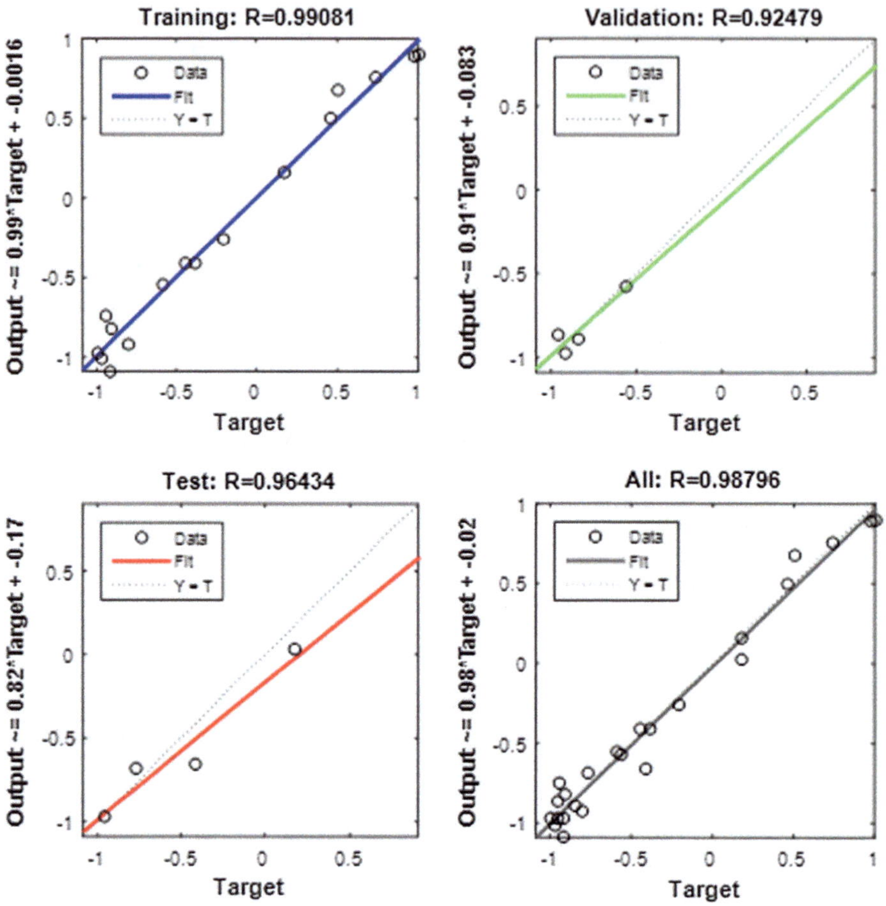

Fig. 8 ANN regression analysis

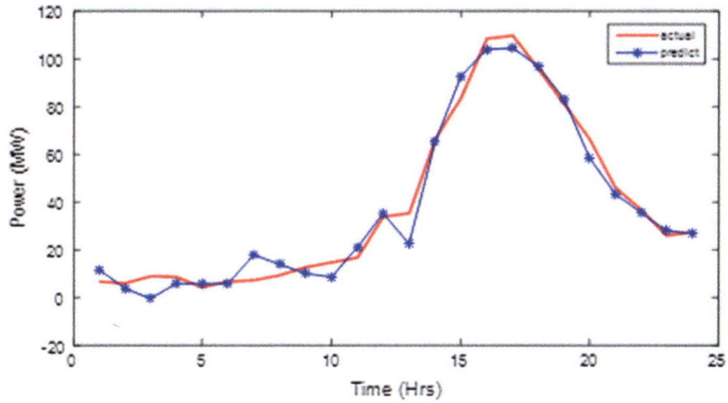

Fig. 9 Comparison of actual and predicted wind power using gradient descent with momentum back propagation

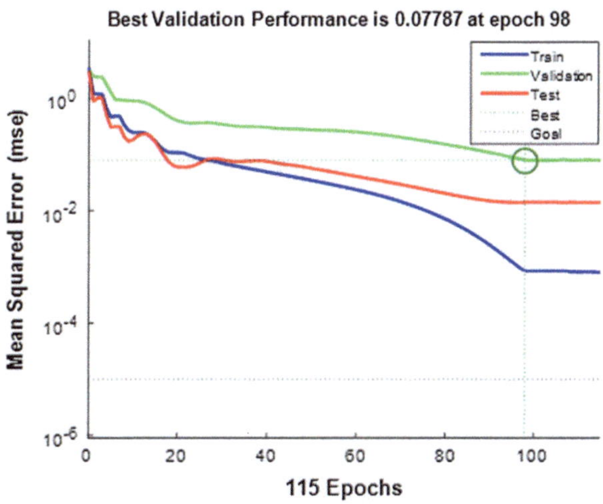

Fig. 10 Performance plot

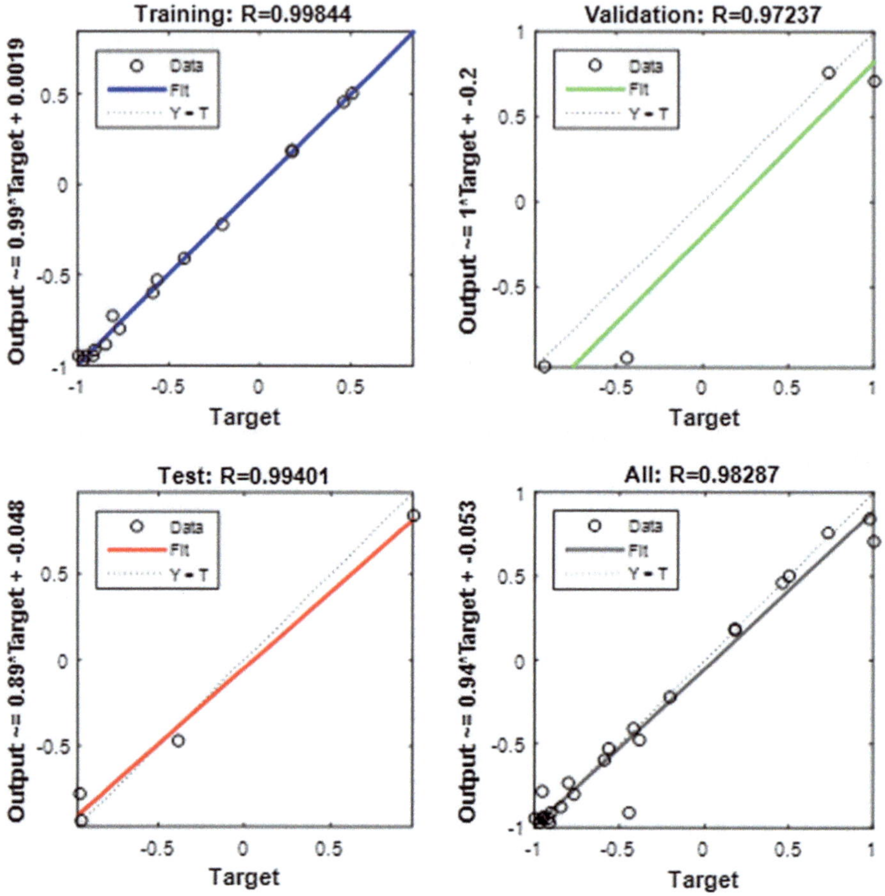

Fig. 11 Regression analysis

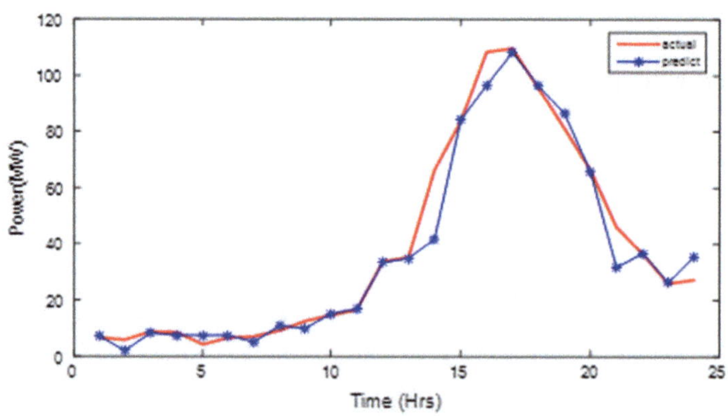

Fig. 12 Comparison of actual and predicted wind power using gradient descent back propagation

Table 1 Comparison of different training algorithm

Training algorithm	MAPE (%)
LM back propagation	13.08
Gradient descent back propagation	18.89
Gradient descent with momentum back propagation	28.75

5 Conclusion

The developed feed forward neural network is capable for providing day-ahead wind energy forecasting. The network is able to forecast on hourly wind power. From Table 1, it is seen that LM back propagation provides accurate forecasting with least deviation between actual and forecasted value. The developed model gives a desirable accuracy. The predicted results show that ANN is an excellent tool, and by using different algorithm, it can provide a better forecasting.

Acknowledgements The author would like to thank SLDC, GETCO for providing weather and wind power data for analysis.

References

1. Ahmed Ouammi, Hanane Dagdougui, Roberto Sacile, "Short Term of Wind Power by an Artificial Neural Network Approach" IEEE 2012.
2. Priyadarshini S, Booma J, Dhanarega A. J, Dhanlakshmi "A very short term wind power forecasting using back propagation algorithm in neural networks", Vol 10, No. 6, April 2015 APRN Journal of engineering and applied sciences.
3. Anirudh S. Shekhawat "Wind Power Forecasting using Artificial Neural Networks.", Vol. 3 Issue 4 April 2014 IJERT.
4. Introduction to Neural Networks using MATLAB 6.0 by S N Sivanandam, S Sumathi, S N Deepa.
5. K. G. Upadhyay, A.K Choudhary, M.M. Tripathi "Short term wind speed forecasting using feed forward Neural Network" Vol 3 No. 5 2011 pp 107–112.
6. Debra Lew, Michael Milligan, Gary Jordan and Richard Piwko "The value of Wind Power Forecasting", 2nd Conference on Weather, Climate and the New Energy Economy Jan 26 2011.
7. Vicky Jain, Alok Singh, Vibhor Chauhan and Alok Pandey "Analytical Study of Wind Power Prediction using Feed Forward Neural Network" ICCPEIC pp 303–306.
8. Neural Network Toolbox, MATHWORK Inc. Mark H Beale, Martin T. Hagan, Howard B. Demuth.

To Find Parameters of Transformer Using Open-Circuit and Short-Circuit Tests

Pooja N. Upadhyaya and Vijay H. Makwana

Abstract The transformer is one of the most important components in power system network. Because of its relatively simple construction, it is a highly reliable piece of equipment. Keeping design principle in mind the parameter of transformer was obtained by open circuit and short circuit tests, without actually loading the transformer, and these two assessed tests give the test results which are used to determine the equivalent circuit parameters. This paper presents the equivalent parameters from test data and significance H.V. or L.V. side in which the calculation to be performed, and then simulated same thing in MATLAB/Simulink. The comparison of simulated results and experimental results of open circuit and short circuit is shown.

Keywords Single-phase transformer · Simulation · Laboratory tests Parameters of transformer

1 Introduction

Transformers are the most vital component of power system. Transformer is a static device which transfers electric power from one circuit to another without changing its frequency [1].

To find out the parameters of transformer, we have to perform some experiments on single-phase transformer in laboratory [2]. Open-circuit test and short-circuit test are such experiments. These tests are performed in laboratory and are replicates via MATLAB/Simulink model. It has enhanced as well as complete understanding of transformer setup under changed condition [3].

P. N. Upadhyaya (✉) · V. H. Makwana
G.H. Patel College of Engineering & Technology, V.V. Nagar, India
e-mail: pooja.upadhyay007@gmail.com

V. H. Makwana
e-mail: vijaymakwana@gcet.ac.in

R. Kher et al. (eds.), *Proceedings of the International Conference on Intelligent Systems and Signal Processing*, Advances in Intelligent Systems and Computing 671,
https://doi.org/10.1007/978-981-10-6977-2_21

With offered Simulink model, we are clear with changed characteristic for transformer operation and we can easily calculate any required data, e.g., efficiency, regulation [1]. The MATLAB simulation as part of laboratory experiment improves laboratory experiment providing the opportunity to confirm results of laboratory experiment and compare them with those obtained by MATLAB simulations. The steady-state operation of the single-phase transformer, connected with the load we can calculate current and power, voltage regulation percentage and efficiency according to the load. This requires the knowledge of equivalent circuit obtained by open-circuit and short-circuit tests [4].

The MATLAB Simulink model of transformer is attempted by using linear transformer model available in Simulink which will work on the linear characteristic of transformer [3]. We have used readings for the no-load test and the short-circuit test.

2 Experimental Setups of Transformer Tests

The parameters of transformer are inspected using an equivalent circuit. The equivalent circuit of transformer referred to its primary side is shown in Fig. 1. In this circuit, R_p and X_p are the primary winding resistance and leakage reactance, respectively, R_s and X_s are the secondary winding resistance and leakage reactance, respectively, R_{o} stands for resistance core loss, and X_{o} stands for magnetizing reactance. N stands for transformer turn ratio [3].

The excitation current is very small compared to the load current of the transformer. In fact, it is very small that under normal condition it causes a completely negligible voltage drop in R_p and X_p [1]. So, the parameters of transformer are found from the open-circuit and short-circuit tests.

The SimPower system toolbox is valuable software to change the simulation model of power system application in MATLAB/Simulink. There are many disadvantages on test of transformer, like testing transformer, testing condition, safety,

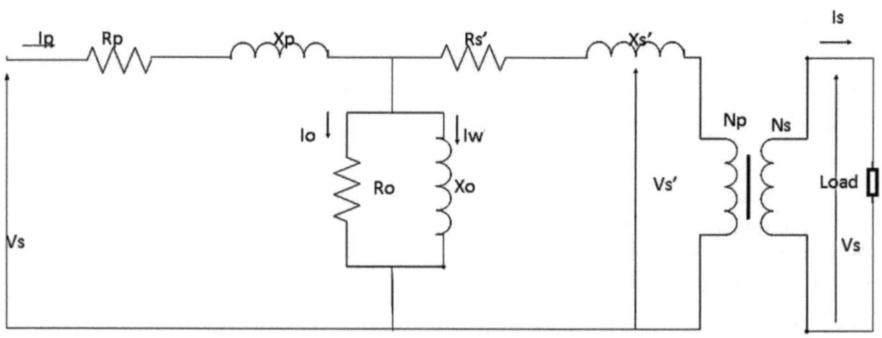

Fig. 1 Equivalent circuit of a transformer referred to primary

and many more. To overwhelm such short coming, we used this Simulink technology to decrease the testing condition [5].

2.1 Open-Circuit Test

As shown in Fig. 2, in open-circuit test the secondary of transformer is kept open circuited, and its primary winding is connected to a full-rated line voltage. The parameters are calculated from short-circuit test reading referred to primary side. The purpose of this investigation is to define no-load loss or core loss which is helpful in finding resistance core loss and magnetizing reactance [6].

When normal voltage is applied to the primary side of transformer, normal flux will be there in the core of transformer; hereafter, normal iron losses will be noted by wattmeter. As the primary no-load current is very minor, copper loss is negligible in primary side and zero in secondary side [1].

The equations used to find out the parameters of transformer are given below:

$$W = V_1 I_0 \cos \emptyset_0$$

$$I_\mu = I_0 \sin \emptyset_0$$

$$I_w = I_0 \cos \emptyset_0$$

$$X_0 = \frac{V_1}{I_\mu}$$

$$R_0 = \frac{V_1}{I_w}$$

The Simulink model of open-circuit test of transformer is shown in Fig. 3.

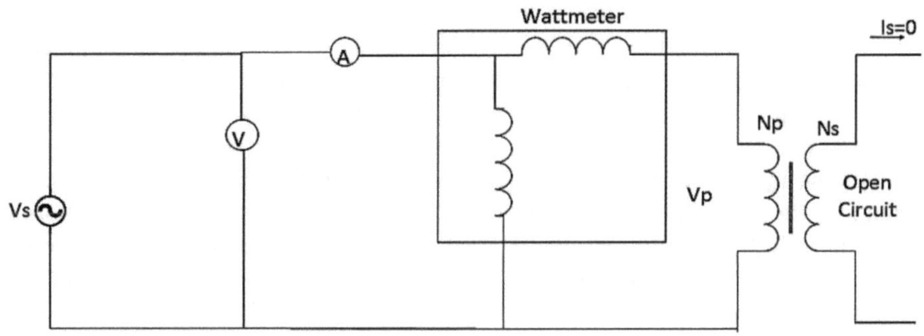

Fig. 2 Experimental setup of open-circuit test on transformer

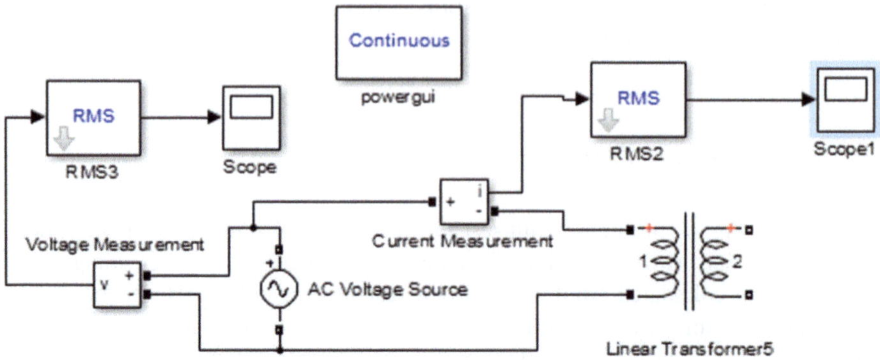

Fig. 3 Simulink implementation of open-circuit test of linear transformer

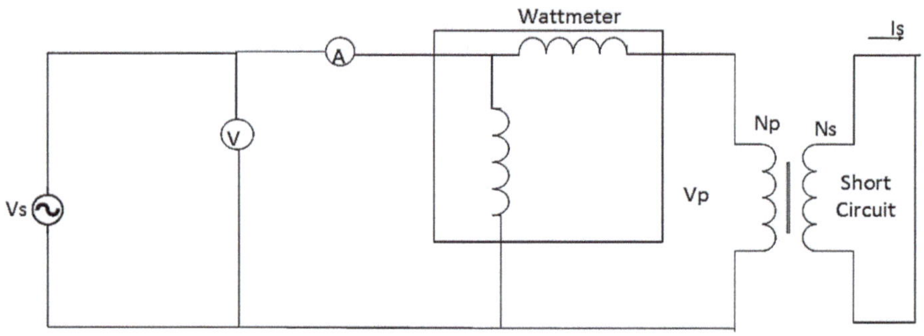

Fig. 4 Experimental setup of the short-circuit test on transformer

2.2 *Short-Circuit Test*

The short-circuit test is carried out to calculate the parameters of transformer. In this experiment, the secondary side of transformer is shorted and the primary side of transformer is connected to lower voltage side as shown in Fig. 4. The parameters are calculated from short-circuit test reading referred to secondary side. The input voltage is adjusted till the current in short-circuited winding is equal to its rated value [6]. The input voltage, current, and power are measured.

Since the input voltage is low at the time of SC test, excitation side has small current. If the excitation current is not taken into consideration, then the total voltage drop in the transformer can be credited to the series element in the circuits [1]. The magnitude of the series impedance raised to the primary side of the transformer is as follows:

$$Z_{01} = \frac{V_{sc}}{I_1}$$

$$W = I_1^2 R_{01}$$

$$R_{01} = \frac{W}{I_1}$$

$$X_{01} = \sqrt{(Z_{01}^2 - R_{01}^2)}$$

It is possible to determine the total series impedance raised to the primary side using this method. To split the R_{eq} into primary and secondary components, the dc test is achieved. In addition, the X_{eq} approximately can be split into the following:

$$X_P = N^2 X_s = \frac{X_{eq}}{2}$$

The Simulink model of short-circuit test of transformer is shown in Fig. 5.

Several measurement blocks are used in Simulink model like current measurement and voltage measurement are used to measure current, voltage, and power.

3 Experimental Setup for Transformer Test

To find out the parameters of transformer, we perform the open-circuit and short-circuit tests on transformer. In this method, the actual load is not used on transformer. But the equivalent circuit parameter of transformer is determined by conducting two tests on transformer.

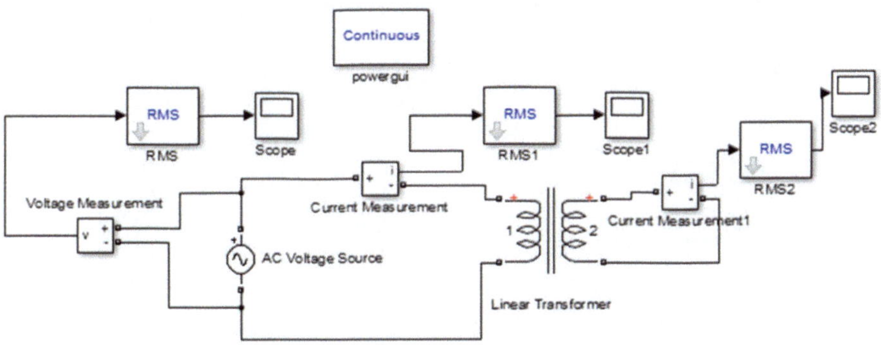

Fig. 5 Simulink implementation of short-circuit test of linear transformer

3.1 Setup for Open-Circuit Test

Procedure

1. Connect the circuit as per the setup shown in Fig. 6.
2. After getting it checked, switch on the supply.
3. Adjust the rated voltage input to the H.V. winding with the help of autotransformer (say 230 V).
4. Take the readings of wattmeter, ammeter, and voltmeter.
5. Switch off the supply.

3.2 Setup for Short-Circuit Test

Procedure

1. Connect the circuit as per the set shown in Fig. 7.
2. First of all, adjust the autotransformer setting for zero output voltage, and then switch on the supply after getting it checked.
3. Increase the input voltage slowly by the means of autotransformer, till rated full-load current circulates in both the windings.
4. Take the readings of all meters for full-load current value.
5. Switch off the supply.

Precautions

1. All the connection should be properly checked.
2. Check all the meters are on zero position; if not all, then set it to zero position.
3. Connection is done by above all the conditions.

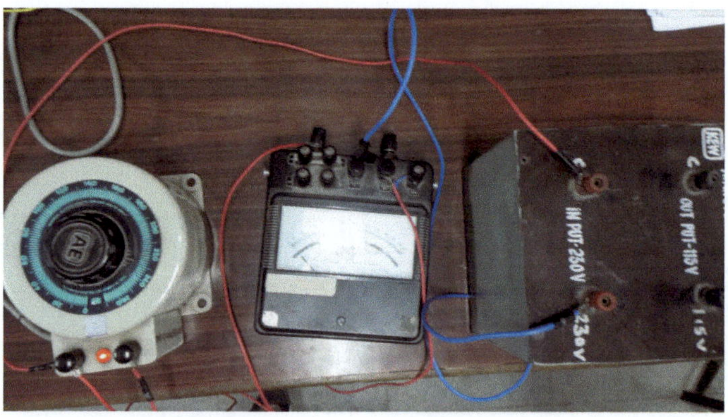

Fig. 6 Setup of open-circuit test

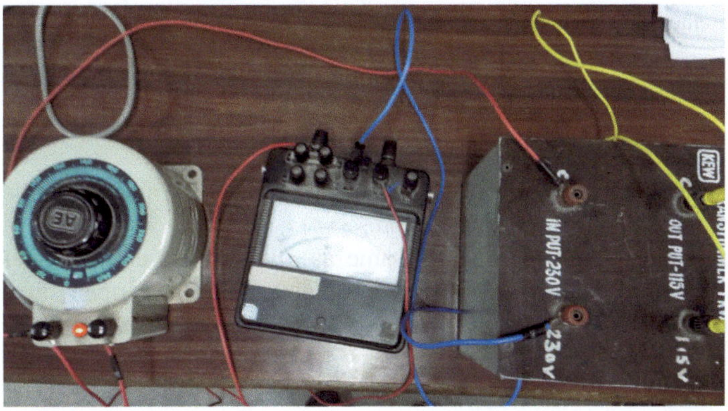

Fig. 7 Setup for short-circuit test

Table 1 Analysis result of open-circuit test

	Experimental results	Simulation results	% Error
V_0 (v)	230	230.17	0.7
I_0 (A)	0.25	0.245	2
P_0 (W)	25	24.494	2.024

Table 2 Analysis result of short-circuit test

Parameters	Experimental results	Simulation results	% Error
V_{sc} (V)	12.35	12.38	0.24
I_{sc} (A)	4.35	4.298	1.19
P_{sc} (W)	50	49.99	0.02

4. While performing SC test, care should be taken to see that autotransformer tapping positions are for zero output and then the input voltage should be increased slowly until the final load current is circulating in the windings.

4 Result Analysis and Discussion

For the transformer tested, the experimental setup of open-circuit and short-circuit tests were performed. The results are given in Tables 1 and 2, respectively. In Table 1, the comparison of open-circuit test between simulated result and experimental result is shown, and Table 2 represents the comparison of short-circuit test

Table 3 Parameters of transformer

R_1	L_1	R_m	L_m
0.04 p.u.	0.01 p.u.	40 p.u.	19.76 p.u.

between simulated result and experimental result. In Table 3, the equivalent parameters were obtained by calculation.

The simulation result was nearly same as obtained by experimental setup in laboratory. The percentage error is shown in Tables 1 and 2 of open-circuit test and short-circuit test, respectively.

4.1 Open-Circuit Test

By performing experiment of open-circuit test on transformer, Table 1 shows experimental result and simulated result.

4.2 Short-Circuit Test

By performing experiment of short-circuit test on transformer, Table 2 shows experimental result and simulated result.

To find out the parameters of transformer, we use the formulas given above. The parameters of transformer like resistance and inductance referred to primary side of transformer and the resistance core loss and magnetizing reactance obtained by mathematical calculation, after performing open-circuit and short-circuit tests on transformer. The result obtained is given in Table 3.

In general, the simulation models were helpful that MATLAB and Simulink can be easily used. Using simulation tool, one can easily find out the parameters and characteristics of transformers under varying condition.

Important Note

If primary is L.V. winding and secondary is H.V. winding, then the transformer is step-up transformer. In short-circuit test, supply is given to H.V. winding and L.V. is shorted. In such case, we connect meter on secondary side of transformer, i.e., H. V. side; for this purpose, the H.V. side acts as primary side of transformer. In such case, the parameters are calculated from the short-circuit test reading, which are referred to secondary side. So, before doing this calculation, it is necessary to find out whether the readings are recorded on transformer primary side or secondary side and accordingly the parameters are calculated.

In step-down transformer, high-voltage side itself is primary side to which supply is given in short-circuit test. So, in that case results of parameters are referred to primary side.

5 Conclusion and Future Work

This paper presented simulation models of transformer performed in MATLAB as well performed test in laboratory to find out the parameters of transformer. Parameters were obtained by performing test experiments in laboratory, and the obtained results were then used for the calculation to find out the other parameters of transformer.

The test performs open circuit, short circuit, presented in this paper, can also be used in the software called MATLAB which gives complete vision for different operation on a single-phase transformer.

In this paper, we have performed the simulation with linear transformer. Further, we can do the same simulation with saturable transformer considering B–H curve of transformer and different characteristics can be obtained.

Acknowledgements Special acknowledgments for G.H. Patel College of Engineering and Technology, Vallabh Vidyanagar, India, for letting us to use electrical machine laboratory which helped us to do experiments to find out the parameters of transformer.

Appendix

The rating of transformer used for this test has following specification:

1 kVA, 230/115 V single-phase transformer, 50 Hz.

References

1. Theraja, B. L., Theraja, A. K., Patel, U. A., Uppal, S. L., Panchal, J. C., Oza, B.,… & Patel, R. M. (2005). A Textbook of Electrical Technology Vol II. *Chand & Co., New Delhi.*
2. Ayasun, Saffet, and Chika O. Nwankpa. "Transformer tests using MATLAB/Simulink and their integration into undergraduate electric machinery courses." *Computer Applications in Engineering Education* 14.2 (2006): 142–150.
3. Laughton, Michael A., and Maurice George Say, eds. *Electrical engineer's reference book.* Elsevier, 2013.
4. A. C. Franklin, and D. P. Franklin, The J & P Transformer Books, 11th ed. New Delhi: Aditya Books, 1983.
5. Song, H., Zhao, F., & He, D. (2012). Simulation Study on Internal Fault of Transformer. *Physics Procedia*, *25*, 459–464.
6. M. G. Say, The Performance and Design of Alternating Current Machines, 3rd ed. NewDelhi: CBSPublishers, 1983.

Parametric Study of Virgin and Aged Transformer Oil

Nikita Rajda and R. R. Patel

Abstract Liquid dielectrics are deemed to be more useful because of their inherent properties. Amongst them, transformer oil is more widely used. It is expected to function as both, a dielectric and a coolant. Due to its interaction with various types of stress while in-service with the transformer oil undergoes consistent degradation. Apart from the degradation due to external factors, the oil also undergoes inherent degradation. In order to gain a broader insight, it is essential to study this inherent degradation. To analyze these characteristics has become even more essential due to the increasing utilization of high voltage. With this aim, this paper assesses the major parameters of the virgin and naturally aged transformer oil. To analyze the dielectric strength, AC breakdown voltage test has been conducted. Density, viscosity, flash, and fire point tests have been conducted over a range of temperature to gauge its physical traits. Weibull analysis has been employed for the statistical evaluation of the AC breakdown voltage data. The tests indicate that the dielectric breakdown strength shows major degradation with age, but the other tested parameters exhibit a minor deterioration.

Keywords Transformer oil · Virgin · Naturally aged · AC breakdown voltage Viscosity · Flash point · Fire point · Weibull analysis

1 Introduction

Transformer oil is typically a highly refined mineral oil that is stable at high temperature which has excellent properties to suppress arcing, serve as coolant and electrical insulation [1]. The transformer oil has an effect on the dielectric properties of paper and pressboard insulations as well. Thermally, the oil removes the heat generated during the operation of transformer by convection. Moreover, size and

N. Rajda · R. R. Patel (✉)
Department of Electrical Engineering, G.H. Patel College of Engineering & Technology, Vallabh Vidyanagar, 388120 Anand, Gujarat, India
e-mail: riteshpatel@gcet.ac.in

R. Kher et al. (eds.), *Proceedings of the International Conference on Intelligent Systems and Signal Processing*, Advances in Intelligent Systems and Computing 671,
https://doi.org/10.1007/978-981-10-6977-2_22

Fig. 1 Virgin and aged transformer oil

weight of the transformer and the current density of the transformer windings depend on the amount of oil and the rate of heat transfer [2].

Oil used in transformers is expected to have high dielectric strength, low viscosity, high flash, and fire point and is required to sustain these properties for a long duration of time. These are the essential electrical, physical, and thermal properties. It is not feasible and economical to replace the oil frequently. Hence, from the performance, safety as well as economic viewpoint, it becomes essential to study its major parameters as well as the changes in them with age.

Keeping these aspects in mind, two types of oil were tested—virgin oil and aged oil. The virgin oil used is TRANSOL GIS: IS 335 manufactured by Savita Oil Technologies Limited. The aged oil is 20 years old, and its manufacturer is unknown. It has undergone a natural degradation without ever coming into contact with a transformer. Figure 1 shows both the types of oil. Visually, it can be stated that the aged oil has undergone severe degradation. Both the types of oil have been sampled directly from the oil container.

2 Experimental Details

2.1 AC Breakdown Voltage Test (50 Hz)

The breakdown voltage is an indicator of the dielectric strength. This test was carried out as per IEC 60156 standard. The test was conducted in oil test kit manufactured by High Voltage, India. Figure 2 shows its schematic diagram. The test cell comprises of spherical brass electrodes at a standard gap of 2.5 mm. A total of three samples each of both the types were taken. Six tests were conducted on each oil sample at an interval of 2 min. The voltage was raised with the help of a knob at an interval of 2 kV/s. Before replacing any oil sample, the test cell was thoroughly cleaned. The mean value of the six consecutive tests was taken as the breakdown voltage strength of the respective sample.

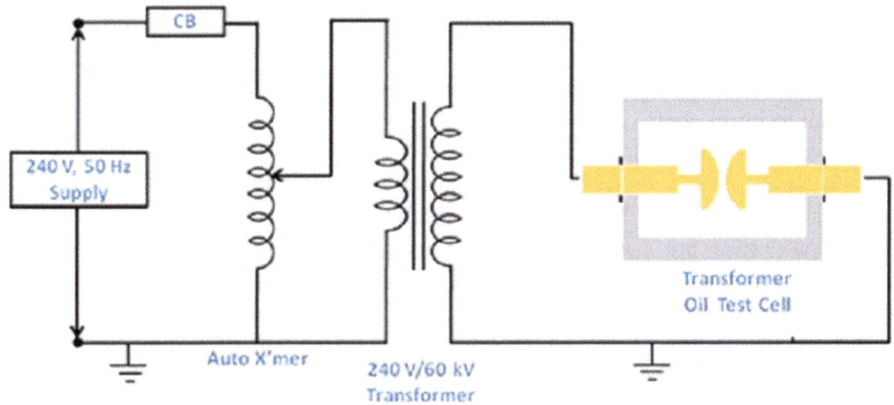

Fig. 2 Schematic diagram of the oil test kit

2.2 *Density Test*

The density of both the oil was measured with the help of 10 ml pycnometers manufactured by Borosil and a high precision weighing scale with four digits of precision manufactured by Mettler Toledo. The pycnometer is also known as specific gravity bottle. The density gives an indication of the suitability of the oil in terms of its thickness at a particular temperature.

2.3 *Viscosity*

Viscosity is an important physical property of oil. It has an influence on both the electrical properties and the heat transfer performance of mineral oil [3]. The kinematic viscosity was measured using Redwood Viscometer-1: Universal manufactured by Raj Instruments, India, as per ASTM D445:04 standard.

It is a capillary type viscometer. The type-1 viscometer is used for fluids having viscosity corresponding to Redwood seconds less than 2000. It makes use of the water bath to heat the oil. After the oil is heated, at a specified temperature, the time taken by the oil to fill 50 ml in a beaker is known as the Redwood second. It is converted to centimeter stokes using a specific formula.

The dynamic viscosity, which describes the force needed to surpass the internal friction between the oil layers, was also calculated using the results obtained from the density test and the kinematic viscosity test.

2.4 Measurement of Flash Point and Fire Point

Flash point and fire point can indicate contamination as well as potential fire hazard. It is vital to know their values since the transformer oil undergoes constant heating under working conditions. The flash point indicates the lowest temperature at which the oil vaporizes to form an ignitable mixture with air. The fire point is the lowest temperature beyond the flash point at which the vapor continues to burn for 5 s [4]. These properties are based on an empirical measurement than a fundamental physical parameter. The test for flash and fire points was conducted in a closed cup type Pensky–Martens apparatus manufactured by Indian Trading. This apparatus makes use of a brass test cup, which is fitted with cover. The oil sample is heated in this cup and stirred at specific rate. The flash point test has been conducted as per ASTM D93:02 standard.

3 Results and Discussion

3.1 AC Breakdown Tests

As can be seen in Table 1, the breakdown voltage value of the virgin oil is nearly 1.85 times that of the aged oil, which indicates that the oil has undergone inherent degradation with time without the effect of any external factors.

In order to obtain broader distribution of the breakdown voltage values at all the probabilities, Weibull analysis, a popular statistical method has been employed. It is appropriate especially for calculating the withstand voltages at low probabilities. Therefore, we use Weibull distribution to calculate the withstand voltages and obtain the shape parameter and scale parameter.

The cumulative distribution function of the two-parameter Weibull distribution is given as follows [5]:

$$F(x) = 1 - \exp[-(x/\alpha)]^{\beta} \tag{1}$$

α = scale parameter calculated at 63.2%,
β = shape parameter,
x = breakdown voltage value.

Table 1 AC breakdown voltage test results

Sr. No.	Virgin (kV)	Aged (kV)
Sample 1	32.5	16.84
Sample 2	29	15.5
Sample 3	29.67	17

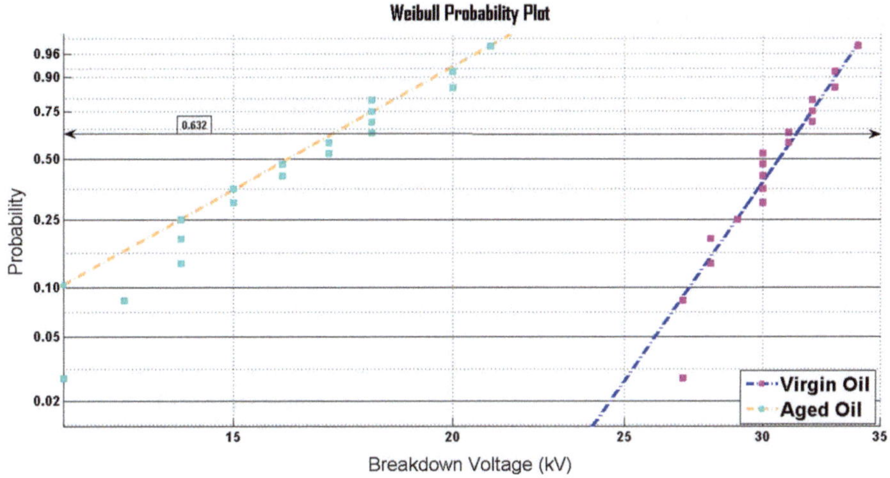

Fig. 3 Comparison of breakdown voltage of virgin and aged oil at Weibull probabilities

Fig. 4 Comparison of Weibull parameters

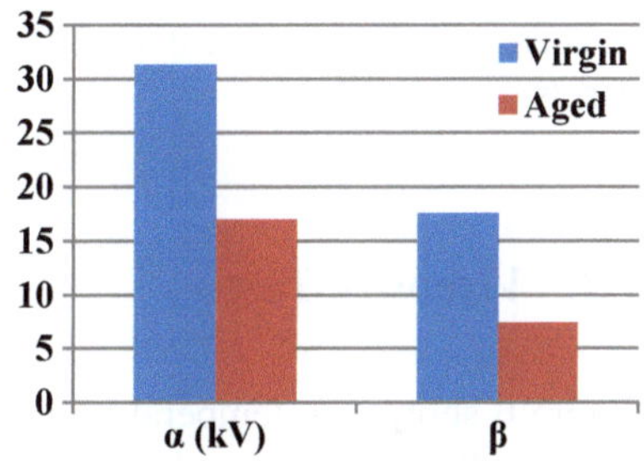

From Fig. 3, it can be seen that the Weibull analysis reaffirms that the aged oil has undergone severe degradation. The virgin oil exhibits superior performance at all the Weibull probabilities.

The shape parameter depicts the homogeneity of the results, which points to the scattering of the data. The virgin oil, as can be seen from the shape parameter depicted in Fig. 4, is more homogeneous in nature, as compared to the aged oil.

The scale parameter in Fig. 4 reaffirms the experimentally obtained values and displays the superiority of the virgin oil over the aged oil. It shows that the aged oil holds a value of 17.5 kV, while the virgin oil holds nearly double value of 31.3 kV

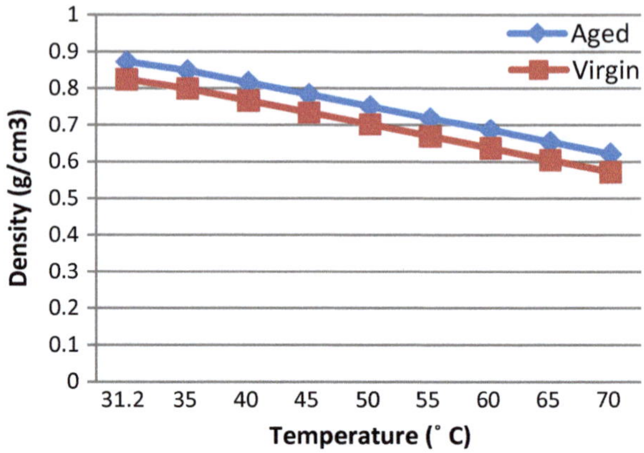

Fig. 5 Changes in density with temperature

3.2 *Density Test*

As can be seen in Fig. 5, there is a definite increase in density with age, which can be attributed to the oxidation of oil and the resulting by-products. Although, it can be seen that the increment in density is minor, which does not lead to disastrous effects.

3.3 *Viscosity Test*

Figure 6 shows the temperature performance of both the types of oil with time in terms of Redwood seconds. A clear nonlinear performance is obtained, indicating the non-Newtonian nature of the oil. The obtained kinematic viscosity of the aged oil is greater than the virgin oil as can be seen from Fig. 7. It is visible that with increase in temperature, nearly after the temperature of 45 °C, the gap between the viscosity values of both the oil decreases, thereby indicating a minor improvement of the viscous characteristics of the aged oil with temperature, which can be attributed to the evaporation of volatile impurities. Figure 8 denoting the dynamic viscosity shows that the virgin oil overcomes the internal friction between the oil layers in a superior manner compared to the aged oil.

3.4 *Flash Point and Fire Point Test*

Table 2 enlists the obtained empirical values for these tests. It can be seen that both the types of oil exhibit high flash and fire points, though the virgin oil is far superior. The minimum requirement of flash point of transformer oil is 140 °C [6].

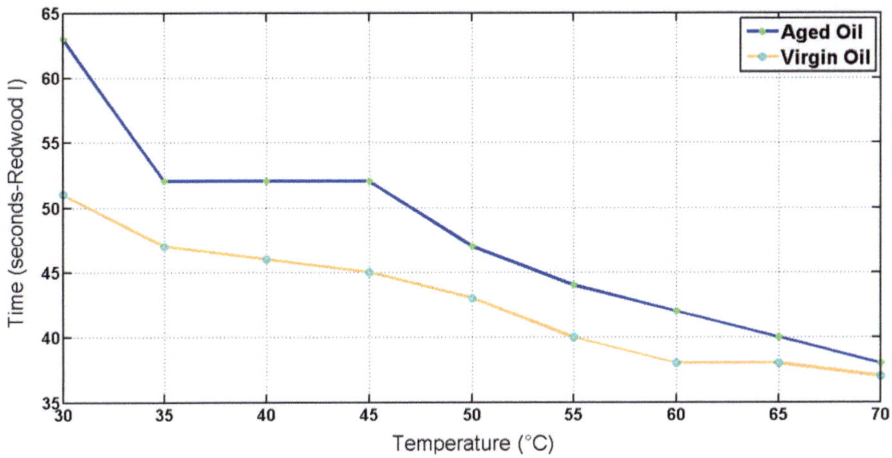

Fig. 6 Redwood seconds versus temperature

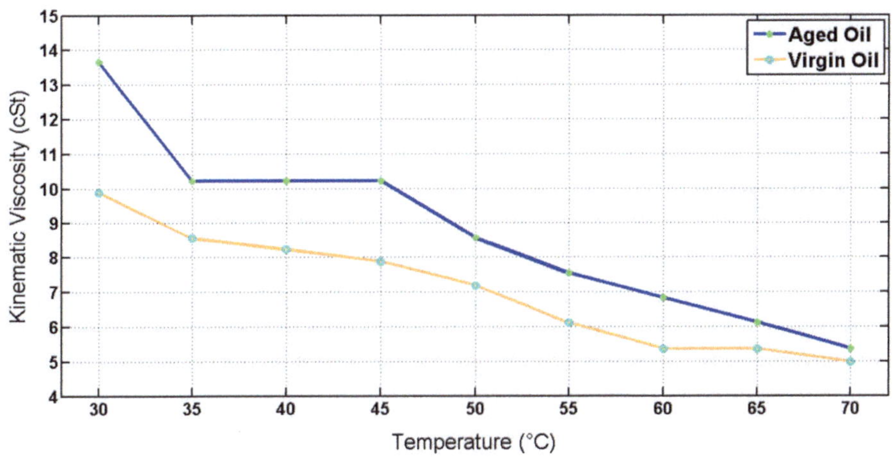

Fig. 7 Kinematic viscosity of oil samples

Even after aging, the 20 years old exhibits sufficient values, this can be attributed to the fact that this particular oil was naturally aged over a period of 20 years and has not been in contact with transformer. Hence, it is devoid of sludge and other such impurities.

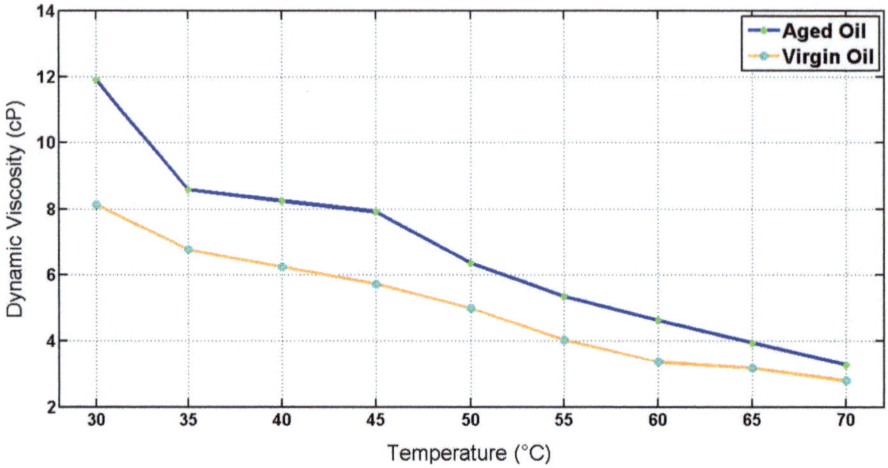

Fig. 8 Dynamic viscosity of oil samples

Table 2 Flash and fire point test results

Oil type	Flash point (°C)	Fire point (°C)
Virgin	164	192
Aged	144	158

4 Conclusions

From this work, it can be concluded that there is a definite inherent degradation of oil with aging. The breakdown voltage values at all probabilities are reduced to a great extent with age, as affirmed by Weibull analysis. Though the density and viscosity values have increased with age, the increment is quite minor and will not have any major impact. The flash and fire points exhibit a sufficient value even after a lengthy span of 20 years.

Thus, it can be inferred that despite the natural aging of transformer oil, the breakdown voltage value and hence the dielectric strength undergo major degradation, while the cooling properties deteriorate only slightly.

Acknowledgements The authors would like to thank Dr. Sunil Chaki of the Department of Physics, Sardar Patel University, Vallabh Vidyanagar for the invaluable inputs and providing the aged transformer oil.

References

1. R. Karthik,T. Sree Renga Raja, R. Madavan, "Enhancement of Critical Characteristics of Transformer Oil Using Nanomaterials", Journal of Arab Science and Engineering, Vol. 38, PP 2725–2733, 2013.
2. Pugazhendi Sugumaran.C, "Experimental Evaluation on Dielectric and Thermal Characteristics of Nano Filler Added Transformer Oil, IEEE International conference on high voltage engineering and application, pp. 207–210, September 2012.
3. Jin, H., T. Andritsch,P.H.F. Morshuis, J.J. Smit. "AC breakdown voltage and viscosity of mineral oil based SiO_2 nanofluids." Electrical Insulation and Dielectric Phenomena (CEIDP), 2012 Annual Report Conference on. IEEE, 2012.
4. Karthik, R., and T. Sree Renga Raja. "Investigations of transformer oil characteristics." IEEJ Transactions on Electrical and Electronic Engineering 7.4 (2012): 369–374.
5. Jian-quan, Zhou, Du Yue-fan, Chen Mu-tian, Li Chen-rong, Li Xiao-xin, Lv Yu-zhen, "AC and lightning breakdown strength of transformer oil modified by semiconducting nanoparticles." Electrical Insulation and Dielectric Phenomena (CEIDP), 2011 Annual Report Conference on. IEEE, 2011.
6. Bureau of Indian Standards, "New Insulating Oils – Specification (fourth revision)", IS 335:1993.

Optimal Placement of Capacitor Using Backward/Forward Sweep Method

Priyanka Parmar and Chintan Patel

Abstract In power system, distribution system is very important part for the consumer side. Consumer demand for the reactive power can cause low power factor, low voltage and high power loss. Simulation is carried out by backward/forward sweep load flow method which is used for the calculation of the voltage and branch current. To improve the voltage regulation, the capacitors are placed at the different nodes of the 15-bus standard radial distribution system. Shunt capacitors are used, to achieve power and energy loss reduction. The achievement of such benefits among other benefits depends greatly on how optimally these shunt capacitors are installed.

Keywords Backward/forward sweep method · Capacitor placement
Genetic algorithm · Power loss

1 Introduction

In distribution systems, the demand of the reactive power is very important. To fulfill that requirement, various objectives are considered. The appropriate placement of capacitor can maintain the voltage profiles by reducing losses. The power capacitor's aims are to generate the proper power supply and reactive power to the consumers and also to reduce the power loss and to improve the voltage regulation. In power system, there are various load flow methods such as Newton–Raphson load flow methods, Gauss–Seidel load flow method and also fast decoupled method, which are used only in transmission system. In distribution system, more preferable method is backward/forward sweep load flow method, because distribution system has high R/X ratio, radial structure of the network and also unbalanced conditions [2].

P. Parmar (✉) · C. Patel
Electrical Department, G H Patel College of Engineering and Technology,
V.V. Nagar Anand, India
e-mail: priyanka1618135@gmail.com

R. Kher et al. (eds.), *Proceedings of the International Conference on Intelligent Systems and Signal Processing*, Advances in Intelligent Systems and Computing 671,
https://doi.org/10.1007/978-981-10-6977-2_23

The backward/forward sweep load flow algorithm is better for the voltage and current calculations. One of the most commonly used methods for improving the voltage profile of the distribution network is the use of shunt capacitor bank with the feeders [4–7]. The size and locations of capacitor banks would have an important impact on the voltage profile enhancement. The objective functions thought in the work include reducing active power loss and improving voltage regulation.

2 Backward/Forward Sweep Method

During the last many years, a variety of approaches based on the forward/backward algorithm are widely used by researchers to get good convergence and fast calculation speed of voltage and current.

This algorithm has advantages of less computation requirements and calculation time compared to the Newton–Raphson method and Gauss–Seidel method.

The network is split into different layers starting from the base bus. This approach is a direct application of Kirchhoff's Current Law (KCL) and Kirchhoff's Voltage Law (KVL). For the calculations of the both sweeps the required data are, active and reactive power and positive sequence impedance model for all the branches. The major steps are included in backward sweep (Fig. 1):

1. Set flat voltage as 1 per unit for the reference of the first iteration.
2. Starting by end node, the branch current is calculated by using KCL using $I_{(i,0)} = (S_i/V_i)^*$
3. Voltage will be updated using previous rated voltage.
4. Same steps are used to calculate for the nodes.
5. Compute the voltage till the reference node.

The major steps are included in forward sweep:

1. By starting with the reference node at rated voltage.
2. Same procedure for calculating voltage and current is performed but in forward direction such as reference node to end nodes.
3. Similarly, the steps are performed till the end node.

By this method, the power loss is calculated. Using voltage and current, the complex power S_{ij} from bus i to bus j and S_{ji} from bus j to bus i is calculated using equation $S_{ij} = V_i I_{ij}^*$.

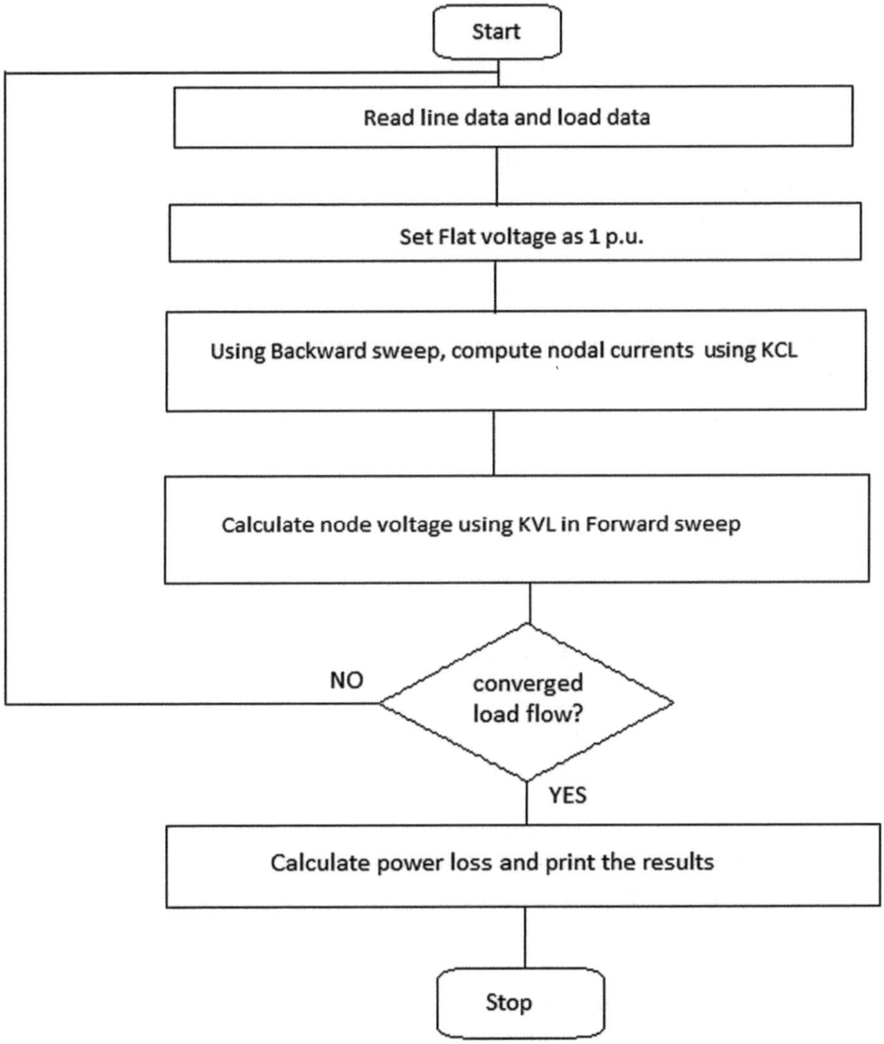

Fig. 1 Flow chart of backward/forward sweep method

3 Problem Formulation

The optimal capacitor placement model can be developed using various constraints [8–13].

1. Minimizing the power loss due to placement of shunt capacitor.
2. Improving the voltage profile.
3. Power flow constraints: The real and reactive power injections from the bus should be within limits.

$$P_{\min} \leq P_G \leq P_{\max}$$
$$Q_{\min} \leq Q_G \leq Q_{\max}$$

Here, the capacitor bank is to be assumed as fixed reactive power capacity Q_0.

Assume X_i is integer variable, which is used to represent the number of capacitor banks installed at ith bus;

The reactive power is

$$Q_{Cj} = X_i Q_0$$

Assume the T_C as the total number of capacitor bank in the given distribution system,

$$\sum_{i=0}^{n} X_i = T_C$$

where X_i is the integer variable.

4 Review on Genetic Algorithm

Normally, genetic algorithm uses the principle of natural evolutional and population genetic to arrive at the high quality of global solution. The traditional optimization techniques require one starting point, GA uses a set of points as the initial conditions. These points are called chromosomes.

Genetic operators are the stochastic transition rules applied to each chromosome during each generation procedure to generate a new improved population from an old population. Genetic operators consist of reproduction, crossover and mutation.

1. **Reproduction**: This process is probabilistic for selecting two parent strings from population of strings on basis of roulette wheel mechanism.
2. **Crossover**: This process is for selecting a random position in the string. This position is called crossover point.
3. **Mutation**: This process is for the random modification of string position by changing position of "0" and "1". Mutation can completely prevent the loss of genetic material through reproduction and crossover.

5 Simulation and Results

The proposed backward forward sweep algorithm is applied to standard 15-bus network. The line data and load data for this system are given in Appendix (Fig. 2).

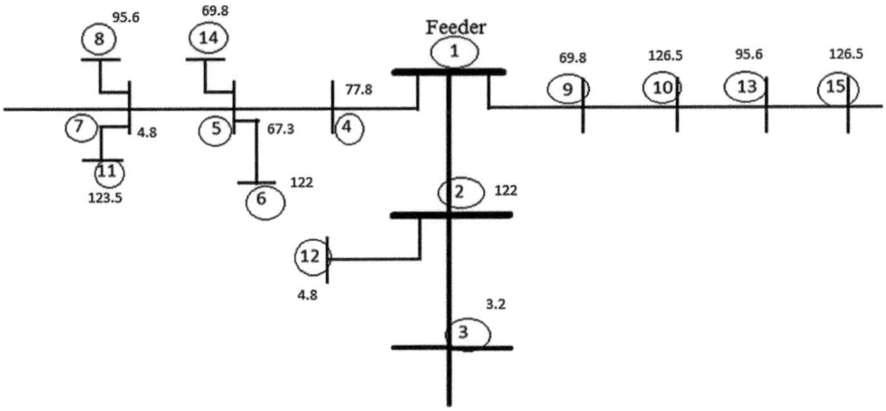

Fig. 2 15-bus standard radial distribution network

Table 1 Voltage magnitude and power loss at all the buses

No. of buses	Voltage (p.u.)	Loss (p.u.)
1	1	0
2	0.9051	0.0925
3	0.9022	0.0001
4	0.9633	0.018
5	0.8995	0.1556
6	0.8349	0.0295
7	0.8519	0.1195
8	0.8083	0.0226
9	0.9671	0.0145
10	0.9266	0.0664
11	0.6542	0.2257
12	0.9019	0.0001
13	0.8401	0.1448
14	0.8168	0.0241
15	0.7453	0.3168

The total active power is 1109.36 kW.

The total reactive power is 842.56 kvar.

The base kV is 11 and base kVA is 100 for the 15-bus radial distribution system.

The voltage and active power losses of 15-bus system are obtained by using the backward forward sweep method. The voltage magnitude at different nodes and power loss of this system are given in Table 1.

It shows the minimum voltages at bus 11, i.e. 0.6542 p.u., 15 i.e. 0.7453 p.u. and 8 i.e. 0.8083 p.u. of the system. The performance of the methodology is shown in Table 2.

Table 2 Performance of methodology

Total active power (kW)	Total active power loss (kW)	Minimum voltage (p.u.)
1109	123.09	0.6542

Table 3 Reduction of losses by adding capacitor banks

Optimal location (No. of bus)	Optimal size (kvar)	Reduction in power loss (%)
11	200	15.64
15	200	23.24
8	200	37.35
11,15	100,100	26.87
15,8	100,100	24.65
11,8	100,100	24.67
11,15,8	100,100,100	38.79

The minimum voltage at the buses can be improved by putting capacitor banks at various possibilities. Various combinations of using capacitors are included. Table 3 shows the capacitor placement locations and sizes. Thus, 200 kvar of capacitor bank is installed at 11th bus. Therefore, the power loss during that instant will be reduced by 15.64% of total power loss.

The elaborated results suggesting the optimal locations and number of shunt capacitor banks are depicted in Table 3. The capacitor banks are optimally allotted to various buses to maintain improved voltage profiles. Comparing the results, it can be observed that the system voltage profile is mostly improved due to capacitor placement. With increasing number of capacitor banks, the least bus voltage will be increased monotonously. Furthermore, it can also be observed from the results that installing capacitor banks at various adjacent buses is likely to provide a better system voltage profile and better reduction of power loss than installing them altogether at a single bus.

6 Conclusion

Using the backward/forward sweep load flow algorithm, the values of voltages at various 15 buses are defined. Voltage regulation of the 15 buses is less at the 11, 15, 8 buses. To improve the performance, capacitor banks are installed. Various sized capacitor banks are used. To improve at single bus 11, 15 and 8, the 200 kvar capacity of capacitor banks are used respectively. And for the better performance, the different two 100 kvar of capacitor banks are used at 11,15 and 15,8 and 11,8 buses respectively. It improved the voltage regulations and higher percentage of the loss reduction for the 15-bus standard radial distribution network. Moreover, capacitor banks could be deputed optimally to have a better voltage profile and reduction of losses.

Appendix

15-bus line data:

Sending end node	Receiving end node	Resistance (Ω)	Reactance (Ω)
1	2	0.0594	0.0226
2	3	0.0643	0.0342
1	4	0.0297	0.0232
1	5	0.0445	0.0125
5	6	0.0198	0.0432
1	7	0.1188	0.0404
7	8	0.0247	0.0273
1	9	0.0297	0.0232
1	10	0.0415	0.0221
7	11	0.148	0.0324
2	12	0.0396	0.0352
1	13	0.1584	0.0194
5	14	0.0495	0.0903
1	15	0.198	0.0226

Active and reactive load data:

Nodes	Active power (p.u.)	Reactive power (p.u.)
1	0	0
2	1.22	0.916
3	0.032	0.024
4	0.778	0.584
5	0.673	0.595
6	1.22	0.916
7	0.0488	0.0366
8	0.956	0.717
9	0.698	0.523
10	1.265	0.949
11	1.235	0.94
12	0.0488	0.0366
13	0.956	0.717
14	0.698	0.523
15	1.265	0.949

References

1. Chang, G. W., S. Y. Chu, and H. L. Wang. "An improved backward/forward sweep load flow algorithm for radial distribution systems." *IEEE Transactions on Power Systems* 22.2,2007.
2. A. D. Rana, J. B. Darji, Mosam Pandya, " Backward/Forward Sweep Load Flow Algorithm for Radial Distribution System", *IJSRD - International Journal for Scientific Research & Development|Vol. 2, Issue 01, 2014.*
3. J. A. Michline Rupa, S. Ganesh, "Power Flow Analysis for Radial Distribution System Using Backward/Forward Sweep Method", *International Journal of Electrical, Computer, Electronics and Communication Engineering Vol:8, No:10, 2014.*
4. Maurizo Delfanti, Gianpietro P. Granelli, Paolo Marannino and Mario Montagna, "Optimal Capacitor Placement Using Deterministic and Genetic Algorithms", *IEEE TRANSACTIONS ON POWER SYSTEMS, VOL. 15, NO. 3, AUGUST 2000.*
5. Mohammad A. S. Masoum, Marjan Ladjevardi, Akbar Jafarian, and Ewald F. Fuchs, *Fellow, IEEE,* "Optimal Placement, Replacement and Sizing of Capacitor Banks in Distorted Distribution Networks by Genetic Algorithms", *IEEE TRANSACTIONS ON POWER DELIVERY, VOL. 19, NO. 4, OCTOBER 2004.*
6. S. Sunisith, K. Meena, "Backward/Forward Sweep Based Distribution Load Flow Method", *International Electrical Engineering Journal (IEEJ) Vol. 5, No.9, 2014.*
7. Minnan Wang and Jin Zhong, "A Novel Method for Distributed Generation and Capacitor Optimal Placement considering Voltage Profiles", *978-1-4577-1002-5/11/2011 IEEE.*
8. Das, D. "Optimal placement of capacitors in radial distribution system using a Fuzzy-GA method". *International Journal of Electrical Power & Energy Systems* 30.6 (2008): 361–367.
9. Sultana, Sneha, and Provas Kumar Roy. "Optimal capacitor placement in radial distribution systems using teaching learning based optimization." *International Journal of Electrical Power & Energy Systems* 54 (2014): 387–398.
10. Reddy, K. Sravan Kumar, and M. Damodar Reddy. "Optimal placement of capacitor in distribution networks using fuzzy and SFLA." *Electrical, Electronics, Signals, Communication and Optimization (EESCO), 2015 International Conference on.* IEEE, 2015.
11. Chiang, H-D., et al. "Optimal capacitor placements in distribution systems. I. A new formulation and the overall problem." *IEEE Transactions on Power Delivery* 5.2 (1990): 634–642.
12. Hocaoğlu, Ulaş Eminoğlu M. Hakan. "Comparison of distribution Systems Power Flow Algorithms for Voltage Dependent Loads".
13. Gnanasekaran, N., et al. "Optimal placement of capacitors in radial distribution system using shark smell optimization algorithm." *Ain Shams Engineering Journal* 7.2 (2016): 907–916.

Improved Model of Modular Multilevel Converter

Rushikesh Chakubhai Kalariya and Mukesh Bhesaniya

Abstract Voltage sourced converter (VSC) exhibits a brilliant open door in different fields of the power system. Newly introduced modular multilevel converters (MMCs) are expected to replace two- and three-level VSC-based configurations for high-voltage direct current (HVDC) transmission systems due to their advantages in terms of performance, scalability, and efficiency. In this paper, construction, basic operation, and control strategies of MMC are discussed. In HVDC transmission systems, due to a large number of submodules in MMC, simulation time in electromagnetic transient (EMT) program is very large. Few researchers have addressed this problem. In this paper, improved model of MMC for simulation is presented. The simulation studies show that the proposed model exhibits high computational efficiency with good accuracy.

Keywords Modular multilevel converter (MMC) · Voltage sourced converter (VSC) · Computationally efficient model (CEM) · High-voltage direct current (HVDC)

1 Introduction

For medium- and high-voltage applications, such as voltage sourced converter-based high-voltage direct current (VSC-HVDC) transmission system, the modular multilevel converter (MMC) has provided best solution as compared to other multilevel VSC-HVDC system [1–6]. In comparison with other multilevel converter topologies, the renowned features of the MMC include (i) it has modular structure to provide any voltage-level requirements, (ii) higher efficiency of the MMC provides better solution for high-power applications, (iii) it has best har-

R. C. Kalariya (✉) · M. Bhesaniya
G.H. Patel College of Engineering and Technology, V.V. Nagar, Anand, Gujarat, India
e-mail: rushikalariya17@gmail.com

M. Bhesaniya
e-mail: mukeshbhesaniya@gcet.ac.in

R. Kher et al. (eds.), *Proceedings of the International Conference on Intelligent Systems and Signal Processing*, Advances in Intelligent Systems and Computing 671,
https://doi.org/10.1007/978-981-10-6977-2_24

monic performance for high-voltage applications, so the passive filter's size can be reduced, (iv) DC-link capacitors are not required or the size is very small in case of the MMC [4].

Toward addressing the technical challenges related to operation and control of the MMC, there has been a great effort over past few years. Voltage sourced converters are based on the insulated gate bipolar transistor (IGBT) combined with control strategies to gain high speed and independent control of active and reactive power by maintaining stable voltage and frequency [2]. For HVDC power transmission system, mainly two topologies are used: (i) two-level and (ii) multilevel neutral-point diode-clamped. Recently developed multilevel topology MMC connects numbers of submodules (SMs) in series to obtain required AC output voltage. MMC employs lower switching frequency to attain lower switching loss. In addition, filter requirement can be eliminated by using large numbers of SMs per phase.

The IGBT switches in MMC can either be modeled in detail, which give better accuracy or using equivalent models which give reduced accuracy but easy to implement in electromagnetic transient (EMT) program [1, 7]. The simulation time with the detailed model is very high; hence, such models are not used for the MMC with a large number of submodules. In [7], the continuous model is presented to reduce the simulation time. In this paper, PWM switching is considered to improve the accuracy of the model.

2 Modular Multilevel Topology

The schematic representation of the three-phase *N*-level MMC is shown in Fig. 1. As shown in Fig. 1, the MMC consists of two arms per leg phase where each consists of N-series-connected submodules (SMs) and a series-connected inductor L_0. The SMs of each arm are controlled to produce required AC phase voltage, and the arm inductor is used to suppress the high-frequency harmonic components of arm current.

2.1 MMC Submodules

As shown in Fig. 1, MMC topology containing series connections of SMs which comprise of a capacitor *C* and two IGBT switches called S1 and S2 (Fig. 2). During normal operation, any of the two IGBT switches S1 or S2 is ON.

There are various types of SMs which are used for MMC construction which are given as follow [4]:

- Half-Bridge circuit
- Full-Bridge circuit

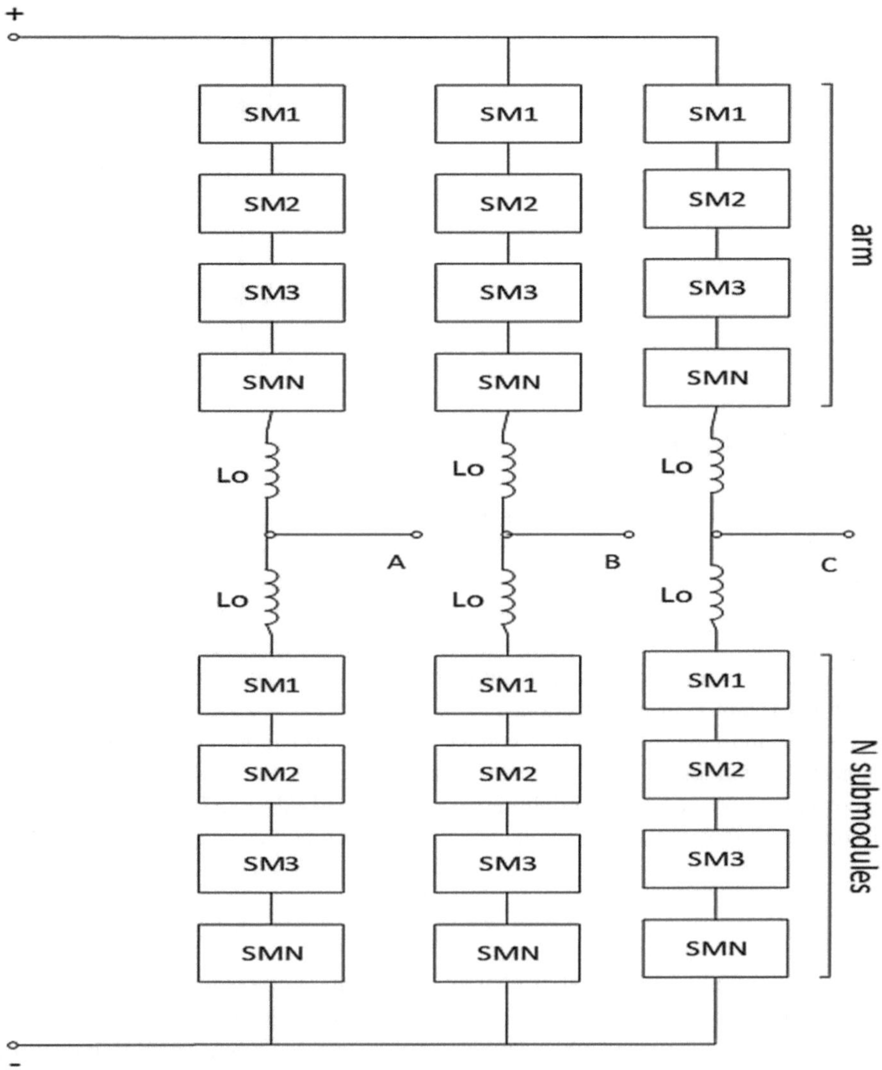

Fig. 1 Basic structure of MMC

- Clamp-double circuit
- Three-level FC (flying capacitor) circuit
- Three-level NPC (neutral-point-clamped) circuit
- Five-level cross-connected circuit

As the Half-Bridge SM has more advantages than other SMs structure which can be used in MMC construction so the Half-Bridge SM mainly used in the MMC construction.

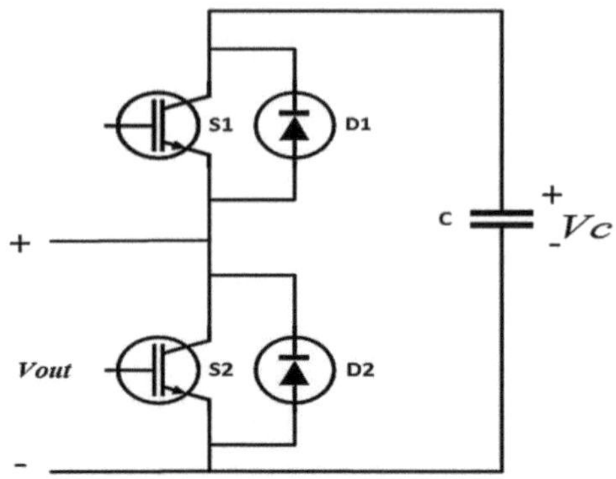

Fig. 2 Structure of submodule

2.2 *MMC Operation*

Figure 3 shows the construction of a single phase MMC leg. In the case of the MMC, *N* submodules are connected in series to form a leg of the converter. The AC system is connected to the midpoint of the leg, and each half of this leg is called arm. In the case of MMC, the energy of DC bus is not stored in a single large capacitor, but it is stored in the each submodules capacitor. For even voltage distribution of voltage across each submodule, various control techniques are used which is explained in next section.

To keep individual submodule voltage constant over time, the total numbers of connected submodules in a leg must remain same [1]. If *N* submodules per arm where all the upper arm submodules are connected and submodules of lower arm are disconnected, then the average voltage across the submodules capacitor is given as:

$$V_{Cx} = \frac{V_{\mathrm{DC}}}{N} \tag{1}$$

If a submodule is inserted from the lower arm, but none of the upper arm submodules is bypassed, there would be $N + 1$ submodules that are inserted in the system and their average voltage becomes $\frac{V_{\mathrm{DC}}}{N+1}$; so the submodules voltage will change. To prevent the change in the submodules voltage, the only combination of *n* submodules from the upper and *N-n* from the lower arm is connected to get desired output voltage [1].

3 MMC Control

As we know that the MMC is a type of voltage sourced converter, the MMC can be controlled by the control strategies used for the control of voltage sourced converter. As we know that the voltage is distributed across submodules and also the

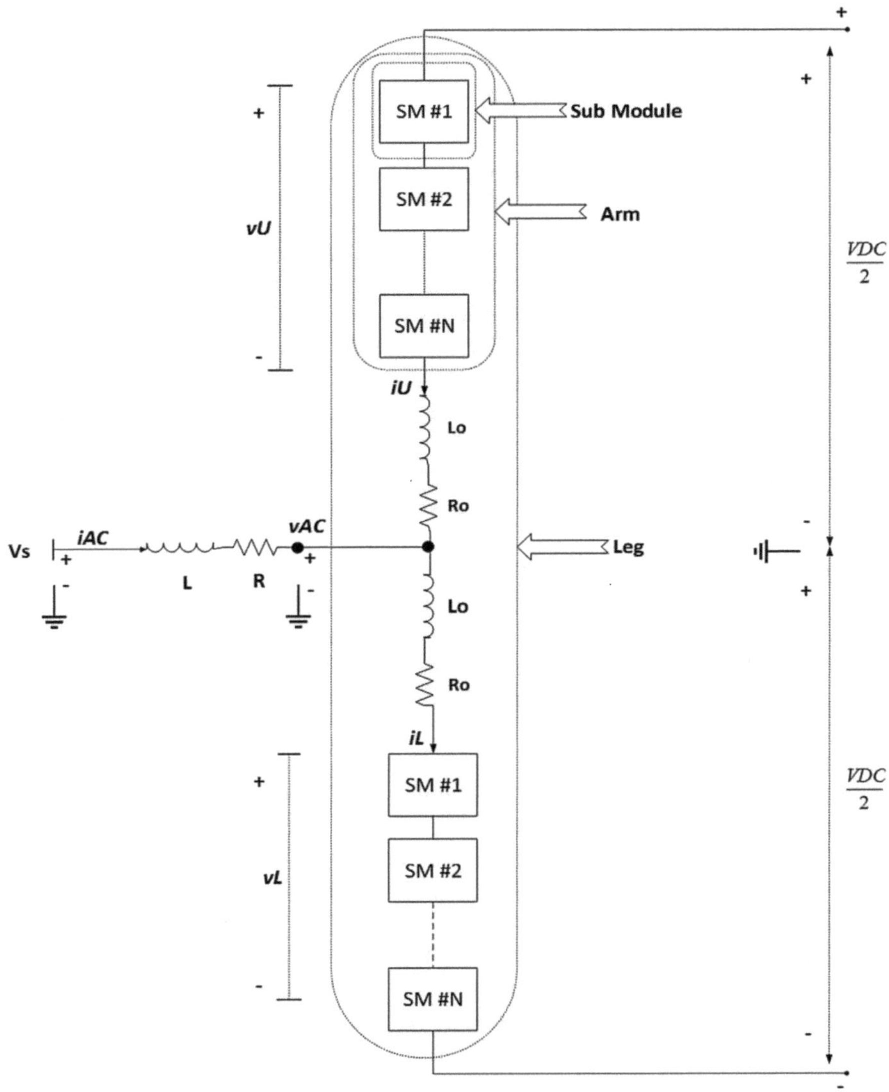

Fig. 3 Single phase MMC leg

inductor present in each phase of the MMC leads us to adopt new control for the MMC. As the voltage is stored in multiple capacitors of submodules instead of a single capacitor as in the voltage sourced converter. So the MMC has to be equipped with the special control to ensure that the voltage is shared equally among all the capacitor of the converter. As a result of unbalancing in capacitor voltage, it will produce the circulating current in the leg of MMC. So the controlling scheme of the MMC can be divided as follow [4]:

- Capacitor voltage balancing
- Switching strategy
- Circulating current control

3.1 Capacitor Voltage Balancing

In the MMC, the submodules are inserted and bypassed out of the system. For capacitor voltage balancing, the submodules are selected which are to be inserted or bypassed. Failure to balancing control results in the disturbance in the output voltage but also results in equipment damage if individual submodule voltage fluctuates outside of the rated values of equipment.

The change in submodule's voltage is dependent on its inserted/bypassed state as well as the magnitude and direction of the arm current of the MMC leg. When the SM is inserted, the voltage of capacitor increases (decreases) if the arm current is flowing into (out of) the SM. And if the SM is bypassed, the voltage of capacitor remains unchanged [4].

According to the state of the SM, the capacitor voltage balance is achieved. If the SM voltage is lower (higher) than the average voltage of arm, the SM should be inserted (bypassed). There are various methods to control capacitor voltage which is classified according to their sorting algorithms used for sorting of the SMs. The sorting method can be given as (i) bubble sorting, (ii) odd–even sorting, (iii) voltage balancing without sorting.

In this paper for the simulation of the MMC, the capacitor voltage balancing is done using bubble sort algorithm. In bubble sort algorithm, an array of unsorted numbers of the SMs is taken and then compared one by one, exchanging their order if the capacitor voltage of first SM has a higher value [1, 2].

Then, the same process is repeated until the array is completely sorted. For example, a small array of four numbers is taken [2, 4, 3, 1]. Table 1 shows the steps required to sort the array [8]. As noticed, the largest item is sorted after the completion of a single iteration. The sorting is said to be completed if an iteration does not produce any change in the order of an array.

3.2 Switching Strategy

The MMC control is designed to determine the number of the SMs to be inserted in each arm at any point of time [9]. The MMC gives the flexibility to decide how many SMs are to be inserted; it requires appropriate techniques. If the current direction is such that will raise the SMs voltage, the SMs should be inserted starting from the lowest voltage and up. If the current direction is reversed, the inserted SMs should be replaced by the highest voltage SMs.

Table 1 Bubble sort algorithm

Initial order	Description	Change in order
	Initial Order	[2, 4, 3, 1]
	First Iteration	
[2, 4, 3, 1]	Compare SM 1 & 2, no exchange	[2, 4, 3, 1]
[2, 4, 3, 1]	Compare SM 2 & 3, exchange	[2, 3, 4, 1]
[2,4,3,1]	Compare SM 3 & 4, exchange	[2, 3, 1, 4]
	Second Iteration	
[2, 3, 4, 1]	Compare SM 1 & 2, no exchange	[2, 3, 1, 4]
[2, 3, 1, 4]	Compare SM 2 & 3, exchange	[2, 1, 3, 4]
[2, 1, 3, 4]	Compare SM 3 & 4, no exchange	[2, 1, 3, 4]
	Third Iteration	
[2, 1, 3, 4]	Compare SM 1 & 2, exchange	[1, 2, 3, 4]
[1, 2, 3, 4]	Compare SM 2 & 3, no exchange	[1, 2, 3, 4]
[1, 2, 3, 4]	Compare SM 3 & 4, no exchange	[1, 2, 3, 4]
	Sorting Completed	

This technique has advantages such as low harmonic distortion and adequate voltage balancing. But it also has disadvantage such as high switching loss. To overcome the above-mentioned shortcomings, other techniques are used such as nearest-level voltage control, sinusoidal pulse width modulation (SPWM), phase shifted sinusoidal pulse width modulation (PS-SPWM), phase disposition sinusoidal pulse width modulation (PD-SPWM) [4].

For the simulation of MMC in this paper, the sinusoidal pulse width modulation (SPWM) is used for switching of SMs. The SPWM is one of the standard techniques used for control of voltage sourced converter (VSC). The SPWM is the simplest techniques used for VSCs. In SPWM, the modulating signal is compared with the carrier waveform which is a triangular wave with higher frequency than the modulating signal which is shown in Fig. 4. If a triangular wave has a value smaller than the required signal, then the resulting command is negative otherwise the command is positive. If the modulating signal has 60 Hz frequency and the triangular carrier signal has 1800 Hz frequency, the modulated signal has 30th harmonic which has magnitude near to fundamental frequency. But after filtering, the modulated signal will trace the modulating signal.

3.3 Circulating Current Control

In the MMC, the main concern is to control the circulating current which flows through the three-phase of the converter. This circulating current does not affect the AC-side voltages and currents. From the previous research, it has been proved that the circulating currents in the MMC are generated by the inner voltage difference

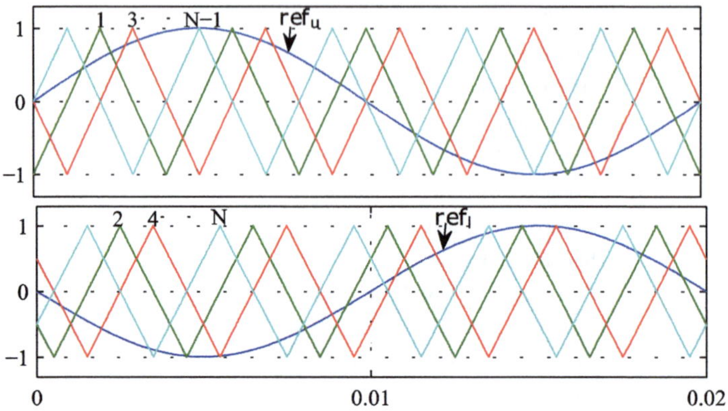

Fig. 4 Sinusoidal pulse width modulation

between the SMs in each phase [5]. The circulating currents are in form of negative sequence with the frequency doubled than the fundamental frequency. As known, the circulating current has no effect on the output voltage and current but it increases the rms value of the arm current which results in higher converter losses. It also increases the ratings of the SMs capacitors which will affect the design of the MMC.

To control the circulating current, various techniques have been suggested such as predictive control technique, open loop control technique, current suppression control technique, active resistance control technique.

The equation of the three-phase inner difference currents is given as below:

$$i_{\text{diff}\,a} = \frac{I_{\text{dc}}}{3} + I_{2f}\sin(2\omega_0 t + \varphi_0) \tag{2}$$

$$i_{\text{diff}\,b} = \frac{I_{\text{dc}}}{3} + I_{2f}\sin\left(2\omega_0 t + \varphi_0 + \frac{2\pi}{3}\right) \tag{3}$$

$$i_{\text{diff}\,c} = \frac{I_{\text{dc}}}{3} + I_{2f}\sin\left(2\omega_0 t + \varphi_0 - \frac{2\pi}{3}\right) \tag{4}$$

where I_{dc} is the total DC current, and I_{2f} is the peak value of the double line frequency circulating current. ω_0 is the fundamental frequency, and φ_0 is the initial phase angle.

From Eqs. (2) to (4), three-phase circulating current can be transformed to two DC components in double line frequency negative sequence rotating reference frame as below.

$$T_{abc/dq} = \frac{2}{3}\begin{bmatrix} \cos\theta & \cos(\theta - 2\pi/3) & \cos(\theta + 2\pi/3) \\ -\sin\theta & -\sin(\theta - 2\pi/3) & -\sin(\theta + 2\pi/3) \end{bmatrix} \tag{5}$$

where $\theta = 2\omega_0 t$. The phase sequence of the transformation is *a–c–b*.

The inner dynamic performance of the MMC is given by the below-mentioned equation.

$$u_{\text{diff}\,j} = L_0 \frac{\mathrm{d}i_{\text{diff}\,j}}{\mathrm{d}t} + R_0 i_{\text{diff}\,j} = \frac{1}{2}\left[U_{\text{dc}} - (u_{pj} + u_{nj})\right] \tag{6}$$

Rewrite Eq. (6) in the *a–c–b* sequence,

$$\begin{bmatrix} u_{\text{diff}\,a} \\ u_{\text{diff}\,c} \\ u_{\text{diff}\,b} \end{bmatrix} = L_0 \frac{\mathrm{d}}{\mathrm{d}t}\begin{bmatrix} i_{\text{diff}\,a} \\ i_{\text{diff}\,c} \\ i_{\text{diff}\,b} \end{bmatrix} + R_0 \begin{bmatrix} i_{\text{diff}\,a} \\ i_{\text{diff}\,c} \\ i_{\text{diff}\,b} \end{bmatrix} \tag{7}$$

Substitute Eqs. (2)–(3) into (7) and multiplying Eq. (5) gives,

$$\begin{bmatrix} u_{\text{diff}\,d} \\ u_{\text{diff}\,q} \end{bmatrix} = L_0 \frac{\mathrm{d}}{\mathrm{d}t}\begin{bmatrix} i_{2fd} \\ i_{2fq} \end{bmatrix} + \begin{bmatrix} 0 & -2\omega_0 L_0 \\ 2\omega_0 L_0 & 0 \end{bmatrix} * \begin{bmatrix} i_{2fd} \\ i_{2fq} \end{bmatrix} + R_0 \begin{bmatrix} i_{2fd} \\ i_{2fq} \end{bmatrix} \tag{8}$$

where $u_{\text{diff}\,d}$ and $u_{\text{diff}\,q}$ are the *dq* components of the inner unbalance voltage.

In this paper, the circulating current control is implemented which is suggested in the reference [4]. Figure 5 shows the structure of the circulating current suppressing controller (CCSC). In CCSC, the upper and lower arm currents are transformed into the double line frequency, negative sequence rotational reference frame. Then, the *dq* components of the circulating current are compared with their respective reference values which are set at zero to reduce the circulating current. The control signals are obtained using the PI controller with cross-coupling compensation. Then, these control signals are transformed to *abc* reference frame by doing inverse transformation.

4 Computationally Efficient Model

As the modules increase in the MMC, it increases the complexity as well as the simulation time. In the proposed model, the IGBTs are not modeled and the MMC behavior is represented using controlled voltage source [1]. Here, in the proposed model, the voltage to be inserted in the arm is derived using the total available capacitor voltages in the arm and the arm current. This reduces the computational complexity of the system model.

The relation between the total available capacitor voltages in the arm and the arm currents is derived by considering the total stored energy in the arms. The total

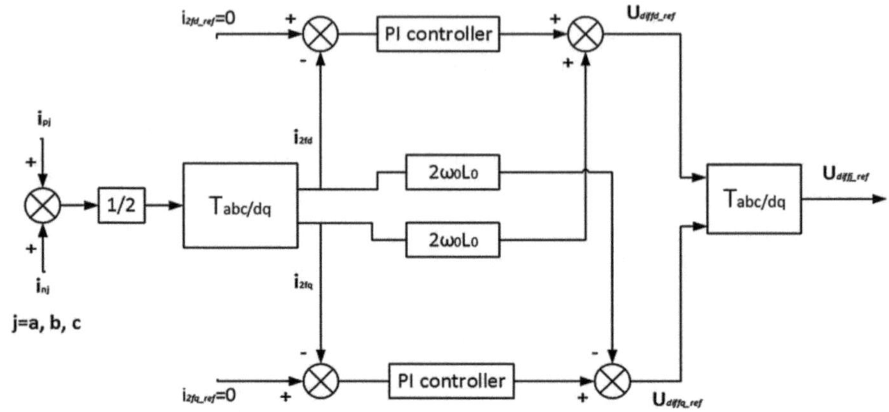

Fig. 5 Circulating current suppressing controller

energy stored in the upper and lower arm of the VSC-MMC with N-series-connected submodules per arm is given by,

$$E_{xu} = \sum_{j=1}^{N} \frac{1}{2} C_{SM} v_{Cxuj}^2 \tag{9}$$

$$E_{xl} = \sum_{j=1}^{N} \frac{1}{2} C_{SM} v_{Cxlj}^2 \tag{10}$$

The line derivative of (9) and (10) will give the instantaneous input power to the upper and lower arm, respectively.

$$P_{xu} = \frac{dE_{xu}}{dt} = C_{SM} \sum_{J=1}^{N} v_{Cxuj} \frac{dv_{Cxuj}}{dt} \tag{11}$$

$$P_{xl} = \frac{dE_{xl}}{dt} = C_{SM} \sum_{J=1}^{N} v_{Cxlj} \frac{dv_{Cxlj}}{dt} \tag{12}$$

If the submodule capacitor voltages in the arm are well-balanced, then the variation in the individual submodule capacitor voltages from the average value of the submodule capacitor voltages will be negligible. Hence, (11) and (12) can be revised as,

$$P_{xu} = C_{SM} \frac{v_{Cxu}^{T}}{N} \sum_{j=1}^{N} \frac{dv_{Cxuj}}{dt} = C_{SM} \frac{v_{Cxu}^{T}}{N} \frac{dv_{Cxu}^{T}}{dt} \tag{13}$$

$$P_{xl} = C_{SM} \frac{v_{Cxl}^{T}}{N} \sum_{j=1}^{N} \frac{dv_{Cxlj}}{dt} = C_{SM} \frac{v_{Cxl}^{T}}{N} \frac{dv_{Cxl}^{T}}{dt} \tag{14}$$

where v_{Cxu}^{T} and v_{Cxl}^{T} are the total capacitor voltages in upper and lower arm, respectively. $\frac{v_{Cxu}^{T}}{N}$ and $\frac{v_{Cxu}^{T}}{N}$ represent the average value of the submodule capacitor voltages in the respective arm. The instantaneous input power to the upper and lower arm can be written as follows,

$$P_{xu} = v_{Cxu} i_{xu} = i_{xu} N_{xu} \frac{v_{Cxu}^{T}}{N} \tag{15}$$

$$P_{xl} = v_{Cxl} i_{xl} = i_{xl} N_{xl} \frac{v_{Cxl}^{T}}{N} \tag{16}$$

By comparing (13) with (15) and (14) with (16), we get

$$\frac{dv_{Cxu}^{T}}{dt} = \frac{N_{xu} i_{xu}}{C_{SM}} \tag{17}$$

$$\frac{dv_{Cxl}^{T}}{dt} = \frac{N_{xl} i_{xl}}{C_{SM}} \tag{18}$$

By integrating (17) and (18), the total capacitor voltages in upper and lower arm are calculated from,

$$v_{Cxu}^{T} = \frac{1}{C_{SM}} \int N_{xu} v_{xu} dt \tag{19}$$

$$v_{Cxl}^{T} = \frac{1}{C_{SM}} \int N_{xl} v_{xl} dt \tag{20}$$

Finally, the inserted capacitor voltages in the upper and lower arms are calculated by,

$$v_{Cxu}^{c} = N_{xu} \frac{v_{Cxu}^{T}}{N} \tag{21}$$

$$v_{Cxl}^{c} = N_{xl} \frac{v_{Cxl}^{T}}{N} \tag{22}$$

From the (21) and (22), the schematic diagram of the modulation block, the generator of the controlled voltage signal, and the proposed model are shown in Figs. 6, 7 and 8.

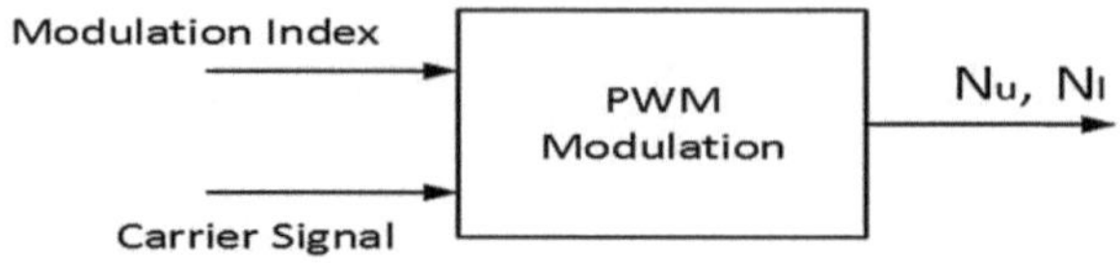

Fig. 6 Modulation block

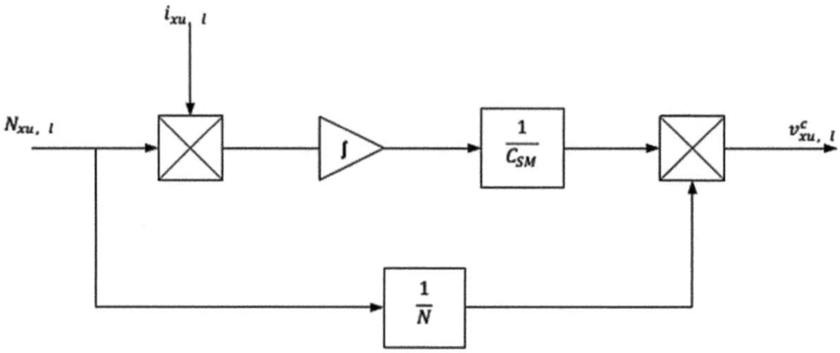

Fig. 7 Block diagram showing the generation of controlled voltage signals of the arm

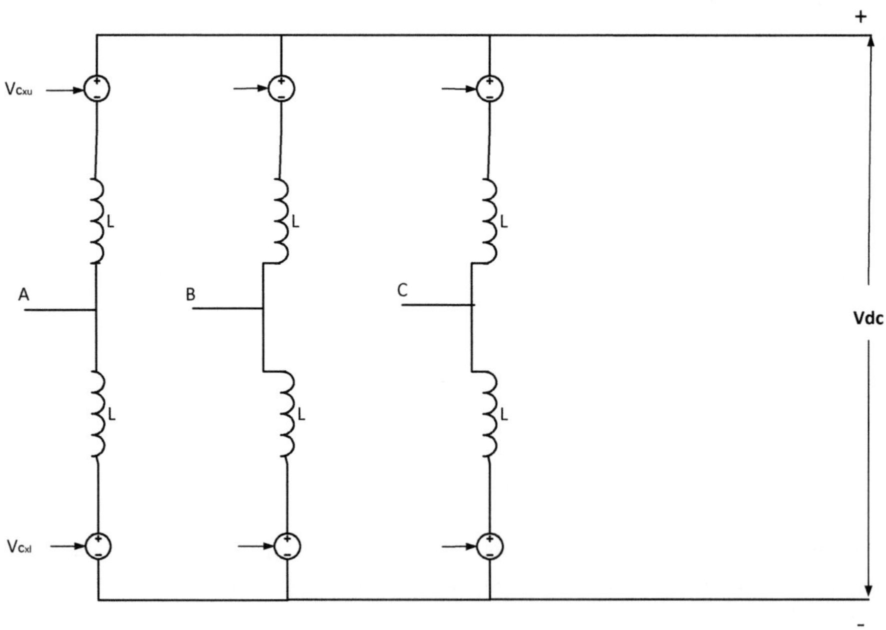

Fig. 8 Schematic diagram of the computationally efficient model

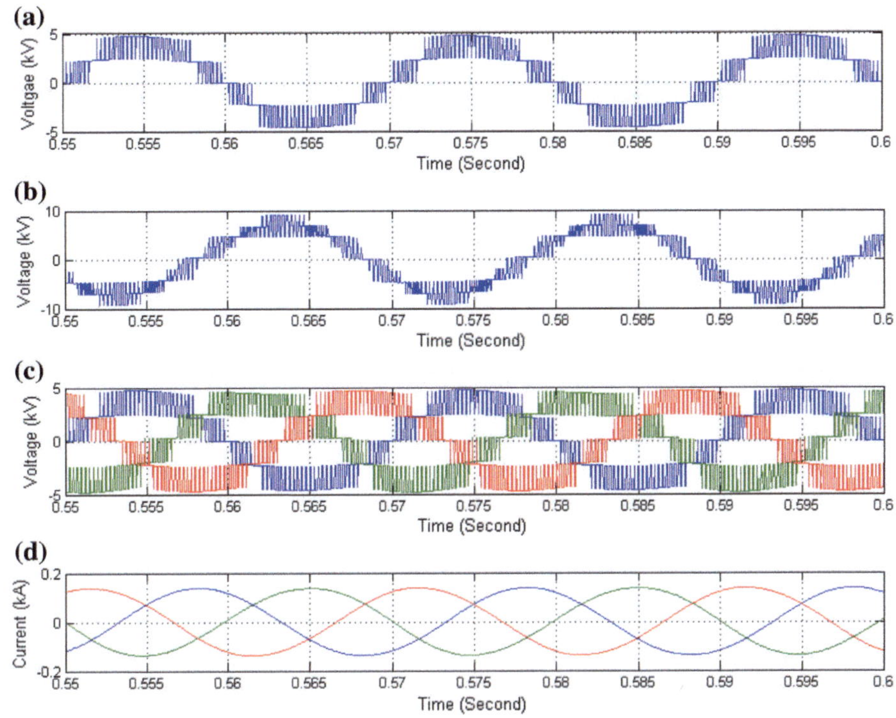

Fig. 9 Output results of five-level DM: **a** output voltage phase voltage, **b** line–line output voltage, **c** three-phase output voltage, **d** three-phase output current

5 Simulation and Results

The simulation of the five-level MMC was carried out using PSCAD/EMTDC, which consisted of 9 kV DC voltage, 1900 μF of submodule capacitance, 3 mH arm inductance with switching frequency of 1 kHz for the R-L load. For the circulating current control, the circulating current suppressing controller (CCSC) was implemented and the results obtained are shown below.

The multilevel output voltage waveform generated by the MMC is shown in Fig. 9. Figure 9a shows the waveform of output phase voltage which has a value of 5.2 kV. The line–line output voltage is shown in Fig. 9b with a value of 9.5 kV. The three-phase output voltage waveform is shown in Fig. 9c. As shown from Fig. 9, as the number of levels increased in the MMC, the output voltage waveform tends toward sinusoidal characteristics. Figure 9d shows the waveform of the three-phase output current of the MMC which has a value of 0.14 kA. From Fig. 9d, it is clear that the output current traces sinusoidal waveform. As discussed in Sect. 3.3, the CCSC was incorporated in the simulation which came into effect after 1 s. From Fig. 10a, it is clear that it contains higher amount of second-order harmonics which was reduced after the CCSC was incorporated. It also reduces the

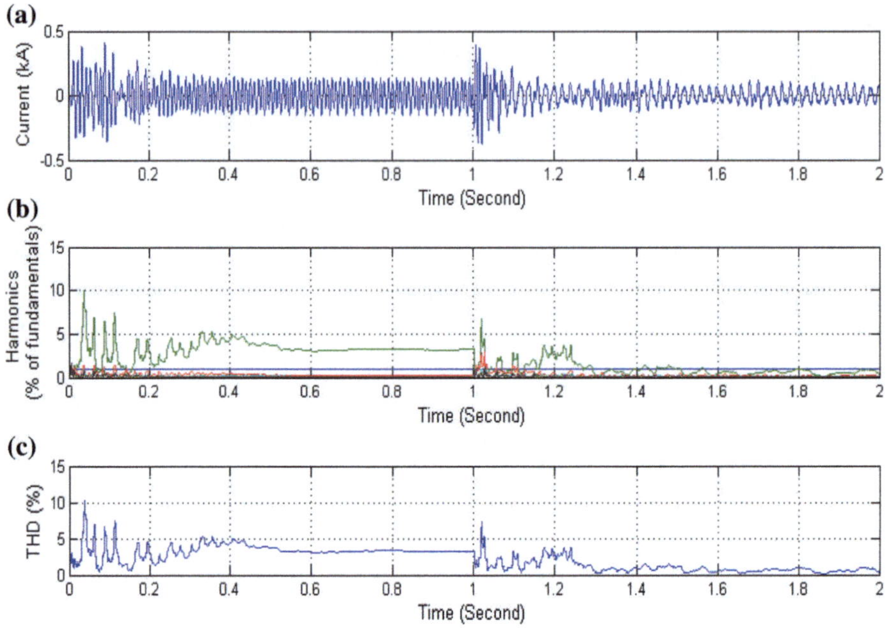

Fig. 10 Circulating current waveform with CCSC: **a** Circulating current of phase-*a*, **b** Individual harmonics of circulating current of phase-*a*, **c** THD of circulating current of phase-*a*

Table 2 Comparison of the computer time

Model	Computer time (s)	
	10	20
Five-level DM	304	607
11-level DM	1296	2570
Five-level CEM	46	81
11-level CEM	48	88

peak to peak value of the circulating current. Figure 10b shows the harmonics of the circulating current. It is clear from Fig. 10b that after incorporation of the CCSC, second-order harmonic was reduced to 0.33%. Figure 10c shows the THD of the circulating current which also reduced after the incorporation of the CCSC.

Another simulation of VSC-MMC is done by the computationally efficient method with PWM signals. This simulation is compared with the detailed simulation of the VSC-MMC. The simulation studies are performed on a 2.30 GHz Intel Core i5-4200U CPU with 8 GB RAM and Windows 8.1 PRO 64-bit OS running PSCAD professional V4.5.

The comparison of the computer time of five-level and 11-level detailed and computationally efficient model is carried out with 10 and 20 s of the simulation time with 20 μs solution time step. The comparison is shown in Table 2.

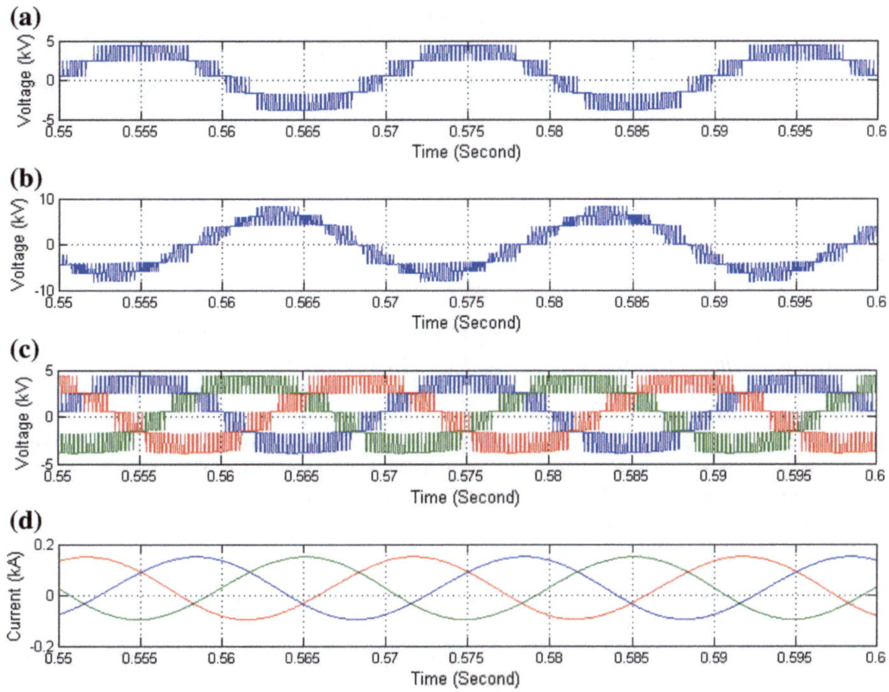

Fig. 11 Output results of five-level CEM: **a** output voltage phase voltage, **b** line–line output voltage, **c** three-phase output voltage, **d** three-phase output current

Figure 11 shows the output voltages and current waveforms of the computationally efficient model. After comparing Figs. 6 and 10, we can say that the computationally efficient model replicates the output voltage and current waveforms of the detailed model. After considering the speed of the simulation by the computationally efficient model, it is clear that the computational efficient model is very accurate and faster than the detailed model of the MMC. From Table 2, it is clear that the computationally efficient model boost up the simulation without loss of the accuracy.

6 Conclusion

In this paper, construction, basic operation, and control strategies of MMC are discussed. Moreover, an improved model of MMC by considering the PWM switching is presented which gives better performance than the continuous model of the MMC. The presented model gives the output results which are comparable to the actual output of the MMC with higher accuracy. The MMC model is analyzed which gives performance advantage over the conventional multilevel VSC-based converter. The main issue related to the MMC is the circulating current which is compensated by implementing the CCSC. The simulation results support the reduction in second-order harmonic in the circulating current. The computationally efficient model represents the faster option for the simulation of the VSC-MMC detailed model which is slower as compared to the other model.

References

1. Peralta, Jaime, et al. "Detailed and averaged models for a 401-level MMC–HVDC system." *IEEE Transactions on Power Delivery* 27.3 (2012): 1501–1508.
2. Lesnicar, A., and R. Marquardt. "A new modular voltage source inverter topology." *Proc. EPE*. Vol. 3. 2003.
3. Debnath, Suman, et al. "Operation, control, and applications of the modular multilevel converter: A review." *IEEE transactions on power electronics* 30.1 (2015): 37–53.
4. Tu, Qingrui, Zheng Xu, and Lie Xu. "Reduced switching-frequency modulation and circulating current suppression for modular multilevel converters." IEEE transactions on power delivery 26.3 (2011): 2009–2017.
5. Marquardt, Rainer. "Modular Multilevel Converter: An universal concept for HVDC-Networks and extended DC-Bus-applications." *Power Electronics Conference (IPEC), 2010 International. IEEE*, 2010.
6. Hagiwara, Makoto, and Hirofumi Akagi. "Control and experiment of pulsewidth-modulated modular multilevel converters." *IEEE Transactions on Power Electronics* 24.7 (2009): 1737–1746.
7. Ahmed, Noman, et al. "A computationally efficient continuous model for the modular multilevel converter." *IEEE Journal of Emerging and Selected Topics in Power Electronics* 2.4 (2014): 1139–1148.
8. Paradis, Dominic, Real-time simulation of Modular Multilevel Converters. Diss., University of Toronto, 2013.
9. Bhesaniya, Mukesh M., and Anshuman Shukla. "Current source modular multilevel converter: Detailed analysis and STATCOM application." *IEEE Transactions on Power Delivery* 31.1 (2016): 323–333.

Closed-Loop Control of Flow Under the Influence of Varying Temperature in Air Blower System Using PLC and VFD

Hiren G. Patel and Mihir Raval

Abstract The main objective of this paper is to design and implement the closed-loop control of mass flow rate in the air blower system when air is under the influence of varying temperature. As mass flow rate of air changes with the temperature, it can be maintained to desired value only by varying the blower speed suitably. For this purpose, we have used mainly programmable logic controller (PLC) and variable frequency drive (VFD). Based on the difference between actual and desired value of flow rate, PLC increases or decreases the speed of air blower operated by three-phase induction motor through VFD to maintain the desired air flow rate under the influence of varying temperature. Obtained results clearly indicate that desired value of flow rate is achieved nicely even under the influence of variation in the temperature.

Keywords Variable frequency drive (VFD) · Programmable logic controller (PLC) · Mass flow rate · Supervisory control and data acquisition (SCADA)

1 Introduction

It is necessary to maintain the mass flow rate accurately in pharmaceutical, petroleum, metallurgy, chemical, building materials, food, machinery, petroleum, and other industries. Recently, more accurate and reliable technics like magnetic field, positive displacement, true mass, and ultrasonics are available for flow measurement [1]. However even with this improved technology, true flow calculations need to take into account temperature, pressure, pipe diameter, and martial purity [2]. Air mass flow rate changes with the temperature. Hence to maintain air mass flow rate to the desire level under the influence of varying temperature is the difficult

H. G. Patel (✉)
Electrical Engineering Department, S. V. National Institute of Technology, Surat, India
e-mail: hpatel.svnit@gmail.com

M. Raval
NiTech Automation, Surat, Gujarat, India

R. Kher et al. (eds.), *Proceedings of the International Conference on Intelligent Systems and Signal Processing*, Advances in Intelligent Systems and Computing 671,
https://doi.org/10.1007/978-981-10-6977-2_25

problem. Here, programmable logic controller (PLC) and variable frequency drive (VFD) are used to maintain the air flow rate to the desired level under the influence of variation in the temperature.

PLC has been widely accepted in industries to control various quantities like pressure, speed, temperature, flow. In an automated system, PLC controller is usually the central part of a process control system. With execution of a program stored in program memory, PLC continuously monitors status of the system through signals from input devices, and based on the execution of logic, it produces the output signal. PLC has several known advantages including, flexibility, reliability, low power consumption, reliability, and ease of expandability [3]. Its flexible configuration will provide users with software tools to quickly build industrial automatic control system. VFD is widely used to vary the speed of three-phase induction motor in industries. It can vary the desired speed by changing the frequency using switching sequence of IGBT [4]. PLC can be programmed to give suitable command to VFD to control the speed of air blower system.

In the literature, authors have reported control of varies quantities using PLC, for example, supervisory control of electrical transmission line was discussed in [5] and dynamic flow controller was discussed in [6]. Temperature control system using fuzzy logic was proposed in [7], and constant pressure irrigation pump was implemented in [8]. But results are not available for the closed-loop control of air flow rate under the influence of varying temperature.

Hence, here we have implemented the closed-loop control of air flow rate using PLC and VFD. Section 2 discusses the design and implementation of air blower system. Control of mass flow rate is discussed in Sect. 3. Obtained results with the implementation are presented in Sect. 4, and concluding remarks are discussed in Sect. 5.

2 Design and Implementation

The air blower system is shown in Fig. 1a, and physically implemented air blower system is shown in Fig. 1b. It consists of three-phase induction motor which is connected with air blower [2]. Air blower blows the air in the pipe structure as shown in Fig. 1a and b. Heaters, sensors for temperature, flow, and pressure are also shown in Fig. 1a, b. Encoder and speed sensor are connected to motor shaft.

Flow rate of air can be easily varied by varying the speed of the three-phase induction motor. The speed of induction motor is directly proportional to the supply frequency and number of poles of motor [9]. Variable voltage variable frequency drive (VVVFD), loosely known as variable speed/frequency drive, varies the speed of the induction motor by changing the frequency and voltage using switching sequence of IGBT [10]. *Yashikava* make *J 1000* series VVVFD is used in the hardware implementation. Air mass flow rate in kg/h is measured using orifice type of flow sensor. Orifice flow sensor works on the principle of differential pressure. By measuring the change in pressure, mass flow can be calculated by flow

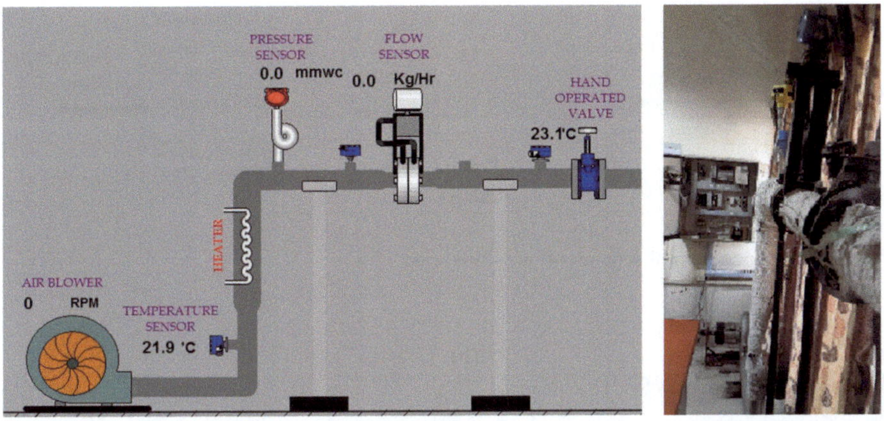

(a) Air Blower system

(b) Hardware implementation

Fig. 1 **a** Air blower system, **b** hardware implementation

calculating device, known as flow totalizer. Depending upon the mass flow rate, flow totalizer gives output in the range of 4–20 mA to the PLC.

Desired value of air flow rate is entered in the supervisory control and data acquisition (SCADA). Voltage across the heaters is controlled using proportionate power control solid state rarely (PPCSSR) and PLC. As the temperature changes, mass flow rate of air in pipe also changes. Thus, the closed-loop control of flow under the variation of temperature would be the difficult problem. But we have successfully implemented the same. Controlling of flow rate using PLC and VFD shall be discussed in the next section.

3 Control of the Mass Flow Rate

Consider the closed-loop control scheme as shown in Fig. 2 [2]. Desired value of the mass flow rate is provided to the controller through SCADA, and actual value of flow rate is provided to controller as a feedback signal. Based on the compression of the actual value of mass flow rate and set value of mass flow rate, error signal is generated. Error signal is used by controller to produce control signal based on the pre-defined logic [2, 11]. Control signal is provided to actuator, and actuator operates the final control element to minimize the error. Desired value of flow rate is achieved when error becomes zero.

Physically implemented control panel of air blower system is shown in Fig. 3a and b. Required value of flow rate is given though the SCADA. *Visual IO* make SCADA is used in the implementation. PLC receives the actual value of mass flow

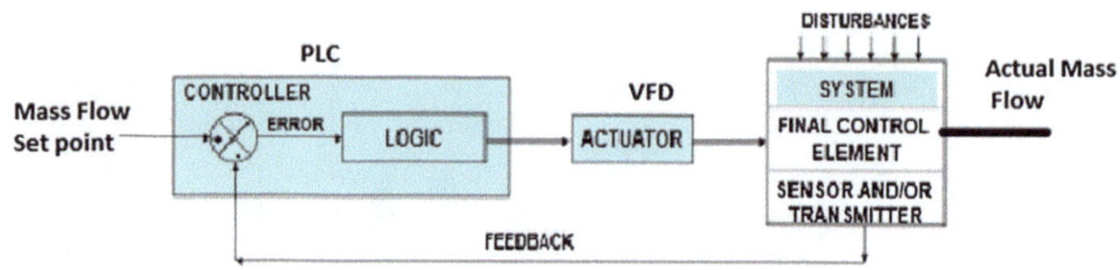

Fig. 2 Close loop control scheme of mass flow rate

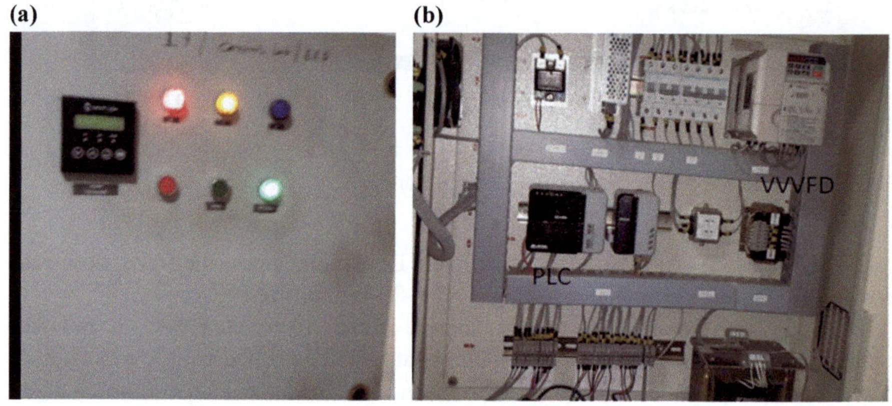

Fig. 3 **a** Control panel (outside), **b** implemented control panel (inside)

rate from the flow totalizer in the range of 4–20 mA. Based on the comparison of set value and actual value of mass flow, PLC produces error signal. Error signal is used by proportional integral derivative (PID) algorithm. PID algorithm is written in the PLC using ladder logic to make error signal low. Proportional gain *Kp*, integral time *Ti,* and derivative time *Td* of PID are varied to get desired value of the flow rate. Process of varying *Kp*, *Ti,* and *Td* to get desired response is known as PID tuning [12].

Hence, with the proper tuning of PID, error signal tends to becomes zero which leads to actual mass flow rate same as that of the set value of flow rate. Output of the PLC is the control signal in the range of 0–10 V. This control signal is given to the VFD, which makes three-phase induction motor to run in the range of 0–50 Hz, i.e., 0–3000 RPM corresponding to 0–10 V depending upon the PID action. Maximum mass flow of 100 kg/h can be achieved with the maximum motor speed of 3000 RPM and similarly 0 kg/h flow with 0 RPM speed of motor. Temperature of air is increased by increasing the % SSR firing. Higher the % SSR firing, higher would be the voltage across the heater and because of higher % SSR firing temperature of air increases quickly.

When the temperature of air increases, air becomes lighter so to maintain the desired value of the mass flow rate motor speed has to be increased. PLC is programmed to handle variation of this temperature effect as well. In a nutshell, mass flow rate, even with the variation of temperature, is maintained because of change in motor speed by the VFD depending upon PLC output signal.

4 Results

Table 1 describes the effect of variation of temperature on mass flow rate. The set value of the mass flow rate is 50 kg/h. Desired value of mass flow rate with output temperature of 24.94 °C is achieved with the motor speed of 1856 RPM. As the temperature of air increases, air becomes lighter so to maintain the desired value of the mass flow rate motor speed has to be increased. This can be easily verified from Table 1. It is clearly seen that to maintain the mass flow rate of 50 kg/h with the output temperature of 34.68 °C, the motor speed increases to 1906 RPM. Similarly, to maintain the mass flow rate of 50 kg/h with the output temperature of 51.20 °C, the motor speed increases to 1994 RPM, and to maintain the desired mass flow rate with the output temperature of 70.42 °C, the motor speed increases to 2098 RPM. Figure 4 is the snapshots from the SCADA and describes the effect of variation of temperature on mass flow rate. The set value of the mass flow rate is 50 kg/h. As described in Fig. 4, with proportional gain Kp, integral time Ti, and derivative time Td equal to 0.75, 5.0, 0, respectively, desired value of mass flow rate with output temperature of 70.42 °C is achieved with the motor speed of 2098 RPM.

The variation of actual value of mass flow rate can be easily seen in Fig. 5. It is clear that actual value of flow rate becomes equal to the desired value of flow rate after few cycles of initial transient. This transient can also be minimized with the proper tuning of PID. Thus, even with the variation in the temperature, mass flow rate is maintained with the corresponding increase in the motor speed, as seen in Fig. 5.

Table 1 Effect of variation in the temperature on mass flow rate

Set value of mass flow (kg/h)	Actual mass flow rate (kg/h)	Output temperature (°C)	Blower speed (RPM)
50	49.91	24.94	1856
50	49.74	34.68	1906
50	50.25	51.20	1994
50	49.36	70.42	2098

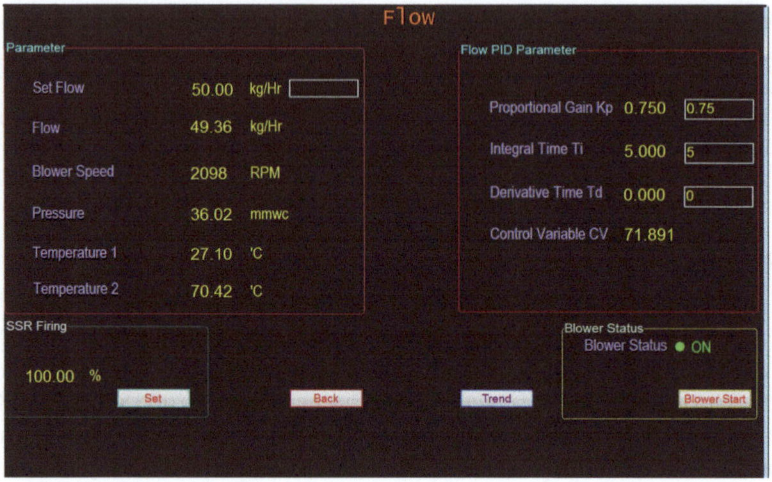

Fig. 4 Effect of variation in the temperature on mass flow rate

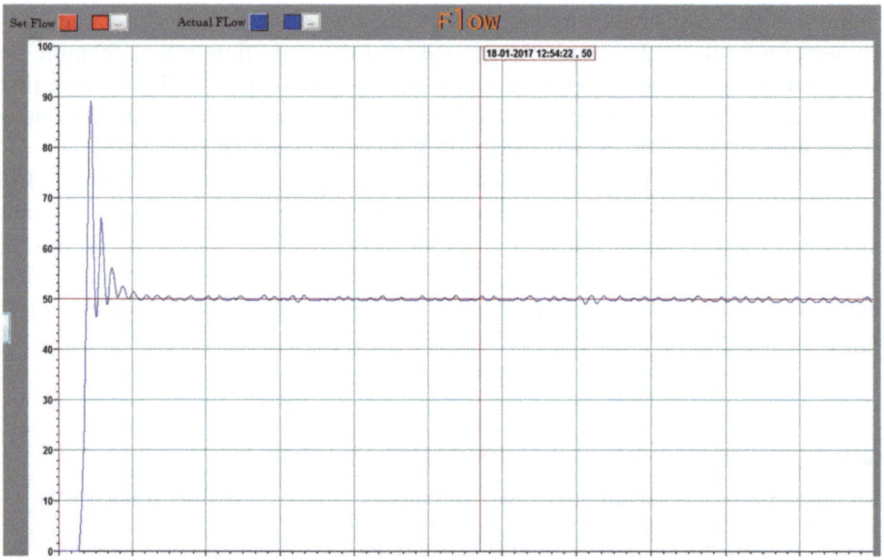

Fig. 5 Tread of mass flow under the influence of varying temperature

5 Conclusion

Close loop control scheme for mass flow rate was designed and implemented successfully using PLC and VFD. Based on the difference between actual and desired value of flow rate, PLC increases or decreases the speed of air blower through VFD to maintain the desired air mass flow rate. Results obtained from the implementation reveal that desired value of flow rate is achieved and maintained nicely even under the influence of varying temperature.

Acknowledgements Authors are thankful to the S.V. National Institute of Technology, Surat, Gujarat, for providing necessary financial support.

References

1. Singh, R. S., Sharma, A., Dubey, H.: PLC based PID implementation in process control of temorature flow and level. International Journal of Advanced Research in Engineering and Technology, 6(1), 19–26 (2015).
2. Patel, H., Raval. M.: Implementation of closed loop control of flow in air alower system using PLC and SCADA. Journal of Control & Instrumentation, 7(3), 39–43 (2016).
3. Webb, J. W., Reis, A.: Programmable Logic Controllers: Principles and Applications. Fifth Edition, PHI Ltd, India (2002).
4. Prakash, R., Singh, R., Kumar, R. R.: Variable Voltage Variable Frequency Speed Control of Induction Motor Using FPGA-Xilinx. International Research Journal of Engineering and Technology, 02(03), 1268–1273 (2015).
5. Prosser, J.: Supervisory control of electric power transmission networks. IEEE Transactions on Power System, 10(2), 1104–1110 (1995).
6. Afzalian, A., Niaki, S.A., Iravani, S. A., Wonham, W. M.: Discrete-event systems supervisory control for a dynamic flow controller. IEEE Transactions on Power Delivery, 24(1), 209–230 (2009).
7. Li, H., Yang, L.: The application of fuzzy control in PLC temperature control System Based on OPC Technology. Seventh International Symposium on Computational Intelligence and Design, pp. 288–291, China (2014).
8. Sungur, C., Çalişir, S., Kaya, S.: Developing an automation system for improving the energy efficiency of constant pressure irrigation pumps. Journal of Irrigation and Drainage Engineering, 142(11), 040160521–7 (2016).
9. Kailaswar, S. V., Keswani, S. V.: Speed control of three phase induction motor by V/f method for batching motion system. International Journal of Engineering Research and Applications, 3(2), 1732–1736 (2013).
10. Deepa, M.: Design of VFD drive for a 3-phase induction motor. International Journal of Innovative Research in Science, Engineering and Technology, 4(1), 18755–18762 (2015).
11. Nagrath, J., Gopal, M.: Control systems engineering. 5th Edition, New Age International Publishers, India (2009).
12. Ogata, K.: Discrete - Time control systems. 2nd Edition, PHI Ltd, India (2005).

Harmonic Analysis and Simulation of a Chemical Plant—A Case Study

Vidhi D. Patel and Sumit K. Rathor

Abstract Numerous kinds of electrical plants exist with different types of load into the system, the reliability of any industrial plant is key important issue, and operation of offender load normally affects either the performance of the system or the whole system that gets collapsed. Harmonics are one of the issues in power system and nowadays needs more focus because of a number of electrical devices exist which generate huge amount of harmonics. Arc furnace is one that considers to be the worst nonlinear load existing in power system, followed by variable frequency drive (VFD), power electronics converter, etc. To run the motors as per requirements, VFD is commonly used facilities in the plant; however, the operation of these facilities presents violent and nonlinear characteristics, which may cause power quality issues. In this study, the problem of harmonics among all power quality issues is the main concern that occurs in power system. For harmonic analysis of 11 kV system of source, transformer and VFD model without filter are modeled and implemented on PsCAD/EMTDC software. FFT analysis at point of common coupling (PCC) shows the harmonic pattern with THD and IHD levels.

Keywords VFD · PCC · Harmonics · Total harmonic distortion
FFT analysis

1 Introduction

At present to make power system and its control efficient, the power electronics-based technology increases frequently. For this reason, in industries the quantity of nonlinear loads is being added to electrical system, and hence, the power quality issues raise in the power system [1]. In this paper, among all power quality issues main concern is harmonics. An attempt has been made in this paper to study

V. D. Patel (✉) · S. K. Rathor
Department of Electrical Engineering, G.H. Patel College of Engineering and Technology, Anand, India
e-mail: patelvidhi3752@gmail.com

R. Kher et al. (eds.), *Proceedings of the International Conference on Intelligent Systems and Signal Processing*, Advances in Intelligent Systems and Computing 671,
https://doi.org/10.1007/978-981-10-6977-2_26

the level, characteristics, and effect of harmonic distortion in chemical plant, which has a good number of power electronics converter like variable frequency drives, arc furnaces as nonlinear loads. And due to this, the harmonics are injected in power system, which may cause worse effect on other main equipment in the industry.

To analyze and reduce these harmonics, harmonic measurement has been done at different points and PCC. Fluke power analyzer has been used to measure these harmonics and total harmonic distortion (THD) at point of common coupling. The analysis has been done based on IEEE Standard 519-2014, after that simulation has been carried out for this chemical plant in PSCAD/EMTDC software. This simulation model is used for FFT analysis of harmonics at different points.

2 Harmonic Sources and Its Effects

There are many harmonic sources like VFDs, UPS, computers, rectiformers, SMPS, arc furnaces in industries [1]. These nonlinear sources create a worse effect in power system, but the main power quality issue is harmonics. It affects so many main power system equipment like transformers, cables, circuit breakers, relays, capacitor banks, bus bars, ac motors. [2]. Effects of harmonics on these main devices are briefly described below.

2.1 Transformer

Harmonics can affect transformers primarily in two ways: (1) core and (2) winding. There are mainly three types of losses in the transformers: (1) hysteresis loss, (2) copper losses, and (3) eddy current losses.

Hysteresis loss. Hysteresis loss increases because voltage harmonics set up hysteresis loop on fundamental loop. So each loop represents higher magnetization power requirement and higher core losses in the transformer.

Copper loss. As harmonic current increases, it may result in the transformer rms current being higher than its capacity [2], and hence, this increased total rms current results in an extra copper loss in conductors.

Eddy current loss. Eddy currents are induced in transformer caused by the magnetic fluxes. This component of the transformer losses increases with the square of the frequency of the current causing the eddy currents [1].

Transformers that are required to supply large nonlinear loads must be derated to handle the harmonics. This derating factor is based on the percentage of the harmonic current in the load and the rated winding eddy current loss. The derating factor is known as "*K* factor" [2].

$$K = I_1^2(1)^2 + I_2^2(2)^2 + I_3^2(3)^2 + \cdots + I_n^2(n)^2 \quad (1)$$

2.2 Cables

When harmonic current flows, it creates additional copper losses in the cables, plus as frequency increases the resistance of the cable increases so current flows from outer periphery of the conductor and it is known as "skin effect" [3]. To handle the nonlinear loads, cables are rated by factor q by which we can know the capability of the cables [2].

$$q = I_1^2(E_1) + I_2^2(E_2) + I_3^2(E_3) + \cdots + I_n^2(E_n) \quad (2)$$

2.3 Protective Scheme

Harmonic current influences the operation of protective devices. Fuses and motor thermal overload devices are prone to nuisance operation when subjected to non-linear load. Electromechanical relay is also affected by the harmonic current. They may be operated faster or slower than their expected time [2].

2.4 Capacitor Bank

Most capacitor bank is used for power factor correction. Large level of voltage and current harmonics present in the system increase the rating of capacitor bank. The reason behind that is as reactance of capacitor is inversely proportional to the frequency, harmonic current of high-frequency flow in capacitor bank behaves as the sink for those harmonic currents [2]. And due to this, overload may occur and capacitor bank may lead to failure. One more serious effect on capacitor bank due to harmonics is "resonance." Resonance conditions are created when inductive and capacitive reactance is equal at any one of the harmonic frequency. If a high level of harmonic voltage and current is present at resonance frequency, more damage is occured in capacitor bank as well as other equipment.

Harmonics also interfere with telecommunication system, especially noise on telephone lines. Induction motor may refuse to start or may run at sub-synchronous speed due to harmonics. It creates problem in excitation panel and may cause generator failure. So if harmonics are beyond its acceptable level, then proper action should be taken to reduce it.

3 Industrial Base Power System

Harmonic measurements are taken in the "Aarti Industries Limited," which is a chemical plant located in Gujarat. The studied system configuration comprises a VFD modeling, load modeling, and generator modeling. The plant model is shown in Fig. 1, in which 11 kV, 50 Hz generator is connected with 11/0.430 kV transformer and 430 V is given to the two VFDs located in plant. VFDs are mainly used to control the induction motor speed.

The harmonic analysis is done at point of common coupling as shown in Fig. 1. All the data measured by fluke meter are shown in Table 1.

4 System Simulation

From the designed circuit parameter, the PSCAD simulation is done as shown in Fig. 2. Here, 11 kV source is used as incomer, and it is given to the 11 kV bus. And it is converted to 11/0.430 kV by transformer. Two subsystems are created for VFDs to connect it with 0.430 kV bus as shown in Fig. 2.

Here, two VFDs in subsystem comprise firstly the rectifier which converts ac power into dc power, then ac capacitive filter removes the dc ripple, and next inverter model again converts dc power into ac power [4]. Induction motor is modeled to connect as a load at the end of VFDs. In the subsystem, whole VFD modeling is shown in Fig. 3.

Fig. 1 Plant model

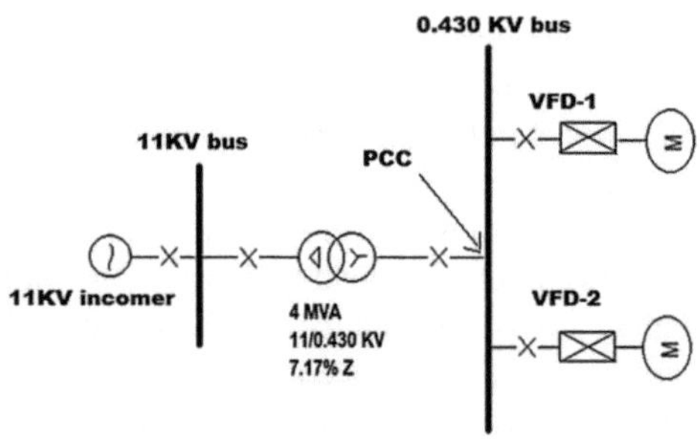

Table 1 Measured data at PCC

	Phase-A	Phase-B	Phase-C
Voltage (L-L, rms) (V)	427.62	428.753	426.557
Current (phase, rms) (kA)	2.9	3.11	2.92

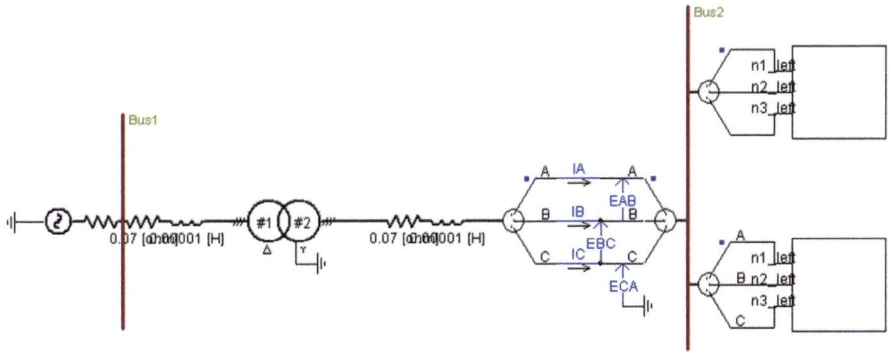

Fig. 2 Simulation of the chemical plant

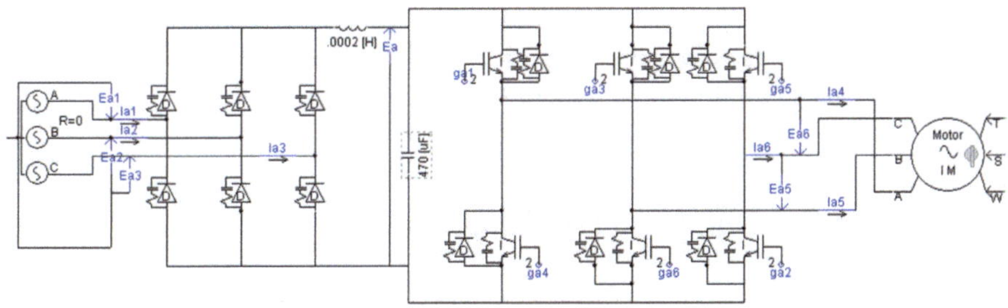

Fig. 3 Simulation of VFD in the subsystem

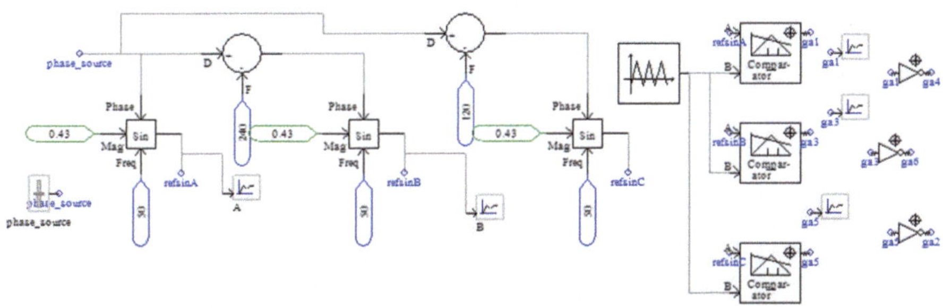

Fig. 4 Simulation of pulse width modulation technique

In subsystem to control the ac output, "controlled inverter scheme" was modeled. The switching signals of the six IGBT switches of the three-legged inverter bridge were generated using a "sinusoidal pulse width modulation" technique as shown in Fig. 4. Here, three sine wave generators were used to generate 50 Hz three-phase sine wave which is compared with 4350 Hz triangle waveform [5]. These three gate signals are given to the upper switch of the inverter, and its invert waveforms are given to the lower switches of the inverter [5].

5 Result and Discussion

To verify the harmonic analysis of the system, a model is developed in PSCAD/EMTDC software as discussed in above chapter. To calculate the total harmonic distortion at a different point, FFT analysis is done in the PSCAD software. For calculation of THD, the FFT analysis block is used [5] and the output of this block is given to one another harmonic distortion block which calculates the total harmonic distortion as well as the individual harmonic distortion.

5.1 Harmonic Analysis at Load End Side

FFT analysis was done at the end of VFDs, and the waveforms of voltage, current, and THD are shown in Figs. 5, 6, and 7. 430 V ac power was converted into 600 V dc by uncontrolled rectifier. And due to controlled inverter, the harmonics were produced in voltage and current waveform.

5.2 At the Point of Common Coupling

At point of common coupling (PCC), two VFDs are connected, and at that point, also voltage and current waveform and its THD were measured which is shown in Figs. 8 and 9.

For this current THD of 18.77%, the individual harmonic graph was shown in Fig. 10. The fifth- and seventh-order harmonics were dominant as shown in harmonic spectrum (Fig. 11) [6]. And this is because of the two VFDs connected at the PCC.

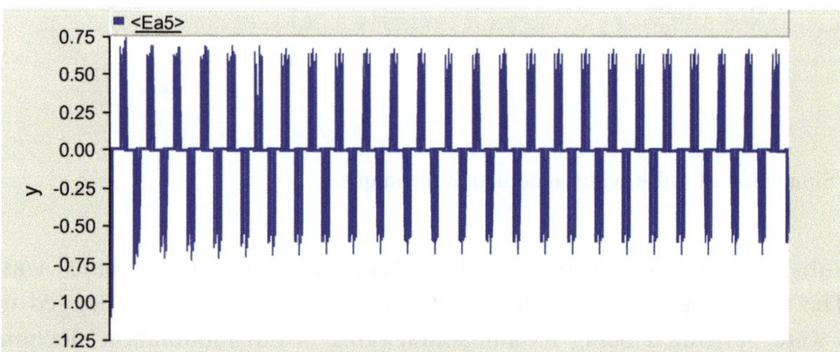

Fig. 5 Distorted voltage waveform at load end side

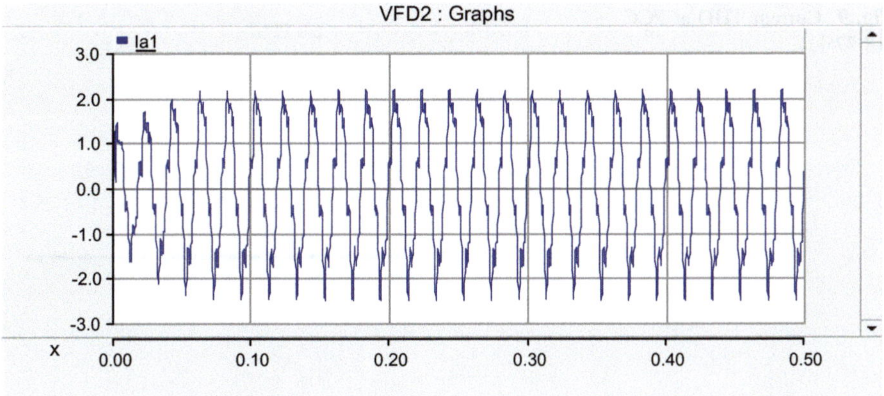

Fig. 6 Distorted current waveform at load end side

Fig. 7 Load-side current THD (14.81%)

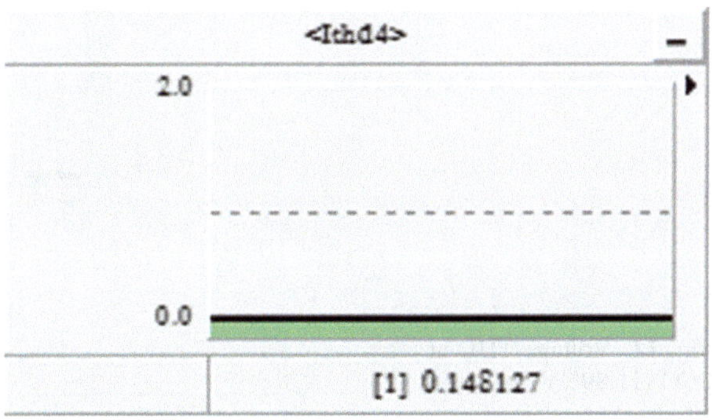

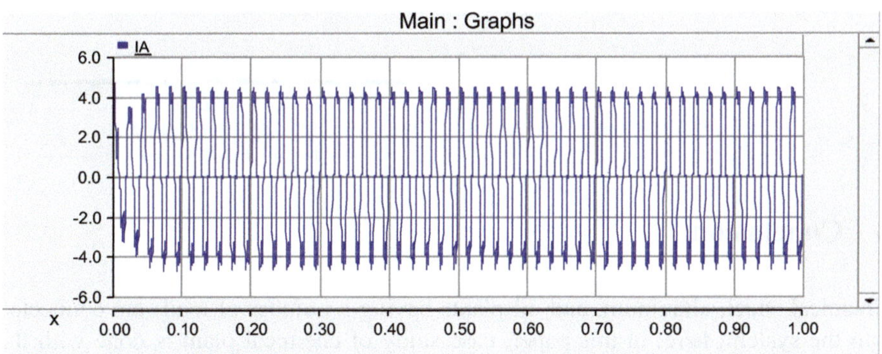

Fig. 8 Current waveform at PCC (2.9 kA)

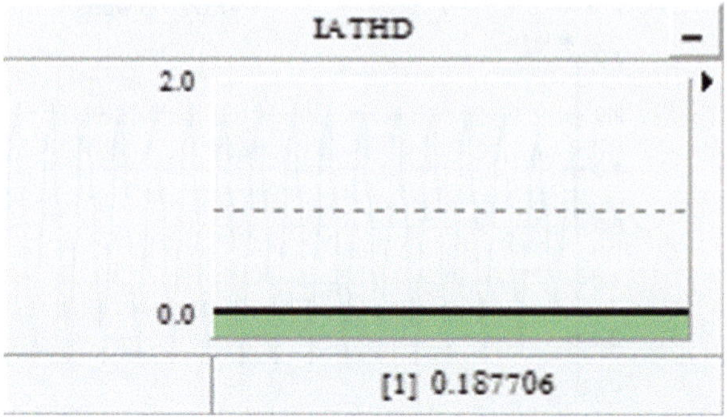

Fig. 9 Current THD at PCC (18.77%)

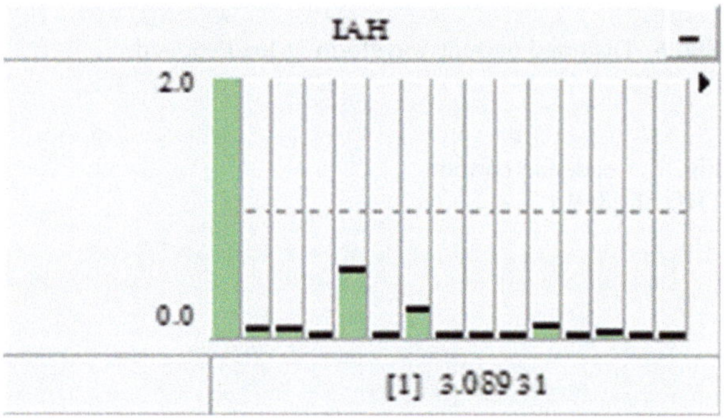

Fig. 10 Individual current harmonic at PCC

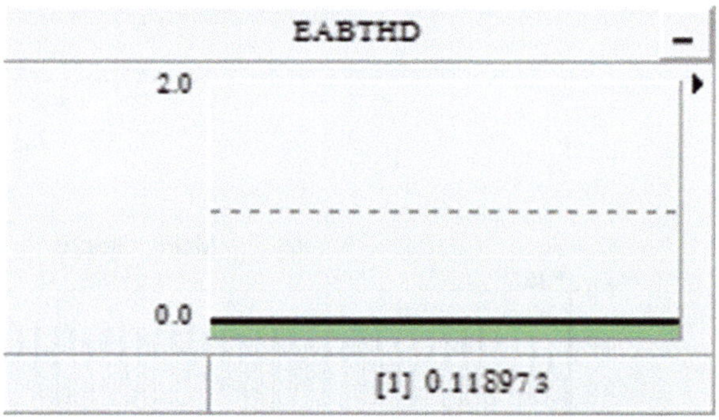

Fig. 11 Voltage THD at PCC (11.89%)

6 Conclusion

Chemical, steel, aluminum, and oil plants having a number of loads are connected into the system; here, in this paper, case study of chemical plant is done with the simulation based on PSCAD/EMTDC that has been modeled for running chemical plant for various types of loads, offender and offending loads. For harmonic analysis, FFT analysis tool is used and total harmonic distortion for voltage and current is calculated. Current THD was low at the end of one VFD, and it is due to the switching of the inverter. But as two VFDs are connected at the PCC, the current THD was increased from 14.81 to 18.77%. And the voltage THD was also increased to 11.89% which is beyond the limit. Active or passive filter is suggested for the same plant for more reliable operation of the plant. Looking at the plant (cost, operation, and configuration) and harmonic signature application of passive filter can mitigate such level of harmonics, which is cost-effective as well with advantage of filter as reactive power compensation in case when high reactive power is demanded.

References

1. Roger C. Dugan, "Electrical Power Systems Quality", 2nd Edition, Tata Mcgraw Hill Publication, chapter 3 and 5.
2. C. Sankaran, "Power quality", CRC Press, 2002, chapter 4.
3. J.P. Srividhya, dr. D. Sivakumar, T. Shanmathi, "A Review on causes, effects and detection technique of harmonics in the power system", International conference on computation of power, energy information and communication (ICCPEIC), 2016 IEEE.
4. M D Singh and K B Khanchandani, "Power Electronics ", 2nd Edition, Tata Mcgraw Hill Publication, 2006.
5. Abdulrahman Kalbat, "PSCAD Simulation of Grid-Tied Photovoltaic System and Total Harmonic Distortion Analysis", 3rd international conference on electrical power and energy conversion, 2013 IEEE.
6. M.B. jannat, J.S. Abdulmalek "Reducing the effects of harmonics produced by electrical arc furnaces in misurata steel plant", environment and electrical engineering (EEEIC),11th international conference, 2012.
7. IEEE Recommended Practice and Requirements for Harmonic Control in Electric Power Systems, IEEE power and energy society, ANSI/ IEEE 519-2014.
8. Mahesh M Swamy, "Harmonic interaction between 1500 KVA supply transformer and VFD load at an industrial plant", Transaction on industry application, 1998 IEEE.
9. D Bentancur, G Perez, A Pardo, A Rondoni, "Study of a Cement Plant Experiencing Power Quality Problems Uruguay", 16th International Conference and Exhibition on electricity distribution, 2001 IEEE.
10. Minghui BAO, Jun XIA, Xiaogen YIN, Manling dong, Hengxin HE, Junjia HE, "Harmonic measurement and analysis in a Morden steel manufacturing facility", power and energy society general meeting, 2010 IEEE.
11. V. Suresh Kumar, P. S. Kannan and V. Saravanan, "Harmonic distortion in a morden cement industry- A case study", Power India conference, 2006 IEEE.

Computational Analysis and Intelligent Control of Load Forecasting Using Time Series Method

Parth A. Joshi and Jatinkumar J. Patel

Abstract Electrical demand forecasting is a nonlinear and complex process for intelligent control of large-scale electrical power system and computational analysis and intelligent control of load forecasting. The research attempts for increasing accuracy of load forecasting. The time series method (TMS) is implemented for the midterm load forecasting for higher accuracy. It reflects the natural growth of the load. Here, the results show that the predication is better for midterm load forecasting (MTLF), which is most important parameter for seasonal planning and intelligent generation control of electric power using autoregressive and Box–Jenkins method by System Identification Tool (SIT) in MATLAB environment. Experimental support for the proposed work has been granted from State Load Dispatch Center (SLDC), Gotri, Gujarat, for forecasting the load. Here, load forecasting is carried out for Uttar Gujarat Vij Company Ltd. (UGVCL), India.

Keywords Load forecasting · Intelligent control · Midterm load forecasting (MTLF) · Time series method (TSM)

1 Introduction

Main difficulty of electrical energy is that it cannot be stored in large-scale according to demand. It is broadly generated by thermal, hydro, atomic, wind, and solar power plants. According to the difference between demand and supply prediction for startup, running and shut down become important factors for smooth functioning of the system every time and some time in case of emergency. If we can predict the required demand of consumers ahead with better accuracy, we can schedule demand consequently to generation [1].

P. A. Joshi (✉) · J. J. Patel
Electrical Department, G H Patel College of Engineering & Technology, V.V. Nagar, Anand, India
e-mail: joshiparth23@yahoo.com

R. Kher et al. (eds.), *Proceedings of the International Conference on Intelligent Systems and Signal Processing*, Advances in Intelligent Systems and Computing 671,
https://doi.org/10.1007/978-981-10-6977-2_27

Table 1 Types of load forecasting [2]

Nature of forecast	Lead time	Application
Short term	0.5–5 h	Operational planning, allocation of spinning reverse, maintenance scheduling, contingency analysis
Medium term	7–30 days	Planning for seasonal peak winter, summer
Long term	30 days to one year	Planning generation growth

This process of predicting active load with actual accordance to load is called **load forecasting**. In India, the load forecasting was the neglected part compared to other countries in previous years. However, after realizing electric power requirement in various states, the government of India conducted first load survey of India in 1958. Nowadays, load forecasting is central process with all phases like generation, distribution, and transmission of the power system. Load forecasting is classified into three categories. They are short term, midterm, and long term [2] (Table 1).

Accurate load forecasting brings toward to the large amount of saving money in electrical utility company. According to Bunn and Farmer, these savings are realized when load forecasting is used to control operation and decision such as unit commitment, dispatch, fuel allocation, and off-line network analysis [3]. The methods for load forecasting are broadly classified as traditional method which includes end use method, econometric method, time series method (BJ, AR, MA, ARMA), regression method. Kalman Filter, Support Vector Machine(SVM). And Artificial Intelligence Method which includes neural network, Expert system, and Fuzzy logic system [4].

2 Load Characteristics

The system load at any given time is mostly a combination of four separate components, i.e., residence load, agriculture load, industrial load, and commercial load. Here, we analyze the load characteristics for the various components at 500–700 MWh with 150 MWh allowable variation. Here, we observed that the load characteristics for UGVCL is consists of 48% (336 MWh), agriculture load 30% (210 MWh), industrial load 13% (91 MWh) load is residential load and 9% (63 MWh) is unusual or any special event load [5]. In general, total system load can be mathematically represented as:

$$T = T_n + T_w + T_s + T_r + T_i + T_a \tag{1}$$

where

T is the total load of the system.
T_n is the normal load of the system which is set for uniform load pattern for each type of day.
T_w is the weather sensatative part which represents the seasonal effect.
T_s is any special event or unusual event of the load.
T_r is the random part of the load which has zero mean white noise.
T_i is the industrial load.
T_a is the agriculture load.

3 MTLF Modeling and Analysis

In this paper, we forecast the midterm load forecasting using time series method. If the historical data are restricted to past value to be forecasted then the procedure is called **time series** [6]. The aim of the time series analysis to discover the pattern from the past data and extrapolate pattern for the future demand curve and the future forecasting is only based on the historical data. Time series is an observation of variable at successive point in the time. The measurement is taken for the span of hourly, weekly, monthly, and yearly. To identify the load pattern, first of all, we construct time series plot. Time series plot is the graphical representation of time and time series variable. The *Y*-axis represents the time which is in the form of hourly, weekly, monthly, and yearly, and *X*-axis represents the time series variable like MW consumption [7]. The reason for choosing time series is that there are many methods that are used in time series method like autoregressive (AR), moving average (MA), autoregressive moving average, Box–Jenkins method (BJ).

A. **Autoregressive Method**

If the load is assumed to be a linear, combination of previous loads then the autoregressive (AR) model can be used for the load profile which is given by Liu et al. in 1996 by,

$$L_t = \emptyset_1 L + (t-2) + \emptyset_2 L(t-2)\ldots\emptyset_p L(t-p) + W_t \tag{2}$$

where L_t is the predicted load at time k, W_t is a random load disturbance. The coefficients $\emptyset_1, \emptyset_2, \emptyset_p$ of the AR can be found out by using least and likelihood method. Also, one well-known algorithm least mean square is used which is represented by Mbamula and El–Hawary in 1993. The algorithm presented by EI–Hawary includes an adaptive AR modeling techniques enhanced with partial autocorrelation analysis [6, 8].

B. **Moving Average Method**

The Moving Average method is used the average of most recent K data value in time series as forecast the next period. The expression for moving average is written as

$$F_{t+1} = (Y_t + Y_{t-1} + Y_{t-2} + Y_{t-k})/k \tag{3}$$

where

F_{t+1}	is the forecasting for term $t + 1$.
Y_t	is the actual value of time series in time t.

Every time a new observation is available so the moving word is used for this method. The new observation is computed by replacing the old observation. As, the average will be change, or new observation become available [7]. This method is the very oldest method of the time series. This method is accurate then autoregressive method because at every interval, we consider a new value to be forecast. Before using "Moving average" to forecast a time series, we have a need to select the order or number of time series values, which has including in time series [9]. If only the most recent data of time series are considered, then the small value of (k) is preferred. And if more past data are considered, then the large value of k is more suitable for time series. The given method has a limitation that it is only applicable for linear system and is not able to accurately predict when other environment constrains are considered.

C. **Box–Jenkins Method**

This method is developed by George Box and Gwilym Jenkins in 1970. This method is the combination of the autoregressive and moving average method [8]. This method is used for both stationary and nonstationary systems. This model can be easily modified for incorporated environmental effects like rainfall, wind velocity, and humidity [10]. Based environmental constrained, has been modified as

$$F_t = \emptyset_1 L + (t-1) + \emptyset_2 L(t-2)\ldots\emptyset_p L(t-p) + W_t + (Y_t + Y_{t-1} + Y_{t-2} + Y_{t-k})/k \tag{4}$$

where

$\emptyset_1, \emptyset_2, \emptyset_p$	are the coefficient of the time series (BJ).
Y_t, Y_{t-1}, Y_{t-2}	are the past data of time series (BJ).
K	is a number of samples.

For the forecasting load using above all methods the mathematically exercise is increase for the data analysis. So to overcome above problem, software-based analysis is highly needed. Dedicated tool developed by MATLAB has been used by authors.

D. **System Identification Tool**

System Identification Tool is mapping the input and output of the system. It takes the current input and output variable of the process as an input and estimates the system parameter. Using the system identification tool, we can estimate linear and nonlinear system are observed.

This is the major advantage of the system identification tool. The network is simulated using MATLAB for different learning rates and different number of training point. The simulated network is implemented to estimate the damping coefficient of the linear or nonlinear variable.

The network is simulated. The system identification determines the mathematical model, block diagram, and flow diagrams. The identification tool is used to estimate nonlinear effect for measurement and control system [11].

4 Result Analysis

Here, estimations made by autoregressive model using System Identification Tool up to fourth order system. Simulated results and best curve fitted are shown in Fig. 1.

The result says that forecasting curve for given mathematical model is indicated by pink color. While black indicts the best fitted values. Suggested methods are used to find the value of time series up to fourth order as per Table 2 input and output of AR method (second order system).

As shown in Fig. 2, the forecasting curve for the given mathematical model is represented by blue color. While the black is presented the best fitted values for the nonlinear AR method to find the value of time series up to second order as per Table 3 the output of the time series for nonlinear AR is linear. The predicted loss function is 13.4529%.

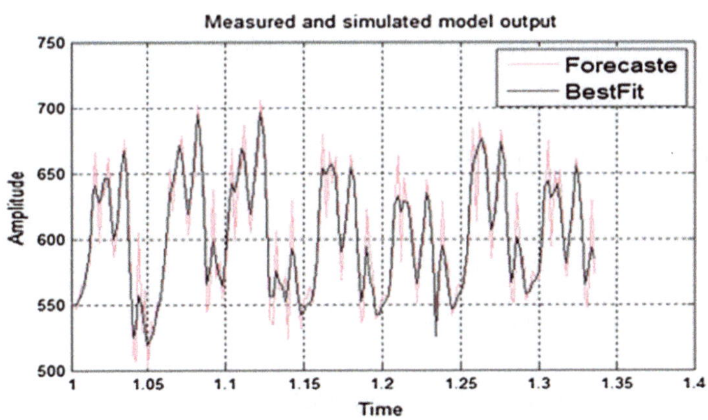

Fig. 1 Load pattern analysis using AR method

Table 2 Input and output of nonlinear AR method (second order system)

% System identification tool constructing nonlinear autoregressive time series %
Name = nlarx1
Nonlinear ARX model with 1 output and 1 input
Inputs: u1
Outputs: y1
Standard regressors corresponding to the orders
$na = 2$, $nb = 2$, $nk = 1$
No custom regressor
Nonlinear regressor
None
Model output is linear in regressors
Loss function: 13.4529%

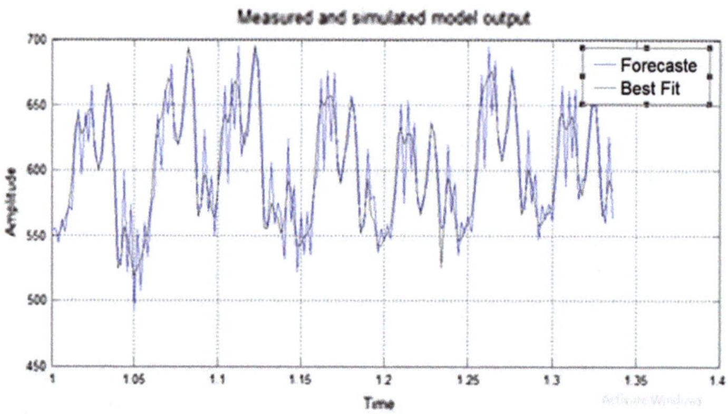

Fig. 2 Load pattern analysis using nonlinear AR (second order system)

Table 3 Input and output of AR method

% Autoregressive with exogenous input (Arx) and the multiple regression model %
Model = Arx441
Discrete-time ARIX model: $A(z)y(t) = B(z)u(t) + 1/(1 - z^{-1})\ e(t)$
$A(z) = 1 - 1.609\ z^{-1} + 0.4974\ z^{-2} + 0.5632\ z^{-3} - 0.28\ z^{-4}$
$B(z) = 1.517\ z^{-1} - 3.402\ z^{-2} + 2.843\ z^{-3} - 0.8046\ z^{-4}$
Name arx441
Sample time: 0.002 s
Parameterization
Polynomial orders: $na = 4$ $nb = 4$ $nk = 1$
Number of free coefficients: 8
Model contains integration on noise channel
Use "polydata", "getpvec", "getcov" for parameters and their uncertainties
Status
Estimated using ARX on time domain data "mydataf"
Fit to estimation data: 99.88% (prediction focus)
FPE: 0.01977%, MSE: 0.01805%

As shown in Fig. 3, the forecasting curve for the given mathematical model is indicated by gray color. While the black is presented the best fittest values. Method is used to find the value series up to fourth order as per Input and output of Nonlinear AR Method (4th Order system). The predicted loss function is 0.03608%. As shown in Table 4, for the input and output of nonlinear AR method (fourth order system), the time series output contains the nonlinearity wavelet. As shown in Table 2, for the input and output of the nonlinear AR model up to second order system with increasing the order of the system, the loss function is reduced,

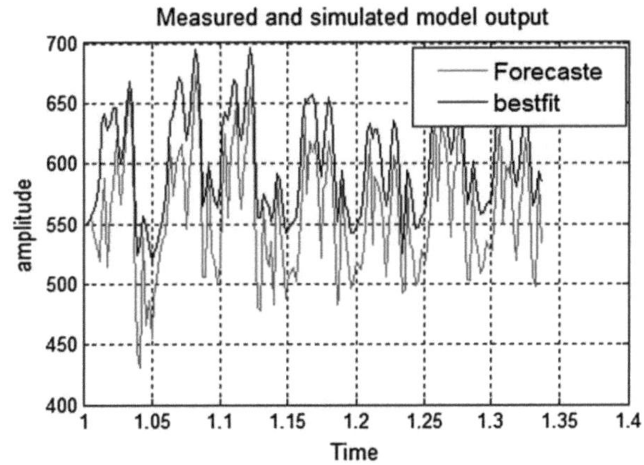

Fig. 3 Load pattern analysis using nonlinear AR (fourth order system)

Table 4 Input and output of nonlinear AR method (fourth order system)

% System identification tool constructing nonlinear autoregressive time series %
Name = nlarx5 Nonlinear ARX model with 1 output and 1 input Inputs: U1 Outputs: Y1 Standard regressor corresponding to the orders $na = 4$, $nb = 4$, $nk = 1$ No custom regressor Nonlinear regressors $Y1(t-1)$ $Y1(t-2)$ $Y1(t-3)$ $Y1(t-4)$ $U1(t-1)$ $U1(t-2)$ $U1(t-3)$ $U1(t-4)$ Nonlinearity: wavenet with nine units Loss function: 0.03608%

but the output of the model is linear in the second order system, while for the fourth order system the output of the model is nonlinear.

As shown in Fig. 4, the forecasting curve of the given mathematical model is indicated by red color. While the black is indicted the best fitted values. Method is used to find the value series up to fourth order as per Table 5. The predicted curve is fit approximate 98.85% to best fit. The mean square error (MSE) is 1.698%. As shown in Table 5, the input and output of Box–Jenkins method, the number of free coefficient of time series using BJ method is 16. This method is suitable for linear and nonlinear system. By using this method, we can observe that the accuracy of this method is accurate compared to other method (Fig. 5; Table 6).

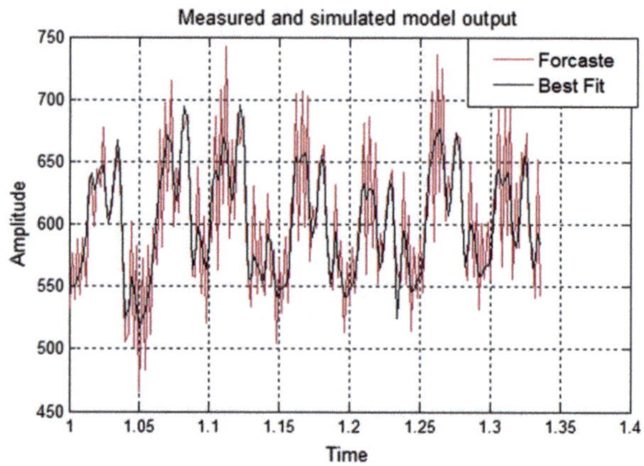

Fig. 4 Load pattern analysis using Box–Jenkins method

Table 5 Input and output of Box–Jenkins method

% Auto-regressive and moving average time series model %
Name = Bj 44441
Discrete-time BJ model: $y(t) = [B(z)/F(z)]\ u(t) + [C(z)/D(z)]\ e(t)$
$B(z) = 1.625\ z^{-1} - 4.412z^{-2} + 4.4\ z^{-3} - 1.609\ z^{-4}$
$C(z) = 1 + 3.508\ z^{-1} + 4.604\ z^{-2} + 2.673\ z^{-3} + 0.5775\ z^{-4}$
$D(z) = 1 - 3.15\ z^{-1} + 3.98\ z^{-2} - 2.326\ z^{-3} + 0.5236\ z^{-4}$
$(z) = 1 - 1.814\ z^{-1} + 0.1927\ z^{-2} + 1.539\ z^{-3} - 0.9123\ z^{-4}$
Sample time: 0.002 s
Parameterization:
Polynomial orders: $nb = 4$ $nc = 4$ $nd = 4$ $nf = 4$ $nk = 1$
Number of free coefficients: 16
Use "polydata", "getpvec", "getcov" for parameters and their uncertainties
Status:
Estimated using POLYEST on time domain data "mydatafd"
Fit to estimation data: 98.85% (prediction focus)
FPE: 0.0001646%, MSE: 1.698%

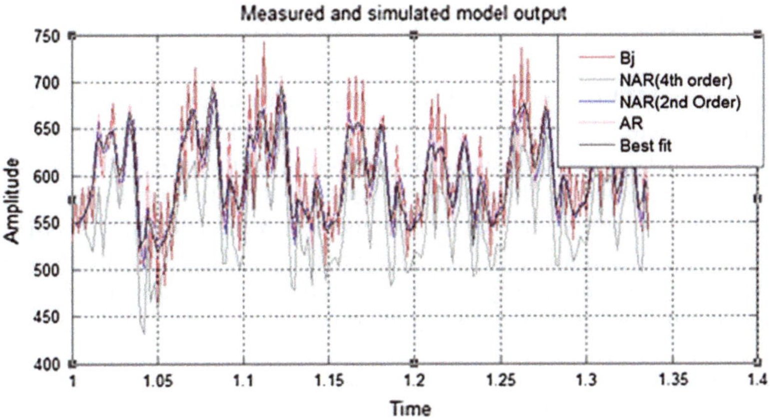

Fig. 5 Load pattern Analysis for all combine methods

Table 6 Comparison of all methods

Sr no	Method name	MSE or loss function (%)
1	AR	0.01805
2	Non linear AR (second order system)	13.4529
3	Non linear AR (fourth order system)	0.03608
4	Box–Jenkins	1.698

After the analysis of the time series methods for electric load forecasting using AR, nonlinear AR, and BJ methods, the BJ method gives better result with linear and nonlinear system. Nonlinear also gives best result for only nonlinear system.

Proposed work plays a significant role in daily operation of electrical utility like fuel resource management and for consideration in balancing the supply and demand of electrical power influence by load forecasting. Obtained such accurate result analysis indicates that it is the base for spot price establishment for entire state electric system and to minimize purchase cost of electricity in recent electricity market. Intelligent control for smart grid integration, such as proposed accurate load forecasting is useful in planning for demand side management, storage, maintains and their schedules integrated control of renewable sources with MTLF benefits consumers with by means of the relationship between demand and price variation with electrical uses pattern based on its prices. i.e. Available Based Tariff (ABT).

5 Conclusion

In given paper, MTLF is presented using time series method in the system identification tool. Here we forecast the amount of (MWh) load using various methods and conclude that autoregressive and Box–Jenkins method provide better accuracy. The mean square errors of these two methods are 0.01805 and 1.698%. Also we understand that in nonlinear autoregressive method, with increasing the order of the system, the loss function is reduced. And we also conclude that the Box–Jenkins method is able to provide nearly precise results for the linear and nonlinear system compared to other method at same sampling time.

Acknowledgements The author(s) of this paper are thankful of State Load Dispatch Center (SLDC), Gotri, Gujarat, India for providing permission and infrastructural support.

References

1. Abu-El-Magd, M. A., and R. D. Findlay. "A new approach using artificial neuralnetwork and time series models for short term load forecasting." Electrical and Computer Engineering, 2003. IEEE CCECE 2003. Canadian Conference on. Vol. 3. IEEE, 2003.
2. Kothari D P and I J Nagrath Moeden Power System Analysis, 4th edition, Tata MCGrew Hill Delhi.
3. Alfares, Hesham K., and Mohammad Nazeeruddin. "Electric load forecasting: literature survey and classification of methods." International Journal of Systems Science 33.1 (2002): 23–34.
4. Prakash, G. L., et al. "Short Term Load Forecasting for Uttarakhand using neural network and time series models." Reliability, Infocom Technologies and Optimization (ICRITO) (Trends and Future Directions), 2014 3rd International Conference on. IEEE, 2014.
5. Annual report 2015 -2016 State Load Dispatch Center, Gujarat Energy Transmission Corporation Limited.
6. Zhang, G. Peter. "Time series forecasting using a hybrid ARIMA and neural network model." Neurocomputing 50 (2003): 159–175.
7. Weron, Rafal. Modeling and forecasting electricity loads and prices: A statistical approach. Vol. 403. John Wiley & Sons, 2007.
8. Shen, Haipeng. "On modeling and forecasting time series of smooth curves." Technometrics 51.3 (2009): 227–238.
9. Chen, J. F., Wang, W. M., & Huang, C. M. (1995). Analysis of an adaptive time-series autoregressive moving-average (ARMA) model for short-term load forecasting. Electric Power Systems Research, 34(3), 187–196.
10. Hagan, Martin T., and Suzanne M. Behr. "The time series approach to.
11. www.mathworks.com.

Cascaded H-Bridge Multilevel Inverter Operating Under Faulty H-Cell Condition

Soniya V. Atodaria and Nilesh V. Shah

Abstract Reliability is an important aspect in the multilevel inverters (MLIs) as they are used for high-power applications and also the number of switch increases with the increase in levels. The efficiency improvement, optimized control strategies, new applications, and fault-tolerant operation are another area of interest of researcher for MLIs. In this paper, the control scheme of cascaded H-bridge multilevel inverter operating under faulty H-cell condition is presented. A modified level-shifted pulse width modulation (LS-PWM) method is used for the control of cascaded H-bridge multilevel inverter operating under faulty conditions. To study the open-circuit and short-circuit fault cases of the H-cell of cascaded H-bridge inverter, two methods of switching have been presented and simulated using MATLAB/Simulink. Neutral shift method is used for achieving balanced line-to-line voltage and line-to-line current. Additionally, to distribute power among H-bridge power cells, rotation scheme in LS-PWM has also been simulated. The simulation results validate the fault-tolerant operation of cascaded H-bridge MLI.

Keywords Cascaded H-bridge multilevel inverter · Level-shifted pulse width modulation · Fault tolerant · Neutral shift

1 Introduction

Multilevel inverters (MLIs) have experienced continuous and increasing growth in terms of application and research. The effort of the researchers and industry has led to a rapid development of different multilevel inverter topologies, modulation techniques, and control strategies. Multilevel inverters are widely used in compressors, blowers, fans, crushers, grinders, etc. [1]. The cascaded H-bridge

S. V. Atodaria (✉) · N. V. Shah
Sarvajanik College of Engineering & Technology, Surat, India
e-mail: atodariasv@yahoo.com

N. V. Shah
e-mail: nilesh.shah@scet.ac.in

R. Kher et al. (eds.), *Proceedings of the International Conference on Intelligent Systems and Signal Processing*, Advances in Intelligent Systems and Computing 671,
https://doi.org/10.1007/978-981-10-6977-2_28

(CHB) MLI has penetrated more industry applications compared to neutral point clamped (NPC) and flying capacitor (FC) [1]. Also, CHB does not have problem of dc-link capacitor voltage balancing. Its losses are also less compared to NPC and FC. Also, CHB uses low-frequency IGBTs compared to high-frequency IGBTs/ IGCTs used in FC. In [2], the comparison between phase-shifted pulse width modulation (PS-PWM) and level-shifted pulse width modulation (LS-PWM) is given, which shows that the LS-PWM gives better output voltage profile compared to PS-PWM. In [3], comparison between different LS-PWM techniques, i.e., in phase disposition (IPD), phase opposite disposition (POD), and alternative phase opposite disposition (APOD) is presented. It shows that IPD gives less THD in the output voltage compared to POD and APOD. IPD also gives more levels in line-to-line output voltage compared to POD and APOD [3].

To increase the reliability of MLIs, several methods have been presented in [4–8]. In [4], space vector modulation (SVM) is used and to bypass the faulty H-cell, SCR is used as fast-acting switch. To make the line-to-line voltage and current balanced, maximum voltage algorithm is used. In [5], phase-shifted pulse width modulation (PS-PWM) control is applied, whereas to bypass the faulty H-cell, contactor is used. To make the line-to-line voltage and current balanced, voltage equivalency triangle is used for which the prior calculations are done and stored in lookup table. In [6], redundant H-cell in each phase is used, which increases reliability only when single cell becomes faulty in any phase. In [7], only one redundant H-cell is used but the method is effective only when there is a single-cell fault condition. In [8], the MLI is used for battery energy storage system and to make the line-to-line voltage and current balanced, neutral shift method is used. In [9], open-circuit fault detection method based on LS-PWM is discussed, which can give information about faulty switch. In [10], fault detection is obtained by the use of artificial intelligence method. The fault-tolerant operation based on LS-PWM is presented in [11, 12]. To bypass faulty H-cell, zero-voltage switching state is used. Thus, reducing the need of extra element/device. The neutral shift method is applied to make the line-to-line voltage and current balanced. But the calculation for offset voltage injection by neutral shift method is given only for single-cell fault condition in the phases. In [13], to overcome limitation of LS-PWM, i.e., unequal power distribution among H-cells, rotation scheme is given.

In this paper, the generalized equations for offset voltage calculation are derived based on the neutral shift method presented in [8], which can be used for fault in more than one cell. To distribute the power equally among the H-cells under faulty condition, rotation technique of carrier wave is used in LS-PWM technique. Also, to achieve balance line-to-line voltage and current, the derived generalized equations of neutral shift method are used. The control strategy using modified LS-PWM technique with the generalized neutral shift equations is verified through simulation results. The simulation result validates the control scheme for different cases of open-circuit and short-circuit fault conditions of single as well as multiple faulty switches in a seven-level CHB–MLI.

2 Neutral Shift Method

Neutral shift method is used to maximize the available output voltage under faulty cell bypassed condition. The seven-level CHB-MLI is shown in Fig. 1 with three cells per phase. In this section, generalized equations for neutral shift voltage are derived. The three-phase ac voltages are given by (1)–(3).

$$v_{an} = mV_{dc}\sin\omega t \tag{1}$$

$$v_{bn} = mV_{dc}\sin\left(\omega t - \frac{2\pi}{3}\right) \tag{2}$$

$$v_{cn} = mV_{\mathrm{dc}}\sin\left(\omega t + \frac{2\pi}{3}\right) \tag{3}$$

where m = modulation index.

The phasor diagram showing relation as in (1)–(3) for cascaded H-bridge MLI under healthy condition is shown in Fig. 2a. Under faulty cell condition, when it is assumed that cell a1 in phase-a gets faulty, without neutral shift, the modulation index of remaining $(N - 1)$ healthy inverter cells would have been increased by 1.5 to balance output line voltage as per (4)–(6). The corresponding phasor diagram is shown in Fig. 2b.

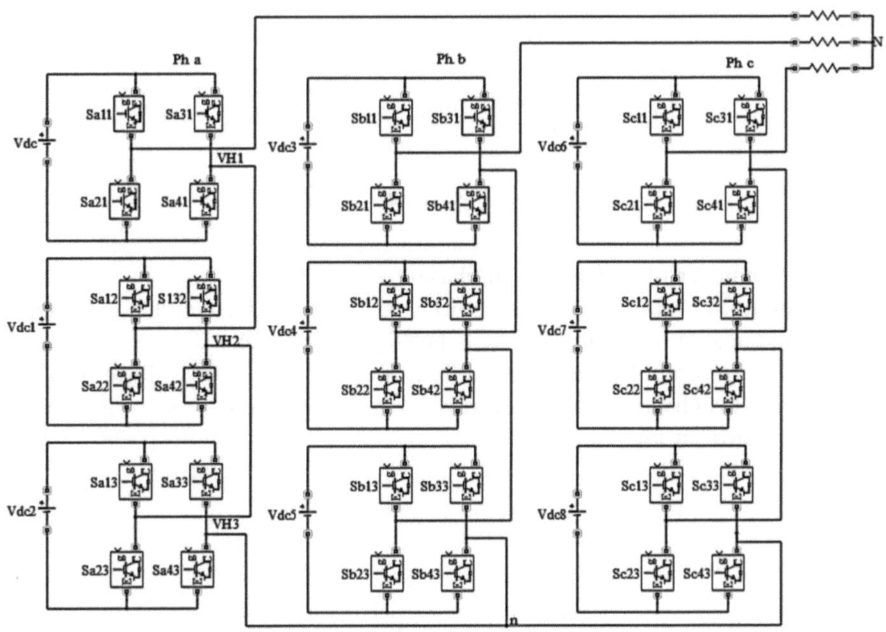

Fig. 1 Seven-level CHB-MLI

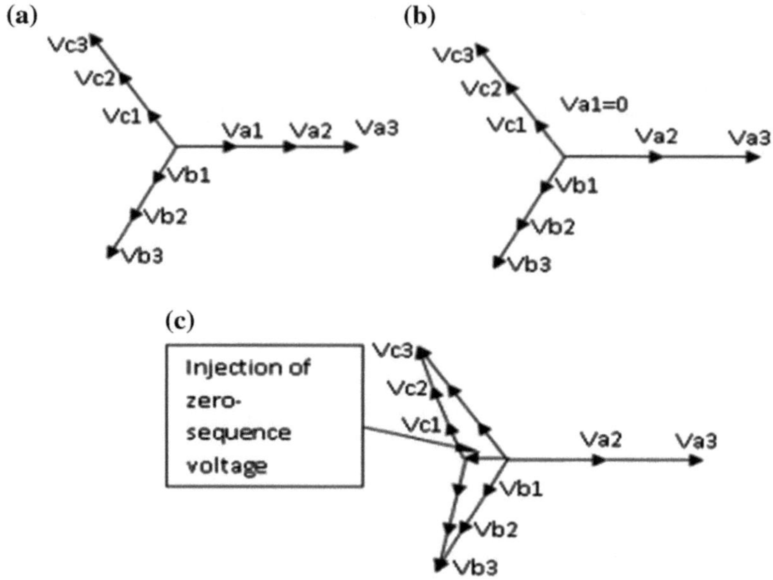

Fig. 2 Phasor diagram of seven-level CHB-MLI with $N = 3$ **a** under normal operation, **b** under cell $a1$ gets faulty and the modulation index of remaining cells v_{a2} and v_{a3} is increased to 1.5 and **c** injection of zero-sequence voltage by neutral shift method

$$v_{an} = \frac{N}{\left(N - n_{fa}\right)} mV_{\text{dc}} \sin \omega t \tag{4}$$

$$v_{bn} = mV_{\text{dc}} \sin\left(\omega t - \frac{2\pi}{3}\right) \tag{5}$$

$$v_{cn} = mV_{\text{dc}} \sin\left(\omega t + \frac{2\pi}{3}\right) \tag{6}$$

where n_{fa} is number of faulty cells in phase-a.

With neutral shift, the burden, i.e., power transferred by the cell that gets faulty, is distributed among remaining $(3N - 1)$ healthy inverter cells. The neutral shift is based on the injection of a fundamental-frequency zero-sequence voltage, $\boldsymbol{v}_{0f}$. The ac voltages of healthy cells can be expressed by (7)–(9), and corresponding phasor diagram is shown in Fig. 2c.

$$v_{an} = \frac{N}{\left(N - n_{fa}\right)} \left(mV_{\text{dc}} \sin \omega t - v_{0f} \sin \omega t\right) \tag{7}$$

$$v_{bn} = mV_{\text{dc}} \sin\left(\omega t - \frac{2\pi}{3}\right) - v_{0f} \sin \omega t \tag{8}$$

$$v_{cn} = mV_{\mathrm{dc}} \sin\left(\omega t + \frac{2\pi}{3}\right) - v_{0f} \sin \omega t \tag{9}$$

where the second term on right, i.e., v_{0f} sin ωt, is zero-sequence voltage.

Let the three-phase line currents be balanced as follows:

$$i_a = \sqrt{2}I \sin \omega t \tag{10}$$

$$i_b = \sqrt{2}I \sin\left(\omega t - \frac{2\pi}{3}\right) \tag{11}$$

$$i_c = \sqrt{2}I \sin\left(\omega t + \frac{2\pi}{3}\right) \tag{12}$$

The input electric power to each converter cell is given by the following:

$$p_{a1} = 0 \tag{13}$$

$$\begin{aligned} p_{an} &= \int_0^{2\pi/\omega} v_{an} i_a \mathrm{d}t \\ &= \frac{\pi\sqrt{2}}{\omega} \frac{N}{(N - n_{fa})} (mV_{\mathrm{dc}} - v_{0f})I, \quad \text{for } n = 2, 3, \ldots, N \end{aligned} \tag{14}$$

$$\begin{aligned} p_{bn}; &= \int_0^{2\pi/\omega} v_{bn} i_b \mathrm{d}t \\ &= \frac{\pi\sqrt{2}}{\omega} \left(mV_{\mathrm{dc}} + \frac{v_{0f}}{2}\right)I, \quad \text{for } n = 1, 2, \ldots, N \end{aligned} \tag{15}$$

$$\begin{aligned} p_{cn} &= \int_0^{2\pi/\omega} v_{cn} i_c \mathrm{d}t \\ &= \frac{\pi\sqrt{2}}{\omega} \left(mV_{\mathrm{dc}} + \frac{v_{0f}}{2}\right)I, \quad \text{for } n = 1, 2, \ldots, N \end{aligned} \tag{16}$$

These powers $(p_{a2}, p_{a3}, \ldots, p_{c3})$ absorbed by the $3N - n_f$ healthy battery units should be equal during the faulty condition, i.e.,

$$p_{an} = p_{bn} = p_{cn} \tag{17}$$

From (10) to (17), the equation for zero-sequence voltage can be derived as given by (18).

$$v_{0f} = \frac{2n_{fx}}{3N - n_{fx}} \tag{18}$$

where n_{fx} is total number of faulty cells in phases x = phase a, b, or c.

If one cell becomes faulty in phase-a and one cell becomes faulty in phase-b, then (7)–(9) are modified as given by (19)–(21).

$$v_{an} = \frac{N}{(N - n_{fa})}\left(mV_{dc}\sin\omega t - v_{0f}\sin\omega t - v_{0f}\sin\left(\omega t - \frac{2\pi}{3}\right)\right) \tag{19}$$

$$v_{bn} = \frac{N}{(N - n_{fb})}\left(mV_{\mathrm{dc}}\sin\left(\omega t - \frac{2\pi}{3}\right) - v_{0f}\sin\omega t v_{0f}\sin\left(\omega t - \frac{2\pi}{3}\right)\right) \tag{20}$$

$$v_{cn} = mV_{\mathrm{dc}}\sin\left(\omega t + \frac{2\pi}{3}\right) - v_{0f}\sin\omega t - v_{0f}\sin\left(\omega t - \frac{2\pi}{3}\right) \tag{21}$$

3 Fault-Tolerant Operation by Modified LS-PWM Method

In this section, the fault-tolerant operation of CHB-MLI is analyzed by two different methods using modified LS-PWM.

3.1 Method-1: For Open and Short-Switching Device Faults

In this method, the faulty cell is bypassed. Hence, inverter gives (L-2) voltage levels in faulty phase. The method of bypassing the faulty H-cell is explained as follows. Considering switch S11 has become faulty, i.e., open-circuited, the switch in opposite leg is made off and the two bottom switches are made on; thus, it gives path to the current and zero voltage. If switch S11 has become short-circuited, immediately the bottom two switches are made off and switch in opposite leg is made on to give path to the current and to give zero voltage.

To allow the reference signal to be modulated linearly with carrier bands, the reference signal is adjusted as shown in Fig. 3. At t = 0.06 s, amplitude modulation index of the reference signal is decreased as given by (22).

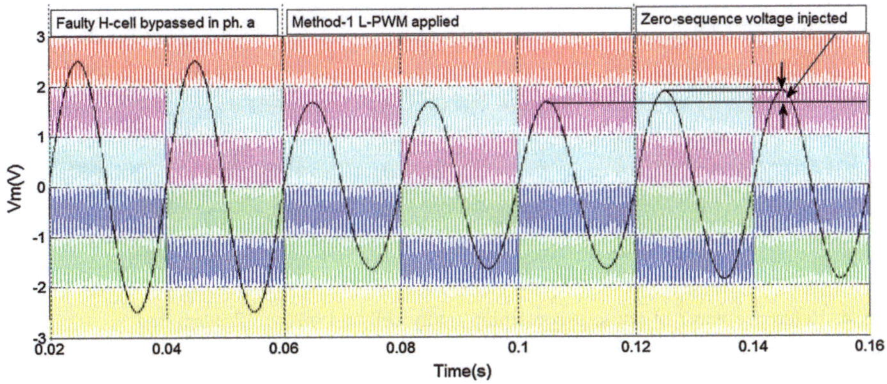

Fig. 3 Reference wave of phase-a containing faulty cell and method-1 is applied for LS-PWM

$$V^{*}_{\text{ref}\_x} = \frac{(N - n_{fx})}{N} V_{\text{ref}\_x} \tag{22}$$

where x represents the phase a, b, or c.

For a seven-level inverter, number of cascaded cell = 3 and with the number of faulty inverter cell $n_{fa} = 1$, the adjusted reference signal $V^{*}_{\text{ref}\_a}$ is given by (23).

$$V^{*}_{\text{ref}\_a} = \frac{2}{3} V_{\text{ref}\_a} \tag{23}$$

At $t = 0.12$ s, the zero-sequence voltage for the neutral shift is injected due to which the $V^{*}_{\text{ref}\_a}$ has been increased as shown in Fig. 3.

3.2 Method-2: Only for Open-Switching Device Faults

In this method, faulty cell is not bypassed. Hence, inverter gives (L-1) levels in faulty phase. Consider that switch S11 has become open-circuited. The reference signal is modified only during the original modulation of S11 as shown in Fig. 4. The carrier Vcr1 corresponding to switching of S11 is made to shift outer-most, and adjusted reference signal is applied as given by (24).

$$\begin{array}{ll} \text{If } V_{\text{ref}\_a} > \frac{2}{3} & \text{then, } V^{*}_{\text{ref}\_a} = \frac{2}{3} V_{\text{ref}\_a} \\ \text{and if, } V_{\text{ref}\_a} < \frac{2}{3} & \text{then, } V^{*}_{\text{ref}\_a} = V_{\text{ref}\_a} \end{array} \tag{24}$$

The faulty cell-1 is responsible for the inner-most modulation level of the negative region. The healthy inverter cells, i.e., cell-2 and cell-3, are responsible for the inner-most and middle level in the positive region and middle and outer-most

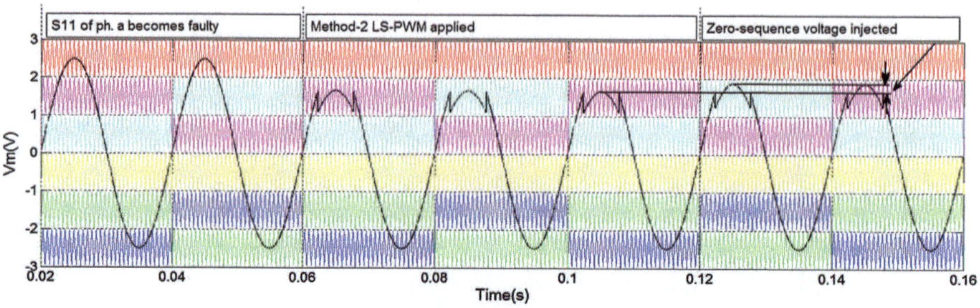

Fig. 4 Reference wave of phase-*a* containing faulty cell and method-2 is applied

Table 1 Simulation parameters for seven-level CHB-MLI

Individual dc voltage source	300 V
Amplitude modulation index, ma	0.833
Reference wave frequency	50 Hz
Switching frequency	1.5 kHz
R-L load	$R = 25\ \Omega$, $L = 4$ mH

level in the negative region. With this modified rotation scheme, the total amount of power delivered by cell-2 and cell-3 is still equal during the two cycles of the fundamental wave.

4 Simulation Results

A seven-level CHB-MLI model has been developed for verifying the control scheme using modified LS-PWM technique for the fault-tolerant operation using MATLAB/Simulink. The parameters used for simulation are given in Table 1.

The simulation is performed for different cases, and the results are discussed as follows:

Case-1: Considering switch S11 in phase-*a* become faulty and method-1 is applied. Figure 5 shows the reference voltages of three phases, phase voltages, line currents, and line-to-line voltages. Up to time $t = 0.06$ s, the faulty H-cell is bypassed as discussed in method-I. At time $t = 0.12$ s, the zero-sequence voltage is injected which gives balanced line currents and line-to-line voltages as shown in Fig. 5.

Case-2: Considering S11 and S12 in phase-*a* become faulty and method-1 is applied. This could be the worst case that out of three cells of phase-*a*, two cells become faulty. Though the possibility of this type of fault is very less, but the simulation results are presented in Fig. 6 to demonstrate the feasibility of the fault-tolerant method for the worst-case situation also.

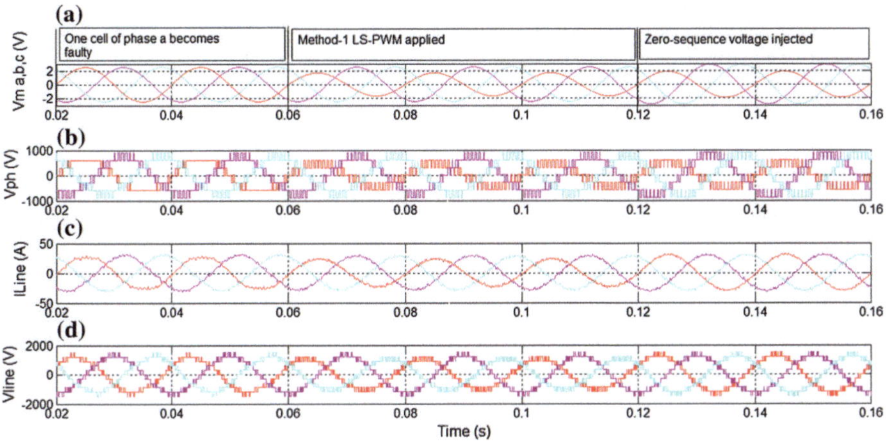

Fig. 5 Simulation results with *Nfa* = 1, *ma* = 0.833 and **method-1 LS-PWM is** applied: **a** Reference waves. **b** Phase voltages. **c** Line currents. **d** Line voltages

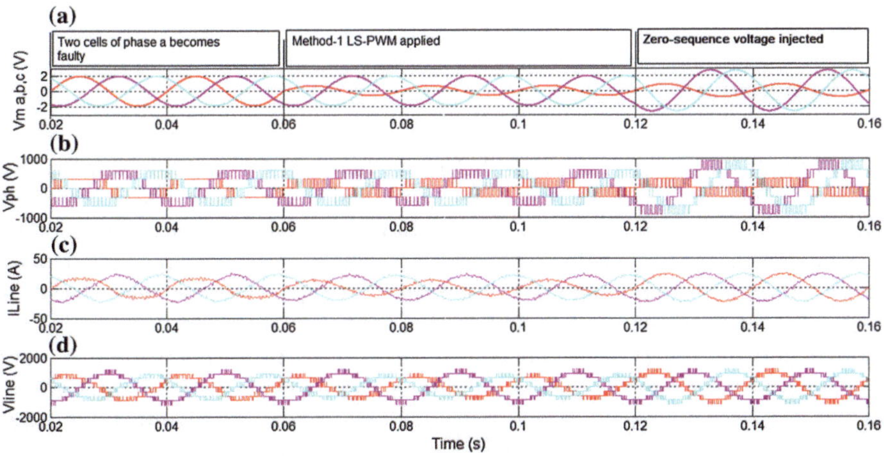

Fig. 6 Simulation results with *Nfa* = 2, *ma* = 0.633 and method-1 LS-PWM is applied: **a** Reference waves. **b** Phase voltages. **c** Line currents. **d** Line voltages

Case-3: Considering switch S11 in phase-*a* become faulty and method-2 is applied. Figure 7 shows the reference waveforms of all three phases, phase voltage, line current, and line-to-line voltage. In method-2, the faulty H-cell is not bypassed; instead, the healthy switches are let to work normally. At t = 0.06 s, the method-2 is applied for control of CHB-MLI under faulty condition and the six-level phase voltage +2 Vdc to −3 Vdc is generated at the faulty phase-*a* as shown in Fig. 7(b). At t = 0.12 s, the zero-sequence voltage is injected that gives balanced line currents as shown in Fig. 7(c) and line-to-line voltages as seen in Fig. 7(d).

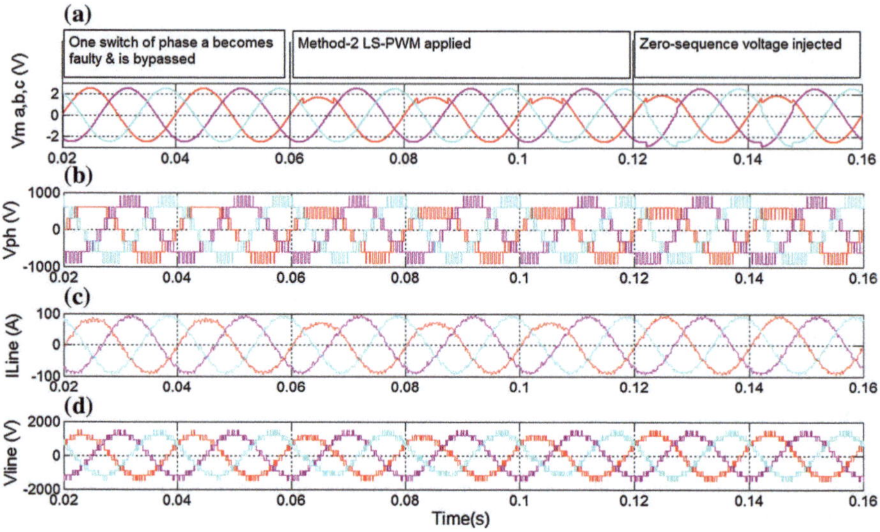

Fig. 7 Simulation results with *Nfa* = 1, *ma* = 0.833 and method-2 LS-PWM is applied: **a** Reference waves. **b** Phase voltages. **c** Line currents. **d** Line voltages

5 Conclusion

In this paper, the control of CHB-MLI under faulty H-cell conditions using two different methods for open-circuit and short-circuit fault conditions is presented. The generalized equations of the neutral shift method for zero-sequence voltage injection are derived, and balanced line-to-line voltage and line current are obtained under different faulty conditions. Method-1 is applicable for the control of CHB-MLI under faulty cell condition to both types of faults, i.e., open and short-switching device faults. Of course, method-1 can be considered as more reliable as it deals with both types of fault, i.e., open-circuit and short-circuit fault. By implementing rotation of the carrier in the presented modified LS-PWM method, equal power distribution among the H-cells can be obtained that improves the lifetime of the switching devices. The simulation results demonstrate the validity of the control scheme for the fault-tolerant operation of CHB-MLI.

References

1. S. Kouro, M. Malinowski, K. Gopakumar, J. Pou, L.G. Franquelo, B. Wu, J. Rodriguez, M. A. Perez, and J. I. Leon, "Recent advances and industrial applications of multilevel converters," IEEE Trans. Ind. Electron., vol. 57, no. 8, pp. 2553–2580, Aug. 2010.
2. B. Wu, "Cascaded H-bridge Multilevel Inverters", in High-Power Converters and ac Drives, John Wiley & Sons, Inc., Hoboken, New Jersey, USA, 1st ed., pp. 127–136, 2006.

3. M. Tamasas, M. Saleh, M. Shaker, A. Hammoda, "Evaluation of modulation techniques for 5-level inverter based on multicarrier level shift PWM", in Proc. of 17th IEEE Mediterranean Electrotechnical Conference (MELECON) 2014, pp. 17–23, 13–16 April 2014.
4. S. Wei, B. Wu, F. Li, and X. Sun, "Control method for cascaded H-bridge multilevel inverter with faulty power cells, in Proc. of Eighteenth Annual IEEE Applied Power Electronics Conference and Exposition, APEC'03, pp. 261–267, 2003.
5. J. Rodriguez, P. W. Hammond, J. Pontt, R. Musalem, P. Lezana, and M. J. Escobar, "Operation of a Medium-Voltage Drive Under Faulty Conditions," IEEE Trans. Ind. Electron., vol. 52, no. 4, pp. 1080–1085, Aug. 2005.
6. W. Song and A. Q. Huang, "Fault-Tolerant Design and Control Strategy for Cascaded H-Bridge Multilevel Converter-Based STATCOM," IEEE Trans. Ind. Electron., vol. 57, no. 8, pp. 2700–2708, Aug. 2010.
7. P. Moamaei, H. Mahmoudi, and R. Ahmadi, "Fault-Tolerant Operation of a Cascaded H-Bridge Inverters Using One Redundant Cell", Power and Energy Conference at Illinois (PECI), IEEE 2015.
8. L. Maharjan, T. Yamagishi, H. Akagi, and J. Asakura, "Fault-Tolerant Operation of a Battery-Energy-Storage System Based on a Multilevel Cascade PWM Converter with Star Configuration," IEEE Trans. Power Electron., vol. 25, no. 9, pp. 2386–2396, Sep. 2010.
9. H. W. Sim, J. S. Lee, and K. B. Lee, "A Detection Method for an Open-Switch Fault in Cascaded H-bridge Multilevel Inverters," in Proc. IEEE ECCE 2014, pp. 2101–2106, Sep. 2014.
10. S. Khomfoi and L. M. Tolbert, "Fault Diagnosis and Reconfiguration for Multilevel Inverter Drive Using AI-Based Techniques," IEEE Trans. Ind. Electron., vol. 54, no. 6, pp. 2954–2968, Dec. 2007.
11. S. M. Kim, J. S. Lee, and K. B. Lee, "Fault-tolerant strategy using neutral-shift method for cascaded multilevel inverters based on level-shifted PWM," in Proc. of 9th International Conference on Power Electronics ECCE, Asia, 2015.
12. S. M. Kim, J. S. Lee, and K. B. Lee, "A Modified Level-Shifted PWM Strategy for Fault-Tolerant Cascaded Multilevel Inverters with Improved Power Distribution," IEEE Trans. Ind. Electron., Jan. 2016.
13. M. Angulo, P. Lezana, S. Kouro, J. Rodriguez and Bin Wu, "Level shifted PWM for cascaded multilevel inverters with even power distribution," in Proc. of IEEE Power Electron. Spec. Conf. (PESC), pp. 2373–2378, Jun. 2007.

Author Index

R. Kher et al. (eds.), *Proceedings of the International Conference on Intelligent Systems and Signal Processing*, Advances in Intelligent Systems and Computing 671,
https://doi.org/10.1007/978-981-10-6977-2

MIX
Papier aus verantwortungsvollen Quellen
Paper from responsible sources
FSC® C105338

If you have any concerns about our products,
you can contact us on
ProductSafety@springernature.com

In case Publisher is established outside the EU,
the EU authorized representative is:
Springer Nature Customer Service Center GmbH
Europaplatz 3, 69115 Heidelberg, Germany

Printed by Libri Plureos GmbH
in Hamburg, Germany